Uncharted Horizons

Unveiling the Dynamic Architecture of Blockchain

*Delving into the World of Mining 4.0
through the Lens of Blockchain Technology*

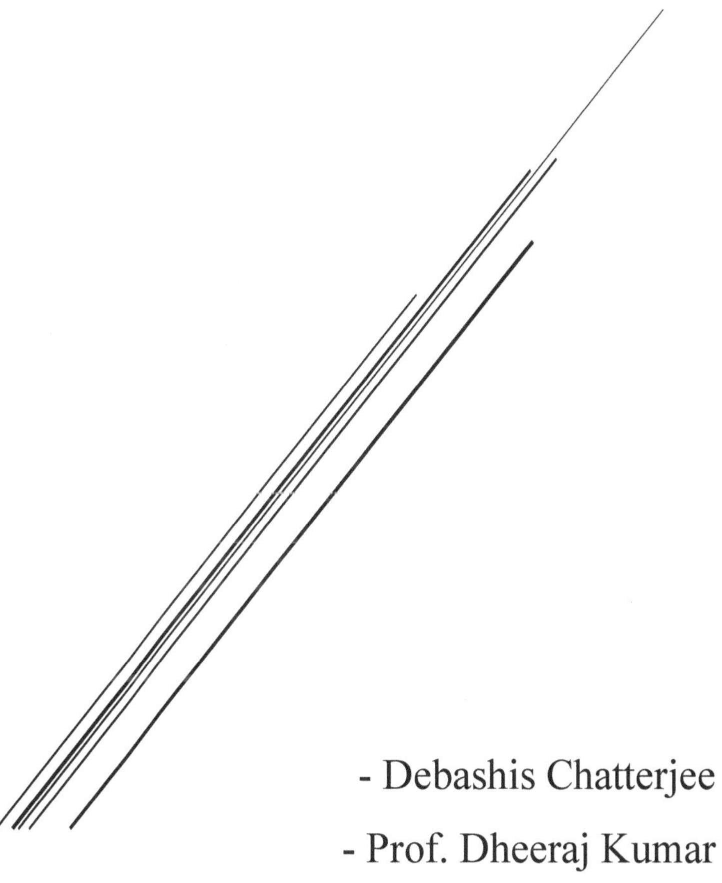

\- Debashis Chatterjee

\- Prof. Dheeraj Kumar

BlueRose ONE
New Delhi • London

BLUEROSE PUBLISHERS

India | U.K.

For permissions requests or inquiries regarding this publication,
please contact:

BLUEROSE PUBLISHERS
www.BlueRoseONE.com
info@bluerosepublishers.com
+91 8882 898 898
+4407342408967

ISBN: 978-93-6261-447-6

Editor: Debashis Chatterjee & Prof. Dheeraj Kumar
Illustrator: Debashis Chatterjee & Prof. Dheeraj Kumar

First Edition: November 2023

Preface

It is with great pleasure and excitement that we, present this book, "Uncharted Horizons: Unveiling the Dynamic Architecture of Blockchain" to the curious minds of students, the innovative spirit of engineers, and the passion of blockchain enthusiasts. Throughout our respective journeys in the realms of technology, we have witnessed the transformative power of blockchain and its potential to revolutionize various industries, including mining.

Within these pages, we embark on a comprehensive exploration of the dynamic and ever-evolving world of blockchain technology. Our aim is to provide readers with a profound understanding of the underlying principles of blockchain and how it can reshape the mining industry to create a more transparent, sustainable, and equitable future.

Blockchain's emergence marked a significant milestone beyond its association with cryptocurrencies. As we began to uncover its vast potential, we realized its profound implications for industries like mining. The decentralized and transparent nature of blockchain offers a unique opportunity to address pressing challenges faced by the mining sector, ranging from ethical sourcing to supply chain management and environmental sustainability.

At the heart of this book lies a deep exploration of blockchain's impact on supply chain management within the mining industry. We delve into how blockchain can foster ethical sourcing of minerals, ensuring that responsible practices are adhered to throughout the supply chain. By recording every transaction on an immutable and transparent ledger, blockchain empowers consumers to make informed decisions about the origin of the resources they consume.

Moreover, we address the challenges and opportunities associated with integrating blockchain technology into the mining ecosystem. As blockchain networks evolve, ensuring standardization and interoperability across various platforms becomes a critical consideration for seamless adoption.

Beyond its industrial applications, blockchain touches the very core of individual empowerment. We are intrigued by the tantalizing idea of granting people control over their personal data, assets, and identities through self-sovereign identity solutions. Blockchain's decentralized nature enables marginalized individuals to access global networks, fostering financial inclusion and social empowerment.

Furthermore, blockchain extends its influence to the realm of employment and social cohesion. By eliminating intermediaries, blockchain can create direct and efficient connections in various fields, revolutionizing traditional work structures. This has the potential to empower freelancers and gig workers by providing them with greater autonomy and control over their careers.

As you traverse through the chapters, we wholeheartedly encourage you to immerse yourselves in the world of blockchain. Grasp its technical intricacies, envision its possibilities, and engage actively with the content. This book is not just a one-sided narrative; we humbly request your feedback, suggestions, and insights to refine and improve the content further. As readers and contributors, you play a vital role in making this work a richer, more accurate reflection of the transformative power of blockchain technology.

While we have endeavoured to present the information with utmost care, we acknowledge that errors or omissions may inadvertently occur. Your assistance in identifying them will enhance the authenticity and reliability of this work, benefiting both present and future readers.

The journey through the blockchain landscape is an enthralling adventure, filled with new discoveries, challenges, and possibilities. We envision a future where blockchain technology fosters a more transparent, equitable, and sustainable world. Together, let us embrace this adventure of knowledge and imagination as we explore the transformative potential of blockchain in empowering mining and beyond.

- Debashis Chatterjee

- Prof. Dheeraj Kumar

Acknowledgements

Embarking on the journey of writing a book about blockchain technology has proven to be an exceptional and transformative experience, one that has far exceeded my initial expectations and led to bountiful rewards. This undertaking has been made possible through the unwavering assistance and expertise extended by TEXMiN, a Section-8 company operating under the esteemed NM-ICPS, DST, Govt of India, situated at the renowned IIT (ISM) Dhanbad. The integral role played by TEXMiN in shaping the content of the book and providing insightful industry perspectives has been monumental, and I hold a deep sense of appreciation for TEXMiN's significant contributions. This collaboration has not only enriched the content of the book but has also seamlessly connected me with an extensive network of field experts. This connection has allowed me to gather fresh perspectives and validate the concepts I have explored. The unwavering support of TEXMiN has indeed been the driving force that has skilfully transformed an abstract concept into a tangible and impactful publication.

I wish to express my deepest gratitude and acknowledge the indispensable efforts of Prof. Dheeraj Kumar, my co-author, and the founding director of TEXMiN, as well as the deputy director of IIT (ISM) Dhanbad. His extensive reservoir of knowledge and profound insights into the intricate landscape of blockchain technology and the mining industry have formed the bedrock upon which the content of this book was meticulously constructed. Prof. Dheeraj Kumar's unwavering dedication to this project has stood as a continuous source of inspiration, propelling the trajectory of this book towards its eventual publication. His steadfast commitment to upholding high standards has driven me to craft a comprehensive and invaluable resource tailored for readers seeking a profound understanding of the subject matter. Prof. Dheeraj Kumar's collaborative ethos and passion for knowledge have played a pivotal role in enriching our discussions and brainstorming sessions, infusing the content with novel ideas and unique perspectives. His unwavering support during moments of challenges and hurdles has provided the impetus for me to overcome obstacles and persevere through every phase of the book's development. The privilege of collaborating with Prof. Dheeraj Kumar throughout this endeavour has been an honour that I deeply cherish. His mastery and guidance in the realm of blockchain technology have undeniably elevated the quality of this book, and his contributions to the advancement of this field demand profound respect.

Furthermore, I extend my heartfelt appreciation to the colleagues at TEXMiN, IIT (ISM) Dhanbad, whose enduring support and encouragement have been pivotal in seeing this project through to fruition. Their insights and steadfast backing have endowed me with the fortitude to navigate the challenges inherent in such an extensive undertaking.

In this journey, I must also acknowledge the foundation of my strength – my loving family, particularly my parents, Swati and Sanjiva Chatterjee, whose unwavering support has been my constant source of strength through every arduous phase. To my sister, Arundhuti, I extend my gratitude for infusing moments of joy into the process. Their unyielding encouragement and love have been my anchor.

Converting an abstract notion into a tangible publication is a demanding pursuit, and I am indebted to my brother-in-law, Debmalya, for his unwavering guidance and support in translating this vision into reality.

Lastly, I wish to extend sincere appreciation to all those who contributed to the book's publication, even if their names are not specifically mentioned in this acknowledgment. Your collective assistance and support have been pivotal in materializing this project within the stipulated timeframe.

I extend my heartfelt thanks to each and every individual who has played a role, no matter how big or small, in this extraordinary journey. Your presence and support have made this endeavour truly remarkable and fulfilling.

-Debashis Chatterjee

Forwarding

Uncharted Horizons: Unveiling the Dynamic Architecture of Blockchain is an exceptional book authored by Debashis Chatterjee and Prof. Dheeraj Kumar, delving into the potential of blockchain technology across various industries, including coal and mineral mining. This comprehensive work offers profound insights into the dynamic architecture and applications of blockchain. Blockchain technology has revolutionized business processes due to its decentralized and secure nature. "Uncharted Horizons" takes readers on a captivating journey, unravelling the mysteries of blockchain and its limitless possibilities, particularly within the coal and mineral mining sector.

This book offers a unique blend of technical expertise and accessibility, making complex concepts understandable for both industry practitioners and newcomers. It demonstrates how blockchain can enhance transparency and traceability in mineral supply chains, automate and secure contractual agreements through smart contracts, and bolster data security and privacy in mining operations.

As the Director of IIT (ISM) Dhanbad and Chairman of HGB, TEXMiN Foundation, I have personally witnessed the positive impact of blockchain in various projects, including those within the coal and mineral mining industry. "Uncharted Horizons" aligns perfectly with our vision of embracing cutting-edge technologies. I wholeheartedly endorse this book as an invaluable resource for employees, researchers, and industry professionals looking to explore blockchain's potential in this sector.

I commend Debashis Chatterjee and Prof. Dheeraj Kumar for their meticulous research and insightful contributions. "Uncharted Horizons: Unveiling the Dynamic Architecture of Blockchain" will inspire readers to navigate the uncharted territories of blockchain within the coal and mineral mining industry, unlocking new horizons of innovation and transformation.

Prof. Jamini Kanta Pattanayak
Director, IIT (ISM) Dhanbad
Chairman HGB, TEXMiN Foundation

Meet The Authors

About Debashis Chatterjee

Debashis Chatterjee is a passionate engineer with a Master's degree in Electronics Engineering, who has amassed an impressive track record spanning over a decade, encompassing industry, government, and academia. His primary focus lies in Research & Development, and he possesses expertise in various cutting-edge domains, including Blockchain, Military Grade Mobile Robotics, Hardware system design, IoT, and hardware test automation. Throughout his career, Chatterjee has made notable contributions to the field. He has invented and patented several electro-mining instruments for the Government of India, showcasing his ingenuity, engineering prowess, and commitment to pushing the boundaries of technological innovation. Additionally, he has contributed to scholarly publications in national and international journals, demonstrating his dedication to advancing knowledge. Moreover, as a provider of competitive intelligence to premier institutes in India, Chatterjee has played a significant role in elevating technological advancements within the country. His expertise extends across multiple domains, notably in robust security protocols and cutting-edge Blockchain-based projects. His commitment to digital security shines through his creation of an innovative AI-driven Authentication Protocol, which detects and prevents fraudulent activities in cryptocurrency portfolios, safeguarding users' investments. A relentless seeker of knowledge, Chatterjee embraces unconventional approaches, exemplifying his passion for technology and its potential to drive progress. Beyond his professional achievements, he fosters collaboration through engaging in meaningful scientific discussions with peers and cherishes the camaraderie of friends. Debashis Chatterjee's contributions continue to reshape the landscape of electronics engineering and technology, making significant strides towards a safer and more advanced digital future. His visionary outlook and dedication to pushing the envelope of technological innovation inspire those around him, leaving an indelible mark on the field and influencing the next generation of engineers and researchers.

For more information on Debashis Chatterjee, follow him on LinkedIn

About Professor Dheeraj Kumar

Professor Dheeraj Kumar is a distinguished figure in the field of Mining Engineering, acclaimed for his significant achievements and contributions at the esteemed IIT (ISM) Dhanbad. As the Founding Director of TEXMiN, as well as Professor and Deputy Director at IIT (ISM) Dhanbad, he brings a wealth of knowledge and expertise to both academic leadership and research. With a Ph.D. from IIT Kharagpur, he has established a strong academic foundation and a passion for research. Throughout his career, he has demonstrated unwavering dedication to teaching and research, earning him recognition as a prominent leader in the field. As Deputy Director of IIT (ISM) Dhanbad, Professor Kumar played a pivotal role in shaping the institution's academic and research landscape. His visionary leadership has fostered the institute's growth as a centre of excellence in mining engineering and geomatics. Under his guidance, the Department of Mining Engineering thrived, and research projects, funded by both national and international agencies, flourished. Professor Kumar's research endeavours span a diverse range of topics, reflecting his versatility and keen interest in addressing critical challenges in the mining industry. From groundwater dynamics to environmental impact assessment, his expertise in geospatial analysis has been instrumental. Notably, his work on coal mining techniques, including the innovative development of Self-advancing Goaf Edge Supports (SAGES), has significantly enhanced safety & efficiency in underground coal mining operations. Professor Kumar's groundbreaking contributions are further complemented by his expertise in blockchain and smart contracts, a revolutionizing force in the mining domain. Through seamless integration of smart contracts into underground coal mining operations, he has ushered in a new era of transparency and trust, effectively streamlining resource management. His prolific publication record in renowned journals attests to his commitment to advancing the frontiers of knowledge in the field of mining engineering and geomatics. Beyond academic excellence, Professor Kumar's mentorship has inspired and nurtured numerous students and researchers, encouraging them to pursue fulfilling careers in mining engineering and related fields. His collaborative and innovative approach to research has cultivated a conducive environment for exploration and discovery at IIT (ISM) Dhanbad.

For more information on Professor Dheeraj Kumar, follow him on LinkedIn

List of Figures

List of Tables

List of Abbreviations

Abbreviations for Different Cryptocurrencies

Although Bitcoin remains the most widely used cryptocurrency, there are several others in circulation. Below is a list of cryptocurrencies and their abbreviations that one may encounter.

- **AUR**: Auroracoin
- **BCC**: BitConnect (inactive)
- **BCH**: Bitcoin Cash
- **BTC** or **XBT**: Bitcoin
- **DASH**: **Dash**
- **DOGE** or **XDG**: Dogecoin
- **EOS**: EOS.IO
- **ETC**: Ethereum Classic
- **ETH**: Ether (also known as Ethereum)
- **GRC**: Gridcoin
- **LTC**: Litecoin
- **KOI** or **COYE**: Coinye (inactive)
- **MZC**: Mazacoin
- **NEO**: Neo
- **NMC**: Namecoin
- **Nxt**: NXT
- **POT**: PotCoin
- **PPC**: Peercoin
- **TIT**: Titcoin
- **USDC**: USD Coin (Stablecoin)
- **USDT**: Tether
- **VTC**: Vertcoin
- **XEM**: NEM
- **XLM**: Stellar
- **XMR**: Monero
- **XPM**: Primecoin
- **XRP**: Ripple
- **XVG**: Verge
- **ZEC**: Zcash
- **THETA:** Theta Token
- **UMA:** UMA
- **UNI:** Uniswap
- **VET:** VeChain

Blockchain & Network Abbreviations

One can utilize a DAO for conducting peer-to-peer transactions or measuring MACD. However, the question arises: what exactly does that entail? Explore the following compilation of frequently used terms and phrases when discussing blockchains.

- ❖ **BIP**: Bitcoin Improvement Proposal
- ❖ **BTM**: Automatic Teller Machine for Bitcoin
- ❖ **DAO**: Decentralized Autonomous Organization
- ❖ **DPoS**: Delegated Proof of Stake
- ❖ **EEA**: Enterprise Ethereum Alliance
- ❖ **EIP**: Ethereum Improvement Proposal
- ❖ **ERC**: Ethereum Request for Comments
- ❖ **EVM**: Ethereum Virtual Machine
- ❖ **FA**: Fundamental Analysis
- ❖ **LN**: Lightning Network
- ❖ **MACD**: Moving Average Convergence Divergence
- ❖ **MoE**: Medium of Exchange
- ❖ **P2P**: Peer to Peer
- ❖ **PoA**: Proof of Authority
- ❖ **PoB**: Proof of Burn
- ❖ **PoD**: Proof of Developer
- ❖ **PoS**: Proof of Stake
- ❖ **PoW**: Proof of Work
- ❖ **SC**: Smart Contract
- ❖ **SegWit**: Segregated Witness
- ❖ **SoV**: Store of Value
- ❖ **TA**: Technical Analysis or Trend Analysis
- ❖ **UoA**: Unit of Account
- ❖ **UTC**: Coordinated Universal Time
- ❖ **WP**: Whitepaper
- ❖ **YTD**: Year to Date
- ❖ **DAG:** Directed Acyclic Graph
- ❖ **DeFi:** Decentralized Finance
- ❖ **DID:** Decentralized Identifier
- ❖ **DLT:** Distributed Ledger Technology
- ❖ **EVM:** Ethereum Virtual Machine
- ❖ **SPV:** Simplified Payment Verification
- ❖ **UASF:** User Activated Soft Fork
- ❖ **UTXO:** Unspent Transaction Output

Technical Abbreviations

Cybersecurity plays a crucial role in the successful utilization of cryptocurrency. It is essential for both developers and users to have a fundamental understanding of technical abbreviations commonly associated with cryptocurrencies. Here are some examples of such abbreviations:

- ❖ **2FA:** Two Factor Authentication
- ❖ **Addy:** Address
- ❖ **API:** Application Programming Interface
- ❖ **ASIC:** Application Specific Integrated Circuit
- ❖ **BFA:** Brute Force Attack
- ❖ **Bech32:** Bitcoin address format (also known as bc1 addresses)
- ❖ **CPU:** Central Processing Unit
- ❖ **BFT:** Byzantine Fault Tolerance
- ❖ **DAG:** Directed Acyclic Graph
- ❖ **DAPP or dApp:** Decentralized Application
- ❖ **DDoS:** Distributed Denial of Service
- ❖ **DEVCON:** Developers Conference
- ❖ **GPU:** Graphical Processing Unit
- ❖ **IPFS:** Interplanetary Files System
- ❖ **PKI:** Public Key Infrastructure
- ❖ **Multi-sig:** Multi-Signature
- ❖ **NONCE:** Number Used Only Once
- ❖ **SHA-256:** Secure Hash Acronym (256-bit)
- ❖ **WWDC:** Worldwide Developers Conference
- ❖ **CLI:** Command Line Interface
- ❖ **CSS:** Cascading Style Sheets
- ❖ **DOM:** Document Object Model
- ❖ **FTP:** File Transfer Protocol
- ❖ **HTML:** Hypertext Markup Language
- ❖ **HTTP:** Hypertext Transfer Protocol
- ❖ **HTTPS:** Hypertext Transfer Protocol Secure
- ❖ **IDE:** Integrated Development Environment
- ❖ **IP:** Internet Protocol
- ❖ **LAN:** Local Area Network
- ❖ **JSON:** JavaScript Object Notation
- ❖ **OOP:** Object-Oriented Programming
- ❖ **PHP:** Hypertext Preprocessor
- ❖ **REST:** Representational State Transfer
- ❖ **SMTP:** Simple Mail Transfer Protocol

Financial Abbreviations

An understanding of the confluence between technology and economics can be gained by referring to a comprehensive list of trading terminologies and financial abbreviations.

- ❖ **AML**: Anti-Money Laundering
- ❖ **ATH**: All-Time High
- ❖ **ATL**: All-Time Low
- ❖ **ALT | Altcoin**: Alternative Cryptocurrency (cryptocurrencies other than Bitcoin)
- ❖ **CEX**: Centralized Exchange
- ❖ **CMC**: Coin Market Cap
- ❖ **DAICO**: Decentralized Autonomous Initial Coin Offering
- ❖ **DCA**: Dollar Cost Averaging
- ❖ **DeFi**: Decentralized Finance
- ❖ **DEX**: Decentralized Exchange
- ❖ **DLT**: Distributed Ledger Technology
- ❖ **ERC-20**: Token standard for Ethereum
- ❖ **ERC-721**: Token standard for NFT (non-fungible tokens)
- ❖ **ETF**: Exchange-Traded Fund
- ❖ **ETP**: Exchange-Traded Product
- ❖ **FIAT**: Government-issued currency (e.g. US Dollar, Euro)
- ❖ **IBO**: Initial Bounty Offering
- ❖ **ICO**: Initial Coin Offering
- ❖ **ITO**: Initial Token Offering
- ❖ **mBTC**: Millibitcoin (0.001 BTC)
- ❖ **MCAP**: Market Capitalization
- ❖ **PnD**: Pump-and-Dump scheme
- ❖ **OTC**: Over the Counter
- ❖ **SATS**: Satoshis (the smallest denomination of a Bitcoin: 0.00000001 BTC)
- ❖ **STO**: Securities Token Offering
- ❖ **TPS**: Transactions Per Second
- ❖ **Tx**: Transaction
- ❖ **TxID**: Transaction Identification
- ❖ **uBTC**: MicroBitcoin (0.000001 BTC)
- ❖ **UXTO**: Unspent Transaction
- ❖ **APR:** Annual Percentage Rate
- ❖ **CAGR:** Compound Annual Growth Rate
- ❖ **CAPEX:** Capital Expenditure

Conversational Cryptocurrency Terms

The world of cryptocurrency has its own unique terminology, much like any other online community. Experienced members may use terms such as FOMO, JOMO, BUIDL, and HODL. Understanding these terms can help newcomers effectively communicate with seasoned users and become more knowledgeable about cryptocurrency.

- ❖ **BUIDL**: "Build" (purposeful misspelling for ironic meaning)
- ❖ **FOMO**: Fear of Missing Out
- ❖ **HODL**: Hold On for Dear Life (purposeful misspelling of "HOLD")
- ❖ **BTD**: Buy The Dip
- ❖ **DYOR**: Do Your Own Research
- ❖ **FUD**: Fear, Uncertainty, Doubt
- ❖ **FUDster**: A person who spreads Fear, Uncertainty, and Doubt
- ❖ **ELI5**: Explain It Like I am 5
- ❖ **JOMO**: Joy of Missing Out
- ❖ **KYC**: Know Your Customer
- ❖ **TLT**: Think Long Term
- ❖ **OCO**: One Cancels the Other
- ❖ **AMA**: Ask Me Anything
- ❖ **REKT**: "Wrecked" (meaning major losses)
- ❖ **TOR**: The Onion Router (one who sends anonymous data)
- ❖ **CT**: Crypto Twitter
- ❖ **ATH:** All-Time High - The highest price a cryptocurrency has ever reached.
- ❖ **ATL:** All-Time Low - The lowest price a cryptocurrency has ever reached.
- ❖ **SATS:** Short for "Satoshis," the smallest unit of Bitcoin (0.00000001 BTC).
- ❖ **MOONING:** When a cryptocurrency's price is rapidly increasing.
- ❖ **REVERSAL:** A trend change in price movement.
- ❖ **BULLISH:** Positive outlook on the market, expecting prices to rise.
- ❖ **SWING TRADING:** Short- to medium-term trading strategy aiming to profit from price "swings" within a trend.
- ❖ **DAY TRADING:** Buying and selling assets within the same day to capitalize on short-term price movements.
- ❖ **HARD FORK:** A significant protocol update resulting in a divergence of the blockchain.
- ❖ **MOONSHOT:** An investment with the potential for substantial gains.

Table Of Content

Introduction

Bitcoin, a novel economic system on the internet, has the potential to transform society by operating based on consensus among networked users, without relying on a central authority. Unlike traditional currencies, Bitcoin does not require participants to trust each other, and algorithmic self-policing rejects fraudulent attempts on the system. Bitcoin is a digital currency that operates in a trustless, decentralized system, utilizing a public ledger known as the blockchain. Since its launch in 2009, Bitcoin has inspired other alternative currencies that share the same fundamental approach with differing tweaks and optimizations. Beyond economics, blockchain technology has the potential to transform humanitarianism, politics, social issues, and science. By creating a decentralized cloud function, blockchain technology can be utilized by organizations such as WikiLeaks and transnational groups like ICANN and DNS services, providing immense benefits. The advantages of Bitcoin and blockchain technology are enormous, impacting society and its operations in all aspects. Blockchain activities can be categorized into three areas: Blockchain 1.0, which focuses on currency; Blockchain 2.0, which includes economic, market, and monetary packages using blockchain technology like stocks, bonds, and smart contracts; and Blockchain 3.0, which encompasses blockchain-based applications beyond finance and markets, including government, health, science, literacy, culture, and art. The mining industry, which has been historically reluctant to adopt technological advancements, could benefit from blockchain technology by enhancing transparency, efficiency, and sustainability in the mining supply chain. Blockchain can provide an immutable record of mineral origin, movement, and ownership, verifying the ethical sourcing of materials. Blockchain-based tracking systems can also provide real-time information on the origin of minerals, reducing the risk of using conflict minerals and improving supply chain transparency. Blockchain technology can increase the efficiency of the mining supply chain by automating processes such as payments, quality control, and logistics. It can also enhance sustainability in the mining industry by providing a transparent and auditable record of the environmental impact of mining activities, reducing their impact on the environment. As a disruptive technology, Bitcoin and blockchain have the potential to transform human activity to the same extent as the World Wide Web. The decentralized and trustless nature of this technology can liberate industries from regulatory and licensing power structures, paving the way for new disintermediated business models. While there are challenges and risks associated with the adoption of Bitcoin and blockchain technology, the potential benefits are vast, and its impact on society is only beginning to be realized. One of the biggest advantages of blockchain technology is its potential to improve supply chain management in various industries. With blockchain, supply chain participants can track goods and products from their origin to their destination with a high level of transparency and security. This can increase efficiency, reduce costs, and prevent fraud, counterfeiting, and other forms of malfeasance. For instance, in the food industry, blockchain can be used to track the origin, quality, and safety of food products, from farm to table. This can help prevent foodborne illnesses, reduce waste, and improve consumer trust in the food supply chain. Similarly, in the coal and mineral mining industry, blockchain technology can bring significant benefits. For example, blockchain can be utilized to track the origin, extraction, and transportation of coal and minerals, ensuring transparency and accountability throughout the supply chain. This can help prevent illegal mining activities, promote ethical sourcing practices, and enhance sustainability efforts. Additionally, blockchain can facilitate efficient and secure transactions between mining companies and their stakeholders, such as suppliers, buyers, and financial institutions. This technology has the potential to revolutionize the coal and mineral mining industry by streamlining processes, improving traceability, and fostering trust among all participants involved. The potential applications of blockchain technology are vast and varied, and its impact on society is likely to be significant in the years to come. However, as with any new technology, there are also challenges and risks associated with blockchain, including energy consumption, regulatory oversight, and security concerns. It will be important for policymakers, industry leaders, and the public to work together to address these challenges and maximize the benefits of blockchain technology.

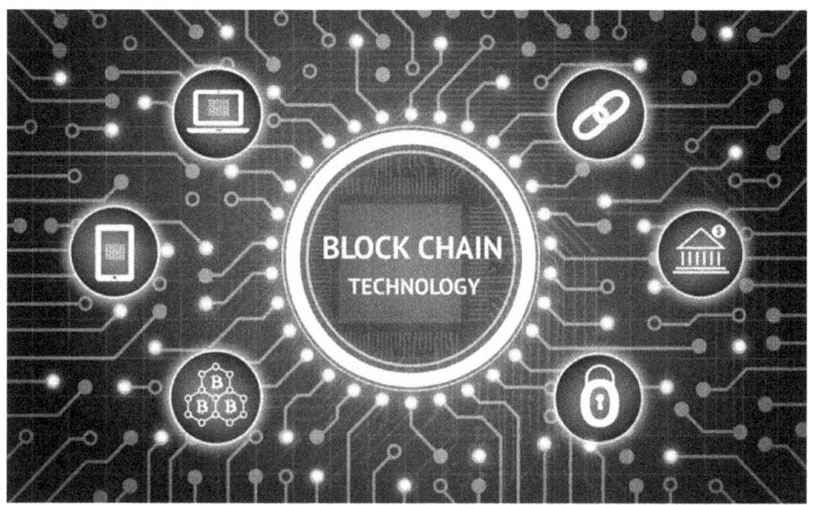

Chapter 1: Blockchain Fundamentals: Decentralization & Security

"Blockchain technology is coming to replace the old, rotten system of governance around the world. Embrace it!"

— Olawale Daniel

Understanding the Basics of Blockchain

A blockchain can be described as a distinctive type of database, also known as Distributed Ledger Technology (DLT). It possesses some unique characteristics, such as rules governing how data can be added, and once data is stored, it is almost impossible to modify or delete it. The data is arranged over time in blocks, with each block built on top of the last one and containing information that connects it to the previous one. By examining the latest block, we can confirm that it was created after the previous one. If we continue tracing back through the chain of blocks, we eventually arrive at the genesis block, which is the first block of the chain. The term "block" refers to computer files that store transaction data in a linear series that creates an unbreakable chain of blocks, hence the name blockchain. All data related to blockchain transactions is accumulated and recorded inside these blocks, and each new block is linked to the previous one through cryptographic techniques. The chain of interconnected blocks stores all transaction data generated since the release of a particular blockchain. The number of blocks verified since the genesis block is known as the block height.

To provide an analogy, imagine a spreadsheet with columns. In the first cell of the first row, you can input any data you want to keep. The first cell's data is converted into a letter identifier, which is then used as part of the next input. For example, the letter identifier KP is used to fill out the next cell in the second row (defKP). If you modify the first input record (abcAA), each subsequent cell would produce a unique combination of letters. When we examine row four, our most recent identifier is TH. As we previously mentioned, it is impossible to go back and delete or alter previous entries because it would be easily detectable and disregarded. If you modify the data in the first cell, a different identifier would be generated, leading to different data in the second block, and so on, resulting in a different identifier in row two, and so on. In essence, TH is a culmination of all the data that came before it. One important feature of blockchain is that it is a decentralized database, which means it is not owned or controlled by any single entity. Instead, it is maintained by a network of computers, known as nodes, that work together to validate and store transactions. When a new transaction is submitted to the network, it is broadcast to all nodes. Each node independently verifies the transaction to ensure that it is valid and that the sender has sufficient funds to complete the transaction. Once the transaction is verified, it is added to a block, along with other verified transactions that have occurred since the last block was added to the chain. Before a new block can be added to the chain, it must be validated by the network. This is done through a process known as consensus, which ensures that all nodes agree on the state of the blockchain. There are various consensus mechanisms used in different blockchain networks, such as proof of work (PoW) and proof of stake (PoS). Once a block is added to the chain, the data it contains is secured through the use of cryptographic hashing. Each block contains a unique hash value that is generated based on the data it contains. This hash value is then used as an input to generate the hash value for the next block in the chain. This creates an immutable, tamper-proof record of all transactions that have occurred on the blockchain. Because of its decentralized nature and cryptographic security, blockchain has many potential use cases beyond just financial transactions. For example, it can be used to create decentralized applications (dApps), supply chain management systems, and even voting systems.

Overall, blockchain technology has the potential to revolutionize the way we store and transmit data, and it is an exciting area of innovation that is still rapidly evolving.

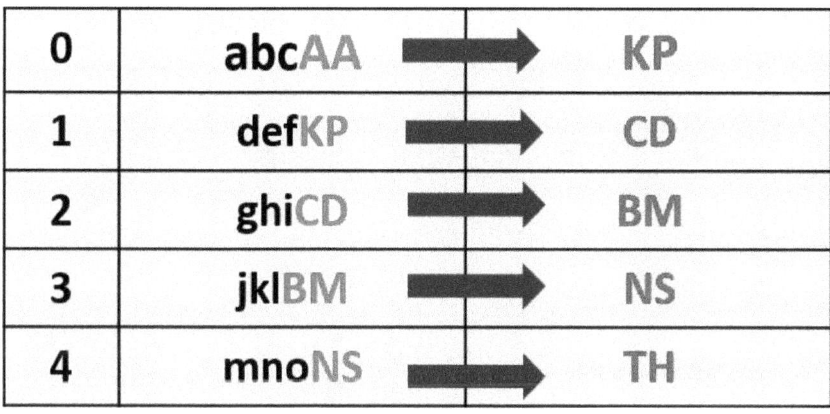

Figure 1.1. A Database where each Entry is Linked to the Previous.

How are Blocks in a Blockchain Network Connected?

As mentioned earlier, the use of two-letter identifiers in our previous analogy was a simplified way of explaining how a blockchain utilizes a hash function. Hashing is a crucial component that holds blocks together in a blockchain network. In essence, it involves taking data of any length and running it through a mathematical function that generates an output known as a hash. Hashing is the process of creating a fixed-length output from a variable-length input using mathematical formulas known as hash functions or hashing algorithms.

The hash functions used in blockchains are unique in that the likelihood of finding two pieces of data that produce the same output (i.e., a hash collision) is incredibly low. Similar to our two-letter identifiers, even a small change in the input data will result in a completely different output hash. An example of a widely used hash function in Bitcoin is SHA256. Even a simple change in the capitalization of letters is enough to entirely alter the output hash, as you can see.

Hashing is a critical component of blockchain technology. It is the mathematical function that takes input data of any length and produces a fixed-size output known as a hash. The hash is unique and provides a digital fingerprint of the input data, meaning any change in the input data would result in a different hash output. The hash function used in blockchains is designed to be a one-way function, meaning it is almost impossible to reverse-engineer the input data from the hash output.

Input Data	SHA256 Output
UNCHARTED HORIZONS	7391fd4b9ac3bb32a50f63ac3956608fa535 4fa03fd663d21d21980272f4146c
Uncharted Horizons	a3741cdd2d79d2029d5508403f41824cdb0 5e0dcdf190f18233cef7ae9cc0e3d
Uncharted horizons	8b63547356ae3d59faaa885ee65b52315f43 f8aa0f3d7ef8a0ae4eb645520dec
uncharted horizons	603d69ebd8cdef1479186bf250a340707a0 9b74fb8b8cd9ddd2c5a1c5040f371
uncharted HORIZONS	dd5cf780062309eba35bd1d20b78add3801 524854572b76e704c02c094d39005
UNCHARTED Horizons	19aec8687c2783b69e7e9fd52a59e6c64e76 99a55d67e119bb4dff3841dd9cfd

Table1.1: SHA256 Hashing Algorithm

0 the hash of the altered block will no longer match the hash stored in the subsequent block, thus revealing any attempts to tamper with the chain. Therefore, the unique property of the hash function ensures that the blockchain is secure and tamper-proof.

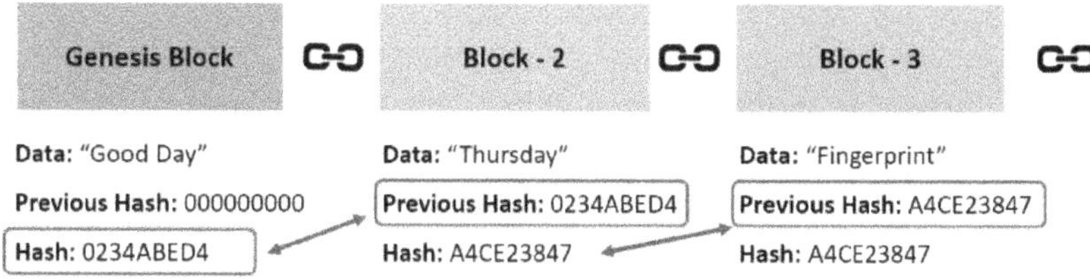

Figure 1.2. Each Block contains an Impression of the Prior Block.

Decentralization using Blockchains Technology

We have already discussed the basic structure of a blockchain network. However, when people talk about blockchain technology, they are not just referring to the database itself. The blockchain ecosystem is built upon several sophisticated technologies such as cryptography and peer-to-peer networking. While blockchains can function as standalone data structures, they are most effective when used as tools for people to interact and coordinate with one another. By integrating diverse technologies and concepts, blockchain technology can act as a distributed ledger that is managed by no one.

Game theory plays a crucial role in the development of cryptocurrencies, and it is one of the primary reasons why the Bitcoin network has managed to thrive for over a decade, despite numerous attempts to disrupt it. The idea behind game theory is that no one has the power to edit the entries outside of the rules set by the system. This means that the ledger is owned concurrently by everyone, and participants reach a consensus on what it looks like at any given time. Essentially, the power is distributed among the network, and no single entity can control it. This is what makes blockchain technology so revolutionary and transformative, as it enables trustless interactions and transactions between individuals without the need for intermediaries.

The Byzantine General's Problem

The challenge facing a system like blockchain is known as the Byzantine Generals Problem, which was first described in the 1980s. The problem arises when isolated participants need to coordinate their actions without a centralized authority. The original dilemma involved a group of military generals surrounding a city and deciding whether to attack it. The generals could only communicate with each other through messengers, and each had to decide whether to attack or retreat. The decision was not important as long as they all agreed on the same course of action. However, if they decided to attack, they would only be successful if they attacked at the same time.

The problem arises when we consider the possibility of one of the generals being malicious and deliberately misleading the others or messages being intercepted and altered. How can we ensure that consensus is reached despite the potential for malicious behaviour or communication failures? This is the same problem that blockchain faces. If there is no central authority overseeing the blockchain and providing users with accurate information, then the users must be able to communicate among themselves to reach consensus.

To overcome the potential for malicious behaviour or communication failures, blockchain mechanisms must be carefully engineered to be immune to such setbacks. A system that can achieve this is called Byzantine fault-tolerant. In other words, the blockchain must be able to withstand the actions of one or more malicious actors without compromising the integrity of the system.

One way to achieve this is through the use of consensus algorithms. These algorithms enable blockchain networks to reach agreement on the state of the ledger without relying on a centralized authority. There are several consensus algorithms used in blockchain, such as proof-of-work, proof-of-stake, and delegated proof-of-stake. Each algorithm has its own advantages and disadvantages, but they all aim to ensure that the network can operate in a decentralized manner while maintaining the integrity of the ledger.

To illustrate the problem mathematically, let us say there are n generals who must coordinate their attack on a city. Each general i has to decide whether to attack or retreat and sends a message mi to all the other generals. The decision of each general is represented by a binary variable $x_i = 1$ for attack and $x_i = 0$ for retreat. The decision of general i is denoted by x_i, and the decision of all other generals is denoted by vector x_{-i}.

However, some of the generals may be traitors who are attempting to sabotage the agreement. Let us say there are m traitors, where $m \leq (n-1)/3$, and the traitors can arbitrarily lie and send different messages to different generals.

A reliable algorithm for reaching consensus among the generals must satisfy the following conditions:

1. Agreement: all loyal generals must decide on the same value.
2. Integrity: if all loyal generals agree on a value, then any loyal general must adopt that value.
3. Validity: the value agreed upon by the loyal generals must be the value initially proposed by a loyal general.

There are various algorithms for solving the Byzantine Generals Problem, such as the Practical Byzantine Fault Tolerance (PBFT) algorithm and the Proof of Work (PoW) algorithm used in Bitcoin. These algorithms use complex cryptographic techniques and game theory to achieve consensus among the generals while accounting for the presence of traitors.

The Byzantine Generals Problem is a fundamental challenge facing blockchain networks that must be addressed in order to ensure the system's resilience to malicious behaviour or communication failures. Consensus algorithms are a key tool in achieving Byzantine fault-tolerance and enabling blockchain networks to operate in a decentralized manner.

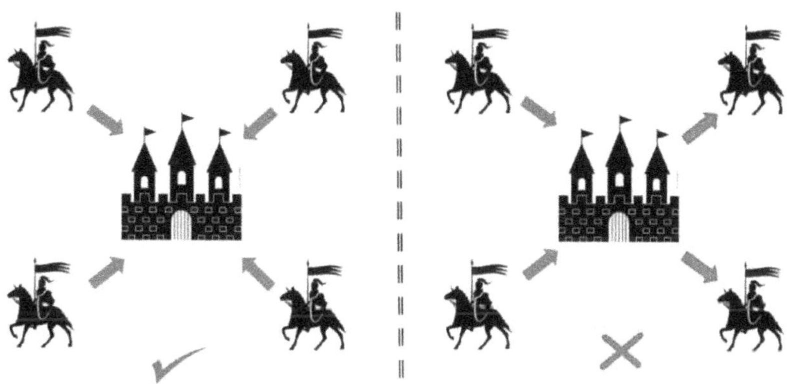

Figure 1.3. The Byzantine Generals Problem

(All generals are successful when attacking (left). When some retreat while others attack, they will be defeated (right))

Why should Blockchain Networks be Decentralized?

Operating a blockchain network independently is possible, but it may not be as efficient compared to more sophisticated alternatives. The true power of blockchain lies in its potential to be utilized in a decentralized environment, where all participants have equal access and control. In this setting, the blockchain is secure from malicious attacks and cannot be deleted, making it a reliable source of truth for everyone. Since all participants have access to the same information, no one has an unfair advantage, and the system operates on a level playing field. This characteristic of a decentralized blockchain network is particularly important in applications where trust is a key concern, such as in financial transactions or sensitive data sharing. By enabling a distributed network of users to maintain and validate the ledger, the blockchain becomes a robust and reliable tool that is resistant to tampering and malicious attacks.

Decentralized Peer-to-Peer Network in Blockchain Technology

In the dynamic blockchain ecosystem, the peer-to-peer (P2P) network plays a pivotal role, akin to the generals in the Byzantine Generals Problem. In contrast to traditional client-server networks that rely on a central server as an intermediary for data exchange, the P2P network empowers participants to engage in direct communication, sending and receiving information directly from their peers. This departure from centralized networks, which often have an administrator overseeing them, bestows upon the decentralized P2P network a heightened level of security and resilience against potential malicious attacks.

To grasp this concept more vividly, consider two scenarios: In the first scenario, participant A seeks to convey a message to participant F, but they are compelled to route it through a central server. In the second scenario, however, A and F are directly connected, without any intermediary intervention. In traditional networks, the server houses all the data that participants require, and any data request necessitates querying the central server. In stark contrast, the blockchain P2P network operates on a different principle – each participant maintains a copy of the entire database on their own computer. This unique feature enables them to access the blockchain even in the face of network disruptions or when a participant leaves the network.

When a new block is appended to the blockchain, the information is rapidly disseminated across the P2P network, allowing every participant to update their individual ledger copies. This decentralized architecture of the P2P network eliminates the presence of a single point of failure and renders it impossible for any single individual or entity to assert control over or manipulate the data. In essence, the P2P network serves as the backbone of the blockchain ecosystem, facilitating participant interaction and coordination without reliance on a central authority, a testament to the true essence of decentralization and security in the digital age.

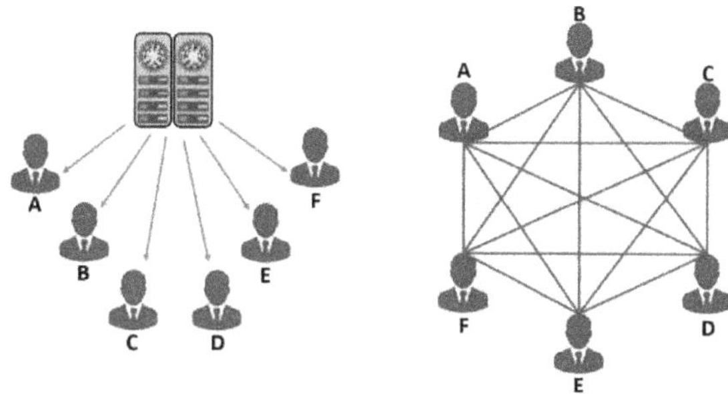

Figure 1.4. Centralized (left) vs. Decentralized Network (right).

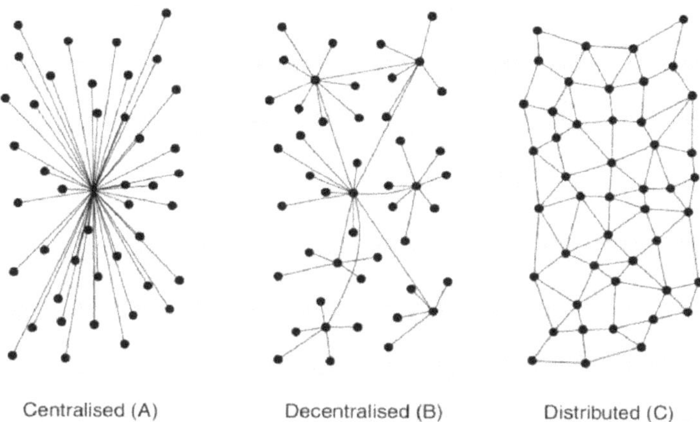

Figure 1.5. Centralized, Decentralized & Distributed Networks.

What are Blockchain Nodes?

In a blockchain network, nodes refer to the machines that are connected to the network, and their main function is to store copies of the blockchain and share data with other machines. Users do not have to worry about managing these processes manually. Instead, they can download and run the blockchain software, and everything else will be taken care of automatically. However, the term "node" can also encompass other users who interact with the network in any way. For instance, in the context of cryptocurrency, a wallet application on a smartphone is referred to as a light node.

Each node stores a complete copy of the blockchain, which is a record of all the transactions that have taken place on the network. Nodes communicate with each other to verify and validate new transactions, add them to the blockchain, and ensure that everyone has an up-to-date copy of the ledger. Nodes are critical to the functioning of the blockchain because they are responsible for maintaining the integrity of the network. Without nodes, the blockchain would not be able to function as a decentralized, secure, and transparent system for recording transactions.

Figure 1.6. Connected Nodes in a Blockchain Network

What is a Public & Private Blockchain Network?

The emergence of Bitcoin laid the foundation for the blockchain industry to grow into what it is today. With Bitcoin proving itself as a valid financial asset, innovators started exploring the potential of the underlying technology for other fields. This led to a wide range of use cases for blockchain beyond finance.

Bitcoin is an example of a public blockchain where anyone can view the transactions on it, and all that is required to participate is an Internet connection and the necessary software. Due to the absence of prerequisites for participation, it is referred to as a permission-less environment. However, there are other types of blockchains, namely private blockchains, that establish rules regarding who can view and interact with the blockchain network. These are referred to as permissioned networks.

Although private blockchains may seem unnecessary at first glance, they have important applications, particularly in enterprise settings. By setting up access rules, private blockchains can provide increased security, privacy, and accountability for businesses that require confidential data management.

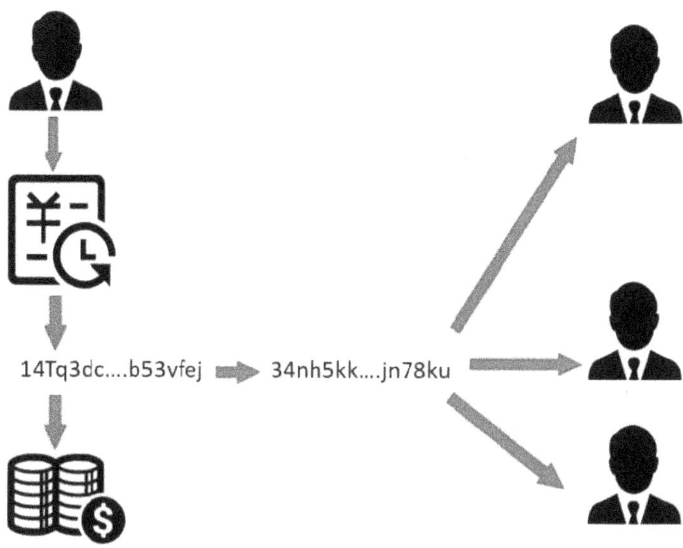

Figure 1.7: Representation of a Blockchain Transactions

Who is the Founder of Blockchain Technology?

The emergence of blockchain technology was marked by the launch of Bitcoin in 2009, which has now become the most widely recognized and popular blockchain network. However, the idea behind Bitcoin was not entirely novel, as its pseudonymous inventor Mr. Satoshi Nakamoto borrowed elements from earlier technologies and proposals. The technology of blockchain incorporates modern hashing algorithms and cryptography, which have been in existence for several decades before Bitcoin's release. It is fascinating to note that the concept of blockchain was first used in the early 1990s to timestamp documents, ensuring their integrity and immutability.

Advantages & Disadvantages of Blockchain Technology

Blockchain networks offer numerous solutions to the challenges faced by various industries, from finance to agriculture. Decentralized distributed networks have numerous advantages over traditional client-server models, but they also come with some trade-offs.

One significant benefit of blockchain networks is their ability to enable the transfer of funds without involving a middleman. This advantage was first mentioned in the Bitcoin white paper published in 2008 by Satoshi Nakamoto. Blockchains have since extended this capability to allow users to send all types of data, which removes counterparties, reduces risk, and decreases network fees.

A public blockchain network is permissionless and decentralized, meaning there is no barrier to entry because there is no central authority. Anyone with an internet connection can interact with other peers on the network. The high degree of censorship resistance is another vital feature of blockchains, making it almost impossible for even well-resourced attackers to compromise the network.

However, blockchain networks are not a one-size-fits-all solution to every technological challenge. While they are optimized for some benefits, they can be inefficient in other areas. One significant obstacle to the mass adoption of blockchain networks is their scalability issues. Since all participants must stay in sync, introducing new data too rapidly can overwhelm nodes and slow down the system. To maintain decentralization, blockchain developers often restrict the speed at which the network can update. This can lead to a long waiting period for users trying to make transactions if many others are trying to do the same.

Another disadvantage of decentralized blockchain networks is that they are not easily upgradeable. While software developers can add new features to their applications, in a network of tens of thousands of users, making modifications can be challenging. Nodes may refuse to interact with modified software if it is incompatible with other nodes in the network. This means that introducing changes to a blockchain network can be a long and complicated process that requires extensive coordination and discussion among ecosystem participants.

one of the most significant challenges faced by blockchain networks is scalability. The decentralized nature of these networks means that every node must keep a copy of the entire blockchain, and every transaction must be processed by every node to maintain consensus. This can lead to a bottleneck effect, where the system can only handle a limited number of transactions per second. Bitcoin, for example, has a capacity of about 7 transactions per second, while Visa can handle up to 24,000 transactions per second.

To address this issue, developers are working on various solutions, such as Sharding, which involves breaking the blockchain into smaller parts to process transactions in parallel, and off-chain solutions such as Lightning Network, which allows for faster and cheaper transactions without requiring every node to process them.

Another challenge faced by blockchain networks is the difficulty of making changes to the system. Since blockchains are decentralized, there is no central authority to make decisions or enforce changes. This means that any modifications must be agreed upon by the majority of network participants. This can lead to long and intensive discussions on forums and community meetings before changes are implemented. In some cases, forks may occur, where a portion of the network splits off to create a new blockchain with different rules and features.

Despite these challenges, the benefits of blockchain networks continue to drive innovation and adoption in various industries. In finance, for example, blockchain-based cryptocurrencies offer a decentralized alternative to traditional banking and payment systems. In agriculture, blockchain technology can be used to track the origin and journey of food products, promoting transparency and food safety. As the technology continues to evolve, it is likely that we will see more use cases and innovations in various industries. *(Ref. No.: 1- 4, 6, 8, 11)*

How New Blocks are Introduced to a Decentralized Blockchain Network through Consensus Algorithms?

The previous section covered a lot of ground, including the fact that nodes in a blockchain network are interconnected and store copies of transactions in the decentralized ledger. Nodes communicate information about transactions and new blocks to each other. However, you might be curious about how new blocks are added to the blockchain network. In a permissionless network where there is no central authority, a robust system is necessary to determine who can add blocks to the blockchain. This system should be designed in such a way that it is expensive for users to cheat, but they are rewarded for acting honestly. Since any rational participant will want to act in a way that is economically beneficial to them, the block creation process must be accessible to anyone. One way to ensure this is by using autonomous programmed protocols that require users to "put some skin in the game" by risking their own money. This enables them to participate in block creation, and if they successfully generate a legitimate block, they will receive a reward. However, if any participant attempts to cheat, the rest of the network will be able to detect this, and they will lose their stake. These mechanisms are known as consensus algorithms because they allow network participants to reach a consensus on which block should be added next.

Understanding the Proof of Work Consensus Algorithm & Block Rewards in Mining

The Proof of Work (PoW) consensus algorithm is widely used in mining as a means of solving a puzzle set by the network protocol. In this algorithm, users sacrifice computing strength to hash transactions and other data contained in a block. However, the hash must be below a certain number to be considered legitimate, which requires miners to continuously hash modified data until a valid solution is found. This process is computationally expensive and requires users to invest in mining hardware and electricity with the hope of receiving a block reward.

The block reward consists of two components: the block subsidy and transaction fees. The block subsidy generates new coins and accounts for the majority of the block reward. Transaction fees are the fees paid by users to have their transactions included in the block. The PoW mechanism is used in many blockchains, with Bitcoin being the first network to use this algorithm. When a miner submits a new block to the network, other nodes can easily verify its legitimacy by running it through a hash function. If the block is not valid, the miner will not receive a reward and will have wasted resources for nothing.

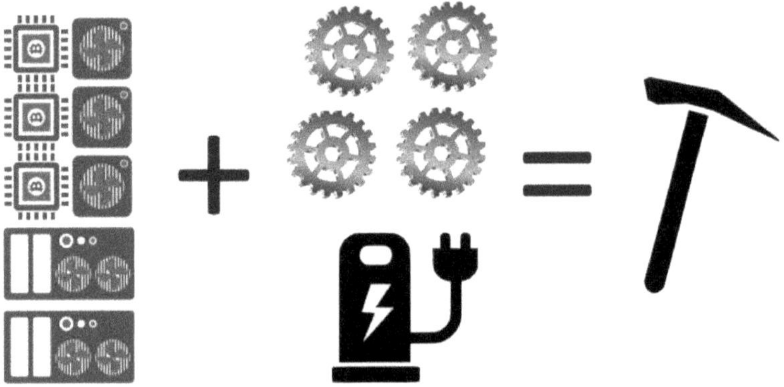

Figure 1.8. Mining in PoW Consensus Algorithm

Pros & Cons of Proof of Work in Cryptocurrency Mining:

Advantages of Proof of Work:

1. Tried and tested consensus algorithm that has secured billions of dollars' worth of value.
2. Permissionless, meaning anyone can participate in the mining competition or run a validating node.
3. Promotes decentralization, as miners compete against each other to create blocks, ensuring that the hash power is never controlled by a single party.
4. Hash rate measures the efficiency and overall performance of a mining machine.
5. Provides a secure and robust system to verify transactions and create new blocks.

6. Encourages miners to participate in the network, which increases its security.
7. Provides an incentive for miners to validate transactions and create new blocks through block rewards.
8. It is resistant to Sybil attacks, where a single entity creates multiple fake identities to control the network.

Disadvantages of Proof of Work:

1. Cryptocurrency mining consumes a large amount of electricity, which is seen as wasteful.
2. Mining puzzle difficulty increases as more miners join the network, requiring users to invest in better mining equipment, which can price out many miners.
3. Possibility of a 51% attack, where one miner or group of miners acquires the majority of the hash power, allowing them to potentially undo transactions and undermine the blockchain's security.
4. Mining can lead to centralization if a few large mining pools dominate the network.
5. High energy consumption contributes to environmental concerns and carbon footprint.
6. Expensive and complex mining equipment is needed to stay competitive.
7. Mining can lead to a concentration of wealth in the hands of a few successful miners.

Proof of Stake (PoS): A More Energy-Efficient Consensus Algorithm for Blockchain Technology

Proof of Stake (PoS) is a consensus algorithm used in blockchain technology that determines which nodes get to add new blocks to the blockchain. PoS works differently than the more widely used Proof of Work (PoW) algorithm. While PoW requires miners to perform complex mathematical computations to validate transactions, PoS uses a different approach.

In PoS, validators are selected to add new blocks to the blockchain based on the amount of cryptocurrency they have staked or held in a digital wallet. The idea behind PoS is that validators who have staked more cryptocurrency have a higher incentive to act in the network's best interest. If a validator acts dishonestly, they risk losing their staked cryptocurrency, which can be a significant financial loss.

Staking is the process by which validators hold a certain amount of cryptocurrency in a digital wallet to participate in block validation. The amount of cryptocurrency required to participate in staking varies from blockchain to blockchain. For example, the amount of cryptocurrency required to stake on Ethereum is 32 ETH.

Once a validator has staked the required amount of cryptocurrency, they are added to the pool of potential block validators. Validators are randomly selected to add new blocks to the blockchain based on a variety of factors, including the amount of cryptocurrency they have staked, how long they have been staking, and their reputation within the network.

When a validator is selected to add a new block to the blockchain, they receive a reward in the form of cryptocurrency. The amount of the reward varies from blockchain to blockchain and is often proportional to the amount of cryptocurrency the validator has staked.

One significant advantage of PoS over PoW is that it is less energy-intensive. PoW requires miners to perform complex mathematical computations to validate transactions, which requires a significant amount of energy. PoS, on the other hand, only requires validators to hold a certain amount of cryptocurrency in a digital wallet, which consumes much less energy.

However, PoS has its drawbacks as well. One concern is that it can lead to centralization, where a small number of validators hold a significant amount of cryptocurrency and have a disproportionate amount of power in the network. Another concern is that PoS may be vulnerable to so-called "nothing-at-stake" attacks, where validators can act dishonestly without risking anything.

Despite these concerns, PoS is gaining popularity in the blockchain community, and many new blockchain projects are choosing PoS over PoW. Ethereum, one of the most widely used blockchain platforms, is currently in the process of transitioning from PoW to PoS in its ETH 2.0 upgrade.

Advantages & Disadvantages of Proof of Stake (PoS) as a Consensus Algorithm for Blockchain Technology

Advantages:

1. Environmentally friendly: PoS has a much smaller carbon footprint compared to PoW, as it eliminates the need for resource-intensive hashing operations.

2. Faster transactions: With no need for arbitrary puzzles set by the protocol, proponents argue that PoS can increase transaction throughput.
3. Staking rewards & interest: Network rewards are paid directly to token holders instead of miners, providing opportunities for passive income through staking funds.
4. Decentralization: PoS incentivizes wider distribution of tokens, as validators must stake a significant portion of their funds to participate in block validation. This can lead to a more decentralized network compared to PoW.

Disadvantages:

1. Relatively untested: PoS protocols are yet to be tested on a large scale, and undiscovered vulnerabilities in implementation or crypto economics may exist.
2. Plutocracy: There are concerns that PoS encourages a "rich get richer" scheme, as validators with a larger stake tend to earn greater rewards.
3. Nothing-at-stake problem: Validators in PoS can work on multiple chains with little added costs, which might cause financial problems during a hard fork.
4. Risk of centralization: Despite its decentralized design, PoS can still lead to centralization if a small group of validators end up controlling the majority of the network's staked tokens.
5. Complexity: Compared to PoW, PoS can be more complex to implement, as it requires careful consideration of factors such as staking rules and penalties for bad behaviour.

Can Blockchain Transactions be Regressed?

Blockchain databases are designed to be extremely robust, with inherent properties that make it incredibly difficult for malicious entities to alter or delete data that has been added to the network. In the case of large blockchain networks like Bitcoin, it is virtually impossible to hack into the network and modify or delete a block. As a result, transactions on a blockchain can be considered to be set in stone forever.

However, it's important to note that there are many different implementations of blockchain, and the most significant difference among them is how they achieve consensus in the network. In some implementations, a relatively small group of participants can gain enough power in the network to effectively reverse transactions. This is particularly concerning for altcoins that operate on small networks with low hash rates due to limited mining competition.

Advantages:

1. Robustness: Blockchain databases are designed to be incredibly resilient, making it very difficult for malicious actors to alter or delete data.
2. Immutable: Once a transaction is added to the blockchain, it cannot be modified or deleted.
3. Security: The decentralized nature of blockchain networks makes them more secure, as there is no central point of failure.
4. Transparency: Blockchain networks are transparent, allowing anyone to view the data that has been added to the network.

Disadvantages:

1. Consensus Mechanisms: The way in which consensus is achieved in the network can affect the security of the blockchain, particularly in smaller networks.
2. Scalability: Some blockchain networks struggle with scalability as they grow larger, leading to slower transaction times and higher fees.
3. Energy Consumption: Proof of Work consensus mechanisms, used by some blockchain networks, consume significant amounts of energy, raising concerns about their environmental impact.

What is meant by Blockchain Scalability?

The term "scalability" in Blockchain refers to a network's ability to accommodate a growing number of participants. While blockchains have desirable properties such as decentralization, censorship resistance, and immutability, they come at a cost. Centralized databases work with substantially better speed and throughput, providing a smoother user experience. However, decentralized blockchain networks face challenges in this regard, making scalability an enormously debated subject amongst blockchain programmers for years.

Various solutions have been proposed, developed, or implemented to mitigate some of the performance drawbacks of typical blockchain networks. But at this point, there is not a clear best approach, and many unique solutions need to be trialed till there are more dependable solutions to the scalability problem. There is a fundamental question regarding scalability: Should the developers focus on enhancing the overall performance of the blockchain itself (on-chain scaling), or should they permit transactions to be executed without bloating the primary blockchain (off-chain scaling)? Both have their own benefits.

On-chain scaling solutions may involve reducing the size of transactions or optimizing how information is stored in blocks. Off-chain solutions involve batching transactions off of the primary blockchain and only adding them later. Some of the most impressive off-chain solutions are referred to as sidechains and payment channels. It is important to note that despite the scalability challenge, a blockchain database is engineered to be very robust, making it extraordinarily difficult for any malicious entity to eliminate or modify blockchain data after it's been registered into a network.

Why should a Blockchain Network Scale?

When comparing blockchain networks to their centralized counterparts, it is necessary to ensure that the blockchain systems perform at least as well as the centralized systems. However, it is more realistic to aim for blockchain systems to perform even better to incentivize blockchain developers and end-users to switch to blockchain-based platforms and applications.

To achieve this goal, blockchain systems need to be faster, more affordable, and simpler to use for both technical and non-technical users compared to established centralized systems. This is not an easy task to accomplish while maintaining the key defining characteristics of blockchains such as decentralization, immutability, and censorship-resistance.

Improving the performance of blockchain systems can involve on-chain scaling solutions such as reducing transaction size or optimizing how information is stored in blocks. Off-chain scaling solutions, such as sidechains and payment channels, can also be implemented to allow for faster and more affordable transactions. However, striking a balance between performance and maintaining the core principles of blockchain remains a challenge that requires ongoing development and experimentation. Ultimately, the success of blockchain-based systems will depend on their ability to provide a user-friendly and reliable alternative to centralized systems.

What is a Blockchain Fork?

Just like any other software application, blockchains also require updates to fix any issues, introduce new policies or remove outdated ones. Since most blockchain software programs are open-source, anyone can suggest new updates to be incorporated into the program that manages the entire network. It is important to note that blockchains are decentralized distributed networks, which means that there is no central authority or organization responsible for making decisions.

When an upgrade is made to the blockchain software, it needs to be efficiently communicated to and enforced by the hundreds of nodes spread out globally. However, there may be instances where participants in the network do not agree on a specific upgrade to be implemented. In such situations, there is not an established decision-making process within the organization to resolve the disagreement, which can lead to the network splitting into two branches: soft and hard forks.

A soft fork happens when the software upgrade is backward-compatible, which means that nodes that do not upgrade can still continue to operate on the original blockchain. On the other hand, a hard fork occurs when the new software upgrade is not compatible with the original version, resulting in a permanent split of the blockchain network into two different versions. In such cases, the network's consensus mechanism is disrupted, and each branch operates as a separate and independent blockchain network.

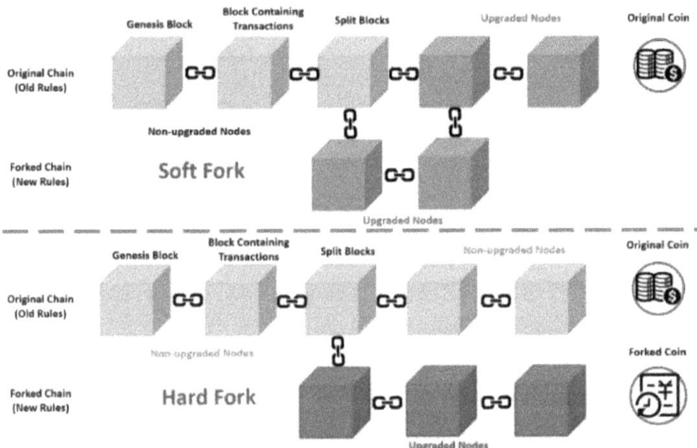

Figure 1.9: Soft & Hard Fork in a Network

Understanding Soft Forks & Hard Forks in Blockchain Upgrades

When there is a consensus among network participants about how a blockchain software upgrade should be implemented, it is a straightforward process. This type of upgrade is called a soft fork. With a soft fork, the new software is backward-compatible, meaning that nodes that have not upgraded can still communicate with nodes that have. However, over time, it is expected that most nodes will upgrade to the new software. A soft fork does not result in a permanent split of the blockchain network. Instead, it is a temporary situation where nodes that have not upgraded to the latest version are still able to function on the network. Soft forks are commonly used to introduce new features, fix bugs or vulnerabilities, or improve network performance.

On the other hand, a hard fork is a more complex process. Once implemented, the new rules may be incompatible with the old rules, leading to a permanent split in the blockchain network. This means that nodes running the new rules cannot communicate with nodes running the old rules. The result is two separate chains: one running the old software and the other running the new software. Each chain operates independently, and they can have different transaction histories and balances. At the time of the hard fork, the balances of the blockchain's native unit are cloned from the old network to the new one. This means that if you had a balance on the old chain at the time of the fork, you will have an equal balance on the new chain as well. Hard forks are typically used to introduce major changes to the blockchain network, such as changing the consensus mechanism, modifying the transaction validation process, or altering the block size limit.

Exploring the Applications of Blockchain Technology

The potential applications of blockchain technology are vast and varied.

Here are a few examples:

Supply Chain Management:

Efficient supply chain management stands as a cornerstone for the smooth operation of numerous businesses, encompassing the intricate process of overseeing goods from the initial supplier to the final consumer. Historically, orchestrating the myriad stakeholders within a supply chain has presented formidable challenges. Nevertheless, blockchain technology emerges as a transformative force, ushering in newfound transparency across a wide spectrum of industries. Through the establishment of an interconnected supply chain ecosystem centered on an immutable, resilient, and dependable database, businesses stand poised to enhance their operational efficiency, thereby gaining a potent competitive advantage. One of blockchain's standout capabilities lies in its capacity to meticulously trace products along the entire journey, from their origin in the manufacturer's facilities to their ultimate destination with the end consumer. This comprehensive visibility affords a profound insight into critical aspects of the goods, including their provenance, quality, and authenticity. By harnessing blockchain's prowess, businesses can transcend the limitations of traditional supply chain management, ushering in an era of unprecedented reliability and trust in the movement of goods.

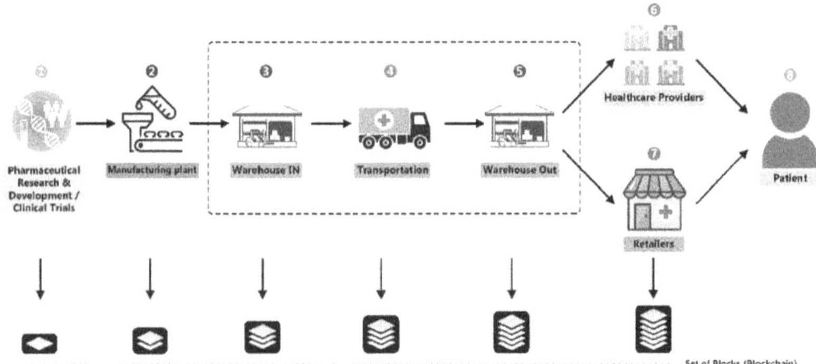

Figure 1.10: Blockchain for supply Chain Management

How Blockchain Technology Can Revolutionize the Gaming Industry

The gaming industry is one of the largest entertainment industries globally and is well-positioned to benefit significantly from blockchain technology. Currently, gamers are typically at the mercy of game developers, who own and control the games they play. In most online games, gamers have to rely on the developers' server space and comply with their ever-changing rules. However, blockchain technology can help to decentralize the ownership, management, and maintenance of online games, providing gamers with more control over the games they play.

One of the most significant challenges in the gaming industry is the inability to own and trade in-game items outside of the game titles. Blockchain technology offers a potential solution by enabling the creation of unique, tradable in-game items known as crypto-collectibles. These items would be stored on a blockchain network and would have real-world value, which could be traded or sold outside of the game. This would create new revenue streams for gamers and developers and provide gamers with actual ownership of their in-game items.

Moreover, blockchain technology can also help to reduce fraudulent activities such as cheating and hacking, as transactions on the blockchain network are immutable and transparent. This would enhance the overall gaming experience for players and help to foster trust between gamers and developers.

Figure 1.11: Blockchain in the Gaming Industry

Blockchain Technology for Healthcare:

Reliably storing medical data is critical for any healthcare network, but the dependence on centralized servers leaves sensitive data vulnerable to security breaches. However, the transparency and security of blockchain technology make it an ideal platform for storing medical data. By cryptographically securing sensitive user information on a blockchain network, patients can maintain their privacy while being able to efficiently and quickly share their medical data with any healthcare institution.

In a currently fragmented healthcare system, if all participants could access a secure, global database, data flow could be significantly faster between them. The use of blockchain technology in healthcare could create a more seamless and efficient exchange of medical data, improving patient outcomes and reducing costs. Moreover, the immutable and transparent nature of blockchain transactions can enhance the security and integrity of medical records, reducing the risk of fraud and errors. *(Ref. No.: 5- 10)*

Blockchain Remittance:

Conventional banking can make sending money across the world a hassle due to the complicated network of intermediaries involved, resulting in high fees and unreliable transaction times. However, blockchain technology can eliminate this ecosystem of middlemen, allowing for cheap and fast transfers globally. While blockchains may sacrifice overall performance for some of their other beneficial properties, several blockchain-based projects are utilizing the technology to enable affordable and near-instant transactions. Additionally, blockchain technology can provide a secure way to manage digital identities on the internet. Currently, a significant amount of personal data is stored on centralized servers and analyzed by advanced autonomous algorithms without the owner's knowledge or consent. Blockchain technology can enable users to take ownership of their data and reveal it selectively to third parties only when necessary, ensuring a smoother online experience without compromising privacy.

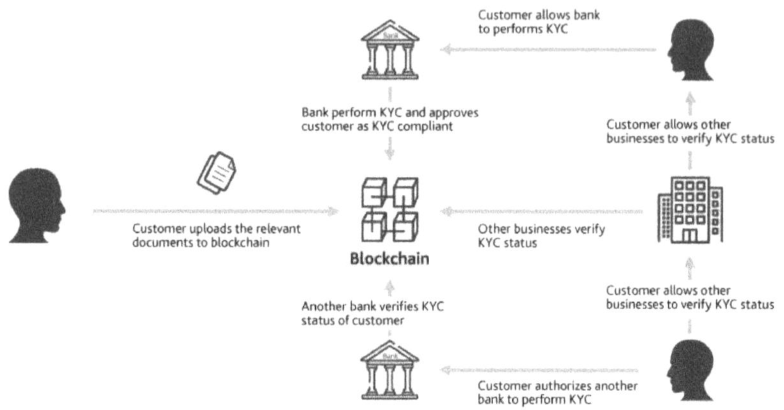

Figure 1.12: Blockchain Remittance

Blockchain & the Internet of Things (IoT)

Mining, manufacturing, and logistics are industries that rely heavily on the supply chain for their success. The use of IoT devices in these industries allows for greater monitoring and optimization of the supply chain. Blockchain technology can further enhance this optimization by adding a layer of transparency and security to the supply chain process.

In the mining industry, IoT sensors can be used to monitor the quality of ore and minerals being extracted from mines. This data can then be recorded and stored on a blockchain, creating an immutable record of the quality of the minerals being extracted. This can provide greater transparency for buyers and investors, and help to combat issues such as fraud and corruption in the industry. In addition, blockchain and IoT can be used to create a more efficient and secure supply chain for the mining industry. This includes tracking the movement of minerals from the mine to the refinery, and then to the manufacturer. By using IoT sensors and blockchain technology, each step of the process can be recorded and monitored, ensuring that the minerals are being transported securely and efficiently.

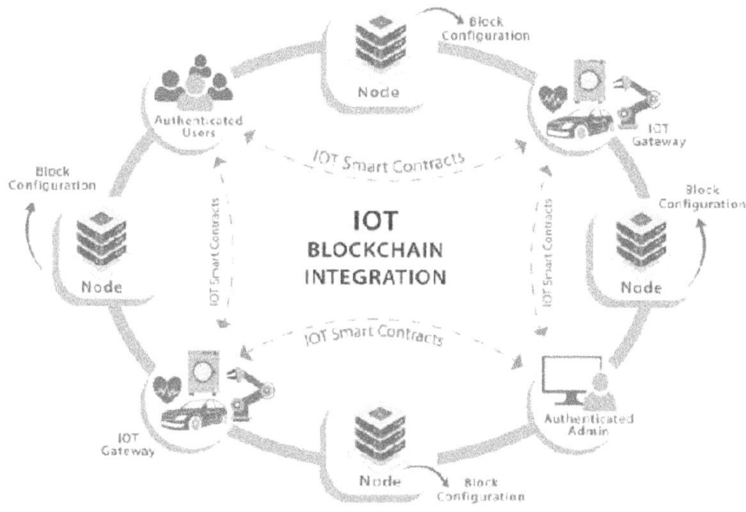

Figure 1.13: Blockchain & IoT

Blockchain for Governance

Distributed networks possess a unique capability – the ability to establish and enforce their own set of rules using computer code. Unsurprisingly, blockchain networks are no exception, holding the potential to revolutionize governance processes at local, national, or even global levels by eliminating the need for intermediaries. This transformative potential extends to addressing a longstanding challenge in open-source development settings: the absence of a reliable mechanism for equitably distributing funding and resources. Blockchain governance introduces a new paradigm where all contributors can actively engage in decision-making, offering a transparent and accountable framework for policy implementation. One of the standout advantages of blockchain governance lies in its ability to incentivize participation and contributions from all network stakeholders. By leveraging cryptocurrencies and tokens, blockchain governance facilitates a fair and inclusive distribution of rewards and funding, fostering an environment where collaboration and innovation flourish.

This approach holds the promise of galvanizing individuals to actively participate in the governance process, driven by the assurance that their contributions will be duly recognized and rewarded, ultimately fostering a more vibrant and innovative ecosystem.

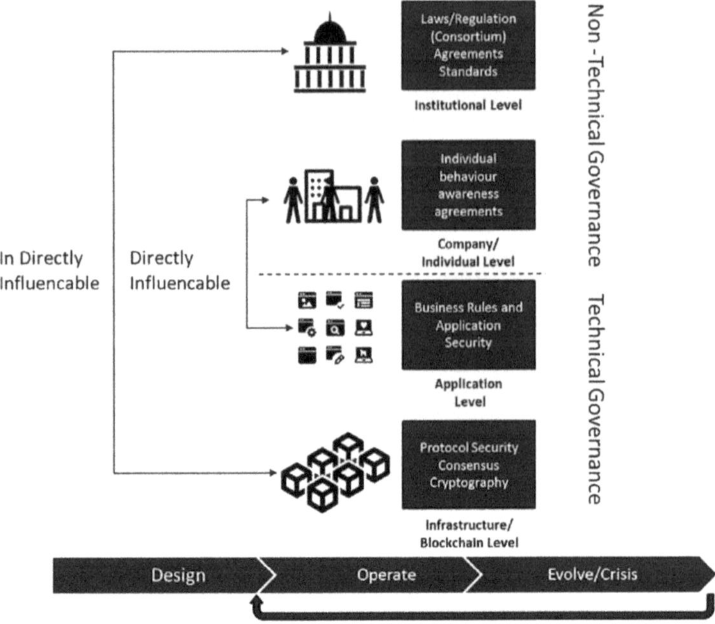

Figure 1.14: Blockchain for Governance

Blockchain for Charity:

Blockchain technology is increasingly being used to enable more transparent and secure donations to charity organizations. One of the main issues faced by charities is the difficulty in tracking donated funds and ensuring that they are used as intended. The use of blockchain technology can provide a solution to this problem by creating a decentralized, transparent ledger that can be easily accessed by anyone. This increases transparency and trust in the donation process and helps to eliminate fraud and corruption. Additionally, blockchain-based charity platforms can offer lower fees and faster transactions, enabling charities to receive funds more quickly and efficiently.

Blockchain for Speculation:

Blockchain technology has created a new era of financial speculation, with its ability to facilitate frictionless transfers between exchanges and provide non-custodial trading solutions. It has also enabled the creation of a growing ecosystem of derivatives products that have made it an attractive playground for all types of speculators. The use of blockchain technology ensures more transparent, secure, and efficient trading, which has led to the creation of a new asset class. Some experts even believe that the global speculative markets will be tokenized on the blockchain, creating a more decentralized and accessible financial system.

Crowdfunding with Blockchain:

Blockchain technology is increasingly being used to facilitate peer-to-peer crowdfunding, which has the potential to revolutionize the way we fundraise for new projects and startups. Crowdfunding platforms have been around for nearly a decade and have been successful in raising funds for many innovative projects. However, they often take a substantial portion of the funds raised as commission fees, and they have their own rules for facilitating settlements among different participants. Blockchain technology can provide a better solution by using smart contracts to create a more automated and secure crowdfunding process. This allows for better-defined agreements and lower fees. Additionally, blockchain-based crowdfunding platforms like Initial Coin Offerings (ICOs) and Initial Exchange Offerings (IEOs) have gained popularity as a way for investors to raise funds in exchange for tokens that represent a share in the future success of a project.

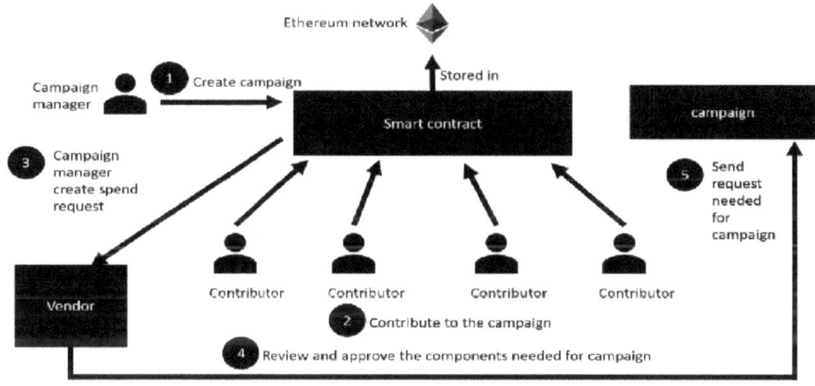

Figure 1.15: Crowdfunding with Blockchain

Blockchain & Distributed File Systems:

Decentralized document storage has many advantages compared to centralized alternatives. Traditional centralized cloud-based servers are vulnerable to malicious cyber-attacks and data loss, while users may face accessibility issues due to censorship from centralized servers. In contrast, blockchain-based data storage solutions work just like other cloud storage solutions. However, when a user uploads a file to a blockchain-based storage unit, it is distributed and replicated throughout numerous nodes. This ensures that the data is not only stored in a decentralized manner but also replicated multiple times, providing increased resilience to data loss.

Furthermore, these decentralized storage systems are incentivized to maintain the integrity of the network by offering storage and bandwidth to the network. If they do not follow the regulations or fail to store and serve files, they are economically punished. Incentives and penalties help to ensure the longevity and security of the network. Users can think of this kind of network as being similar to Bitcoin, with the primary purpose of the network being to enable censorship-resistant, decentralized document storage facilities.

The Inter Planetary File System (IPFS) is an example of an open-source protocol that is paving the way for this new, more permanent, and distributed web. While IPFS is not a blockchain network, it applies some concepts of blockchain technology to enhance security and efficiency. By leveraging the blockchain's strengths, IPFS aims to provide a more resilient and censorship-resistant decentralized storage system. Overall, distributed file systems based on blockchain technology have the potential to provide a more secure, decentralized, and resilient alternative to centralized storage systems.

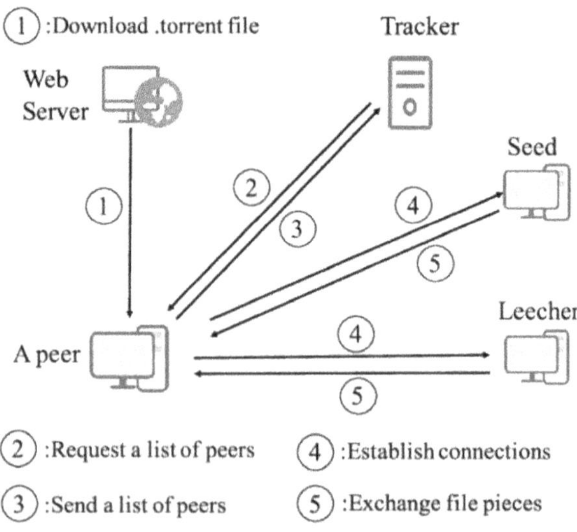

Figure 1.16: Mechanism of file sharing in BitTorrent

(Mechanism of file sharing in BitTorrent, which can help to understand the Blockchain-based distributed file system IPFS)

Blockchain & Money Transfer

In essence, a blockchain is a decentralized digital ledger that consists of a collection of data records. The data is organized into blocks, which are arranged chronologically and secured using cryptography. The first iteration of a blockchain network was developed in the early 1990s when Stuart Haber, a computer scientist, and W. Scott Stornetta, a physicist, used cryptographic techniques to secure digital files in a series of blocks and prevent data tampering. Their work inspired many other computer scientists, engineers, and cryptography enthusiasts, ultimately leading to the creation of Bitcoin in 2008 as the first decentralized digital cash system, or cryptocurrency. Although blockchain technology predates cryptocurrencies, it was not until the introduction of Bitcoin that its full potential was realized. Since then, interest in blockchain technology has been steadily growing, and cryptocurrencies have gained mainstream recognition. Although blockchain technology is primarily used for recording cryptocurrency transactions, it can be used for various types of digital data and applied to a broad range of use cases. The oldest, safest, and most extensive blockchain network is Bitcoin's, which was meticulously designed and developed with a balanced combination of cryptography and game theory.

How Blockchain Transactions Occur in a Decentralized Network

A blockchain network, in the context of cryptocurrencies, is essentially a distributed database that maintains a chain of blocks that store a list of previously confirmed transactions. The network consists of numerous nodes, which are computer systems spread across the globe. Since each participant in the network maintains a copy of the blockchain data and communicates with others to confirm that they are all on the same page or block, the network operates as a decentralized database or ledger. The decentralized nature of the network makes it impossible for any single entity to control or manipulate the data. This decentralized approach to transactions and data management is what makes Bitcoin a unique and ground-breaking innovation. It is a decentralized digital currency that is not subject to borders or censorship. In addition, most blockchain networks are considered trustless since they do not require any form of trust. Bitcoin's network is maintained and controlled by a community of participants acting as nodes, and there is no single authority in control. Mining is an integral part of almost every blockchain network and is based on hashing algorithms. Bitcoin uses the SHA-256 algorithm (Secure Hash Algorithm 256 bits). This algorithm takes an input of any length and generates an output of a fixed length (64 characters or 256 bits). The output produced by the algorithm is referred to as a "hash," which is constantly fabricated from 64 characters. One of the main features of hash functions is their deterministic nature. This means that the same input will always produce the same output, regardless of how many times the process is repeated. However, if a small update is made to the input, the output will change entirely. This one-way function property of hash functions makes it almost impossible to reverse calculate the input from the output. A malicious actor can only guess what the input is, but the probability of guessing it correctly is exceptionally low. As a result, Bitcoin's blockchain is highly secure and is one of the primary reasons why it has gained immense popularity over the years. Having gained an understanding of how the algorithm operates, we can now examine how a blockchain functions using a straightforward example of a basic transaction. Suppose that an individual named "Mr. A" residing in India wishes to transfer funds to another individual named "Mr. B" residing in Israel. The traditional way to perform such a transaction would necessitate the involvement of a third-party intermediary, as follows:

Figure 1.17: Block Chain Simplifies Money Transfer

In the traditional system, when a person "Mr. A" from India wants to transfer money to a person "Mr. B" in Israel, they have to rely on a third-party service, such as a bank or a money transfer company, to facilitate the transaction. In this process, Mr. A initiate the transfer request with the third-party service, and the service verifies the identity of Mr. B as a person in Israel. Once the verification is complete, the third-party service deducts a commission fee and transfers the money from Mr. A's bank account in India to Mr. B's bank account in Israel, which can take more than three days.

However, the emergence of blockchain technology is disrupting this traditional money transfer system. In a blockchain-based system, Mr. A and Mr. B can transact directly with each other without the need for a third-party intermediary. The transfer is executed through a decentralized network of nodes that verify the transaction and record it on a secure and tamper-proof digital ledger.

In a blockchain-based system, Mr. A and Mr. B can also enjoy faster transaction speeds and lower transaction fees since there is no intermediary to deduct commission fees. Additionally, the use of smart contracts in blockchain can enable the automation of the entire transaction process, making it more efficient and transparent.

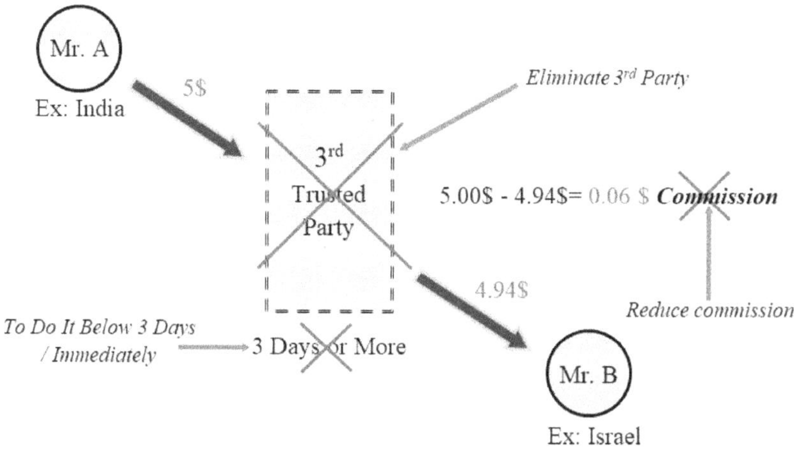

Figure 1.18: Blockchain Tries to Eliminate the 3rd Party

(Reduce Intermediate Commission, Reduce Time to Transfer Money)

1. Blockchain technology aims to eliminate the need for a third-party intermediary in the money transfer process.
2. It seeks to increase the speed of money transfer, making it immediate instead of taking several days.
3. Blockchain technology also strives to make money transfers cheaper by reducing or eliminating the commission fees charged by third-party intermediaries.
4. The decentralized nature of blockchain allows for peer-to-peer transactions, making it possible for individuals to transact directly with each other without the need for intermediaries.
5. The use of smart contracts in blockchain can automate the transaction process, making it more efficient and transparent.

The concept of Open Ledger

Assuming there is a network of four individuals - Mr. A, Mr. B, Mr. C, and Mr. D - who want to exchange money with each other's bank accounts, let us see how a blockchain-based system would work from the genesis:

1. At the genesis, Mr. A has $10 in their account.
2. When Mr. A want to transfer $5 to Mr. B, a new transaction is added to the blockchain ledger, recording the transfer as "Mr. A -> Mr. B; 5$". This transaction is then linked to the existing transactions on the blockchain ledger.
3. When Mr. B wants to transfer $3 to Mr. D, another transaction is added to the blockchain ledger, recording the transfer as "Mr. B -> Mr. D; 3$". This transaction is also linked to the existing transactions on the blockchain ledger.
4. Finally, if Mr. D wants to transfer $1 to Mr. C, the transaction is added to the blockchain ledger as "Mr. D -> Mr. C; 1$". This transaction is also linked to the existing transactions on the blockchain ledger.

In this way, the blockchain-based system enables peer-to-peer transactions without the need for a third-party intermediary. The transactions are recorded on the decentralized ledger, which is secure, transparent, and tamper-proof, ensuring the integrity of the transactions. Additionally, the use of smart contracts can automate the transaction process, making it more efficient and reducing the need for manual intervention. Overall, a blockchain-based system provides a faster, cheaper, and more secure alternative to traditional money transfer methods.

The concept of the "Open Ledger" is essentially a sequence of transactions that forms the basis of the "Blockchain" network. This ledger is open to the public, which means that anyone can access and view the ledger. This allows individuals to monitor the money, check how much each person has, and determine the validity of transactions.

For instance, if Mr. A attempt to transfer $15 to Mr. C's account, the network will immediately detect that the transaction is invalid. This is because Mr. A only had $10 at the genesis and transferred $5 to Mr. B. The blockchain's open ledger ensures that any invalid transaction is rejected, preventing it from being added to the existing chain.

Although the open ledger provides a centralized place to manage the ledger, it creates a new problem of centralized control, which goes against the goal of the

Blockchain network. This leads to the next principle of the Blockchain network, the "Notion of Distributed Ledger."

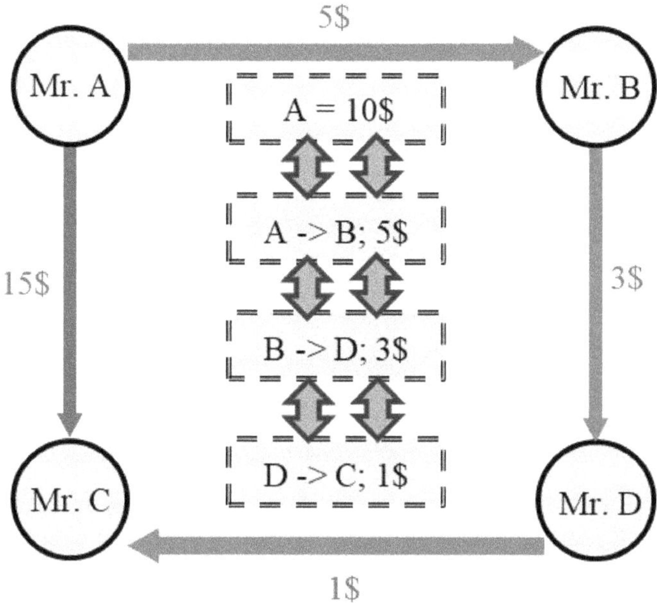

Figure 1.19: Concept of Open Ledger

The Concept of Distributed Ledger

The concept of an "Open Ledger" is fundamental to understanding how blockchain technology operates. It refers to a ledger, or a record of transactions, that is completely transparent and accessible to the public. In an open ledger system, anyone can view the entire transaction history, including the amounts of money or assets involved and the parties involved in each transaction. This transparency is a core feature of blockchain technology and is what distinguishes it from traditional centralized systems. However, it's important to note that while the concept of an open ledger promotes transparency, it doesn't necessarily guarantee decentralization. This is where the idea of a "Distributed Ledger" comes into play. Here is a more detailed explanation of the "Open Ledger" concept in blockchain:

Transparency: In a blockchain network, every transaction is recorded in a series of blocks, and these blocks are linked together in a chain.

Validation: The open ledger allows participants to validate transactions collectively.

Security: The openness of the ledger contributes to the security of the blockchain.

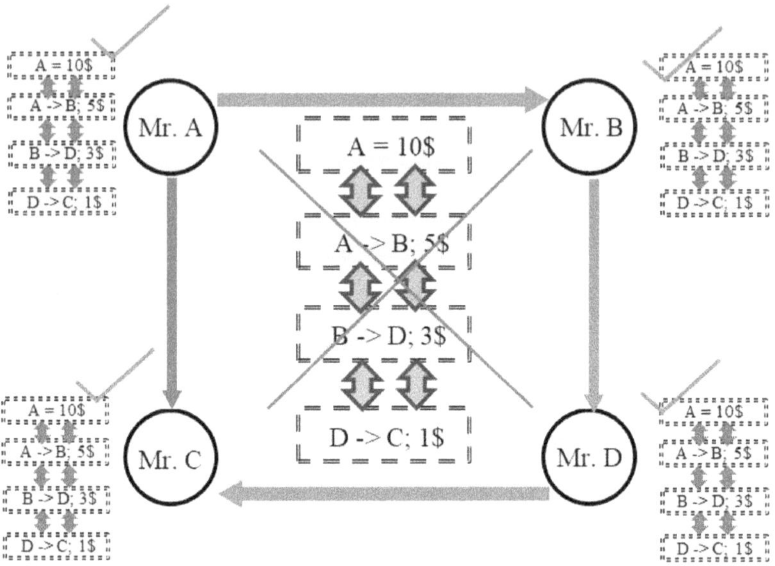

Figure 1.20: Distributed Open Ledger

Distributed Ledger Technology (DLT) refers to the distribution of the "Open Ledger" across all the nodes in a network. For example, each person in the network can have their own personal copy of the ledger, including a chain of events from past and present. When everyone has their own copy of the ledger, the centralized Open Ledger becomes distributed and useless. However, using DLT creates another problem. With multiple identical copies of the ledger in the network, the interface needs to ensure that all the ledgers are synchronized accurately with each other so that all the participants in the network can see the same ledger. To achieve this, special nodes in the Blockchain network known as "Miners" are responsible for validating and updating transactions in the ledger. The notion of a distributed ledger is a key principle of blockchain technology because it eliminates the need for a centralized entity to manage and control the ledger. This decentralization enhances security, resiliency, and trust in the system since no single point of failure or control exists. Transactions are validated and recorded by a network of participants who collectively maintain the ledger, ensuring its integrity and reliability.

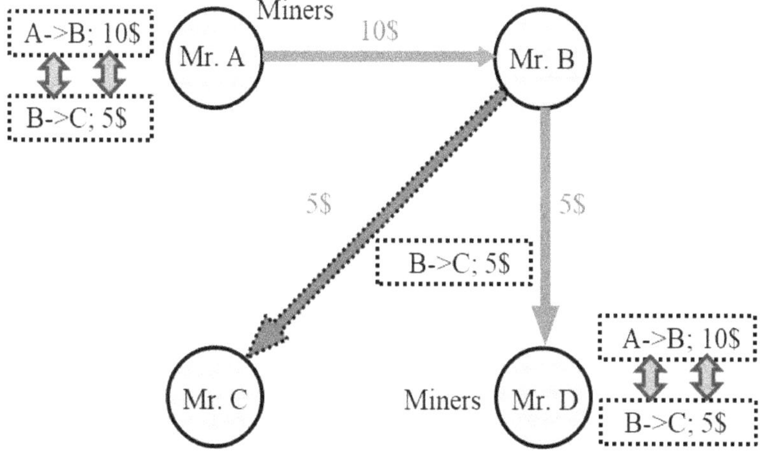

Figure 1.21: Concept of Miners

For instance, if "Mr. B" wants to transfer 5$ to "Mr. C", he will broadcast this transaction to the network for everyone to see. This becomes an invalidated transaction, and it will not be updated in the ledger until it is validated by a Miner. Miners compete to solve complex mathematical problems to earn a reward. In this example, "Mr. A" and "Mr. D" are the miners, and they will work together to validate the transaction and update the ledger. Once validated, the transaction will be added to the ledger chain and distributed across the network for all participants to see. Miners in a Blockchain network engage in a competition to verify transactions, which involves solving a complex mathematical puzzle and adding the validated transaction to the ledger. The miner who solves the puzzle first is rewarded with a financial incentive, typically in the form of the cryptocurrency used in that particular Blockchain network. In the case of a Bitcoin-based Blockchain, the reward would be in the form of Bitcoin.

Allow me to explain what it means to win the competition in the blockchain network. To be the first to validate a new transaction and add it to the ledger, a miner must do the following:

1. Validate the new transaction by checking if the sender has sufficient funds to make the transfer. Since the ledger is open, this is a simple task.
2. Find a unique key that will enable the miners to lock the new transaction to the previous transaction and update the ledger.

3. Solve a complex mathematical puzzle using computational power and time. The search for the key is random, and the miner that solves it first and adds the transaction to the ledger receives a financial reward. In the case of Bitcoin-based blockchains, this reward is in the form of Bitcoin.

How Ledgers are synchronized across the network.

Assuming that a miner, "Mr. D," has successfully solved the complex mathematical puzzle and added the transaction to his ledger, the next step is to make sure that all the other nodes in the network are aware of this update. "Mr. D" will broadcast the new transaction and its security key to the entire network so that other miners can download the validated ledger and add it to their copies.

Once all the miners have updated their ledgers, they will verify that the transaction has been successfully added and validated by checking the security key. At this point, any miner who was attempting to solve the same transaction will realize that it has already been validated, and there is no point in continuing to work on it. Instead, they will look for other transactions to validate and earn a reward for their efforts.

It is worth noting that this process of adding transactions to the ledger is ongoing and continuous. As new transactions are added to the network, miners will continue to compete to solve the mathematical puzzles and validate them. This ensures that the ledger is always up-to-date and accurate, and all nodes in the network have the same copy of the ledger.

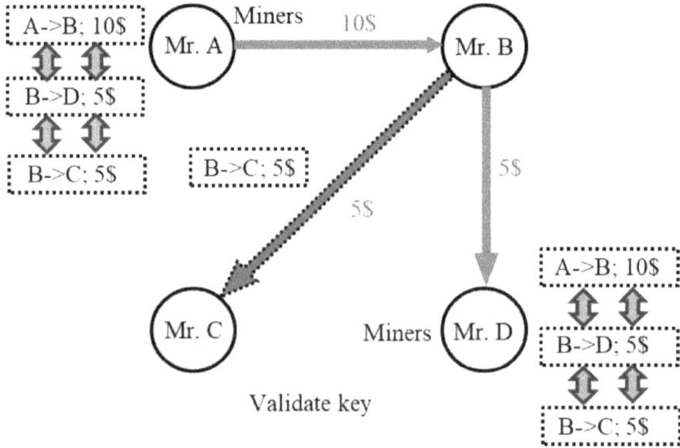

Figure 1.22: Synchronizing the Ledger

Benefits of Distributed Ledgers

1. Highly transparent, secure, tamper-proof, and immutable
2. The need for a third party is eliminated
3. Inherently decentralized
4. Highly transparent

Distributed Ledger Technology (DLT) offers various benefits, some of which include transparency, security, immutability, and decentralization. These benefits are achieved by creating multiple identical copies of the ledger and distributing them across all nodes in the network, allowing each participant to hold a personal copy of the ledger. This eliminates the need for a centralized third party to manage the ledger, as every participant can hold their own copy of the ledger and make updates to it.

Furthermore, Distributed Ledger Technology has great potential to transform the way governments, institutions, and corporations operate. It can be used by governments to collect taxes, issue passports, and maintain records of land registries, licenses, social security benefits, and voting procedures. The transparency and security of the ledger make it ideal for these purposes, as it ensures that records cannot be tampered with or manipulated.

Despite its potential benefits, Distributed Ledger Technology is still in its early stages of adoption, and there is still much to explore in terms of how best to use it. However, one thing is clear: the future of ledgers is likely to be decentralized. With its ability to provide transparency, security, and decentralization, Distributed Ledger Technology has the potential to revolutionize the way we keep records and conduct transactions.

Understanding the Security Mechanisms of Blockchain Technology

Blockchain technology is a secure way of managing data through the use of cryptographic techniques and mathematical models. It is commonly associated with cryptocurrencies, as it prevents duplication or destruction of decentralized digital cash. However, the potential applications of blockchain technology extend beyond cryptocurrencies. One of the most important aspects of blockchain technology is its immutability and security. The technology provides robust security mechanisms that make it difficult for attackers to tamper with the data stored on the blockchain. These mechanisms include cryptographic hashes, digital signatures, and consensus algorithms. The use of blockchain technology is being explored in various contexts, including charity donations, medical databases, and supply chain management. Blockchain's ability to create an immutable and transparent record of transactions can help in reducing fraud, increasing efficiency, and improving accountability. Despite its many benefits, blockchain security is still a complex and evolving subject. It is crucial to understand the fundamental principles and mechanisms that provide robust security to these revolutionary systems. In conclusion, the security mechanisms of blockchain technology are essential to its success in providing secure and tamper-proof data management solutions.

The Principles of Immutability & Consensus

The bedrock of security in a blockchain network hinges on two paramount principles: consensus and immutability. Consensus embodies the ability of nodes within a distributed blockchain network to arrive at a unanimous decision regarding the network's accurate state and the validity of transactions. This consensus process is typically governed by consensus algorithms. In contrast, immutability encompasses the blockchain's inherent capacity to thwart any alterations to transactions once they have undergone verification and become part of the immutable blocks. While these transactions often pertain to cryptocurrency transfers, they also encompass non-financial applications like decentralized file sharing. Collectively, consensus and immutability form the robust foundation that underpins data security within blockchain networks. Consensus algorithms ensure that network rules are diligently adhered to, fostering unanimity among all participants regarding the network's present status. Simultaneously, immutability safeguards the integrity of data and transaction information, fortifying their resistance to unauthorized modifications following validation and inclusion within new blocks.

The intertwined dynamics of consensus and immutability are pivotal attributes of blockchain networks, actively safeguarding the security and integrity of both data and transaction information. They operate in concert to avert illicit alterations to the network and to ascertain unanimous agreement among all participants regarding its current condition.

The Immutable Nature of Blockchain: Why Old Blocks Cannot be Manipulated or Deleted

The security of blockchains relies heavily on the use of cryptography. In particular, advanced cryptographic hashing functions are essential to ensure data security. Hashing involves inputting data of any size into a pre-programmed algorithm, which returns an output of fixed length called a hash. Regardless of the size of the input, the output will always be the same length, and any change to the input data will result in a different hash output. Hashes are used as identifiers for data blocks within blockchains. The hash function of each block is generated before syncing to the hash of the previous block, which creates a sequence of connected blocks that cannot be altered.

These hash identifiers play a crucial role in ensuring that the blockchain network is secure and immutable. The hashing algorithm is also used in consensus algorithms that validate transactions during block synchronization in a network. For instance, the Bitcoin blockchain employs the Proof of Work (PoW) consensus algorithm, which utilizes a hash function called SHA-256. Secure Hash Algorithms (SHA-256) generate a hash that is 256 bits or 64 characters long.

Apart from safeguarding secure transactions on a decentralized ledger, cryptography also plays a vital role in securing cryptocurrency wallets. Asymmetric or public-key cryptography is used to generate paired public and private keys that allow users to acquire and send payments securely. Private keys are used to generate digital signatures for transactions, which authenticate the ownership of the money being sent. Asymmetric cryptography ensures that only the private key holder can access funds stored in a cryptocurrency wallet, thus maintaining those funds secure until the owner decides to spend them, as long as the private key is not shared or compromised.

Crypt Economics:

Besides cryptography, another critical concept in maintaining blockchain network security is crypto-economics, which is based on game theory. Game theory is a mathematical model that utilizes rational actors and predefined policies and rewards to simulate decision-making in various situations. Crypto-economics focuses specifically on modelling the behaviour of decentralized nodes on a distributed blockchain network. In short, it is the study of the economics of blockchain network protocols and the potential consequences that their design may have on participants' behaviour.

The security of blockchain systems through crypto-economics is primarily based on the notion that these systems offer more incentives for nodes to behave positively than to adopt malicious or faulty behaviours. This is evident in the Proof of Work (PoW) consensus algorithm used in Bitcoin mining. Satoshi Nakamoto designed Bitcoin mining to be a resource-intensive and expensive process. PoW mining requires a significant investment of time and money, making it challenging for malicious or inefficient nodes to participate. In contrast, honest and efficient miners receive significant block rewards, which acts as an incentive for them to maintain the integrity of the blockchain network. Dishonest or inefficient nodes are quickly expelled from the network.

This balance of risks and rewards also provides security against potential cyber-attacks that may undermine consensus by placing the majority hash rate of a blockchain network into the hands of a single centralized group or entity. A targeted attack on a particular network is known as a 51 percent attack and can be extremely destructive if successfully executed. However, due to the competitiveness of PoW mining and the magnitude of the Bitcoin network, the probability of a malicious node gaining control of the majority of nodes is exceptionally low. The cost associated with the computing hardware/energy needed to attain 51 percent control of a massive blockchain network is also very high, providing a significant disincentive to make such a massive investment for a small potential reward.

This property of blockchain networks is referred to as Byzantine Fault Tolerance (BFT), which is essentially the ability of a distributed system to continue working normally even if some nodes become compromised or act maliciously. As long as the cost of setting up a majority of malicious nodes remains prohibitive, and higher incentives exist for honest activity, the blockchain network will thrive without significant disruption.

It is worth noting that small blockchain networks are more vulnerable to cyber-attacks since the entire hash rate committed to these systems is significantly lower than that of Bitcoin.

Fundamental Blockchain Security

Blockchain technology relies on fundamental principles, namely cryptography, decentralization, and consensus, to underpin the security and dependability of transactions conducted within its framework. Within a blockchain system, data is organized into blocks, each containing one or multiple transactions. These blocks are intricately linked together using cryptographic methods, forming an unchangeable chain that virtually prevents any tampering with or alteration of prior transactions. Furthermore, every transaction housed within the blocks undergoes scrutiny and validation through a consensus mechanism engaging all network participants. This meticulous process guarantees the legitimacy and accuracy of each transaction, making it impossible for any single user to manipulate or modify the transaction history.

Blockchain's capacity for decentralization is another core facet, allowing participation from members across a widely distributed network. This decentralization ensures the absence of a single point of failure, granting the network resilience and the ability to function smoothly even if certain nodes become compromised or go offline.

It is important to note, however, that not all blockchain technologies offer uniform levels of security. Variations exist; for example, some blockchains might employ consensus mechanisms susceptible to attacks, while others may lack robust encryption, rendering them more vulnerable to hacking. Hence, careful assessment of the security attributes of different blockchain systems is crucial before considering their deployment in critical applications.

Types of Blockchain Networks & Their Security Characteristics

Blockchain networks can have different security features depending on their type and design. These differences include access controls, membership, consensus mechanisms, and decentralization levels. One of the most significant distinctions in blockchain networks is whether they are public or private. Public blockchains, like Bitcoin and Ethereum, are open to anyone who wants to join and are accessible through the internet. These networks rely on anonymous users, which means that participants are typically identified only by their public keys. Public blockchains use a consensus mechanism called Proof of Work, where miners compete to validate transactions and create new blocks in exchange for rewards.

In the realm of private blockchains, stringent identity verification mechanisms are employed to control membership and access rights, creating a tightly-knit consortium of trusted organizations. This exclusive arrangement results in the formation of a confidential "business network" where access is limited to verified participants, ensuring a high degree of security and privacy. Within permissioned networks, consensus is achieved through a specialized process known as Selective Endorsement, wherein only authorized users are empowered to validate transactions. These networks boast robust identity and access controls, rendering them ideal for sectors and applications where regulatory compliance, data confidentiality, and meticulous access management are pivotal. Another way to differentiate blockchain networks is based on whether they are permissionless or permissioned. Permissionless networks, like Bitcoin and Ethereum, have no restrictions on who can participate and validate transactions. Anyone can run a node and participate in the consensus mechanism, which makes these networks highly decentralized. However, they may also be more vulnerable to attacks and have lower transaction throughput.

Permissioned blockchains, on the other hand, are limited to a select set of users who are granted identities using certificates. These networks offer greater control over membership and access privileges, which makes them ideal for private applications that require higher levels of security and compliance. Consensus in permissioned networks is achieved through a variety of mechanisms, such as Byzantine Fault Tolerance, Practical Byzantine Fault Tolerance, or Raft.

When choosing a blockchain network, it is essential to consider the specific business goals and security requirements. Public networks are generally more open and decentralized, making them suitable for applications that prioritize transparency and immutability. Private networks, on the other hand, offer greater control over membership and access, making them preferable for compliance and regulatory purposes. Permissioned networks can be either public or private, and they provide additional layers of security and control over data access and processing. *(Ref. No.: 12- 14)*

In summary, the security of blockchain networks varies depending on their design and the types of controls and consensus mechanisms they employ. Public blockchains offer greater decentralization but are less secure and have lower throughput. Private blockchains are more secure but offer less transparency and openness. Permissioned networks offer greater control over access and processing but require more identity and access controls. When designing a blockchain application, it's essential to consider the specific security requirements and choose the appropriate network type to achieve those goals.

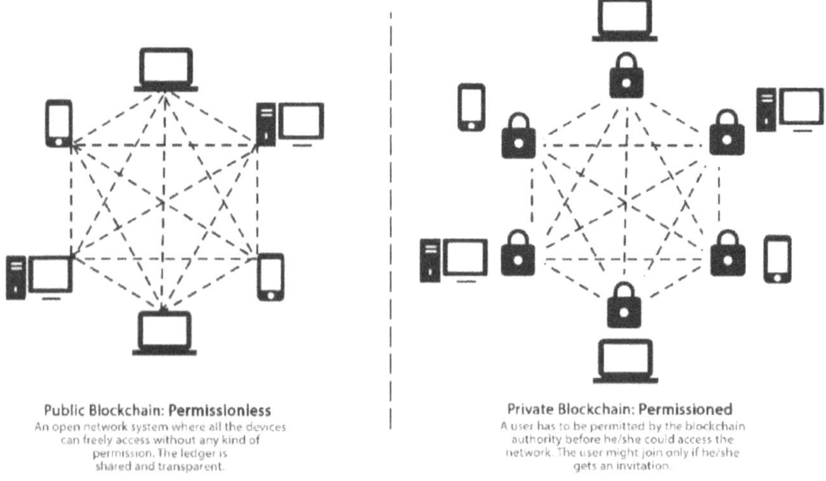

Figure 1.23: Public vs Private Blockchain Network

Cyberattacks & Forgery

Although blockchain technology is known for providing a secure and tamper-proof record of transactions, it is not immune to cyberattacks and frauds. Despite the fundamental security features of blockchain, bad actors can still exploit the known vulnerabilities in blockchain infrastructure and attempt to carry out attacks. This is why it is crucial to understand the various ways in which hackers and fraudsters can threaten blockchain networks. In the following sections, we will explore some examples of how blockchain networks can be attacked and manipulated for malicious purposes.

How Fraudsters attack Blockchain Network

Blockchain networks face threats from cybercriminals who can exploit vulnerabilities in their infrastructure to execute various types of attacks and frauds. The most common types of attacks include phishing, routing attacks, Sybil attacks, and 51% attacks.

Phishing Attacks

Phishing is a type of social engineering attack where attackers try to trick victims into revealing their login credentials or other sensitive information by impersonating a legitimate entity. In the context of blockchain networks, fraudsters can send emails or messages to wallet owners, pretending to be a trusted source, and trick them into revealing their private keys or other credentials. These phishing emails or messages usually contain fake hyperlinks that lead the victims to a fake website that looks identical to the real one, where they are asked to enter their login credentials.

Once the attackers have obtained the login credentials, they can steal the victim's digital assets or tamper with the blockchain network. Phishing attacks can result in significant losses for both the users and the blockchain network. Users may lose their digital assets, and the blockchain network's security and integrity can be compromised, leading to a loss of trust in the network.

To protect against phishing attacks, users should always be vigilant and verify the authenticity of any messages or emails they receive. They should avoid clicking on suspicious hyperlinks or downloading attachments from unknown sources. Additionally, blockchain networks can implement security measures such as two-factor authentication and anti-phishing filters to protect against phishing attacks.

Routing Attacks

One of the ways in which blockchain networks can be attacked is through a routing attack, where hackers intercept data as it is being transferred to internet service providers. Participants of the network may not be able to detect this attack, as everything appears normal on the surface. However, the attackers can extract sensitive data and even cryptocurrencies behind the scenes, putting the security of the blockchain network at risk. This type of attack exploits vulnerabilities in the routing infrastructure of the internet, making it a difficult challenge to overcome.

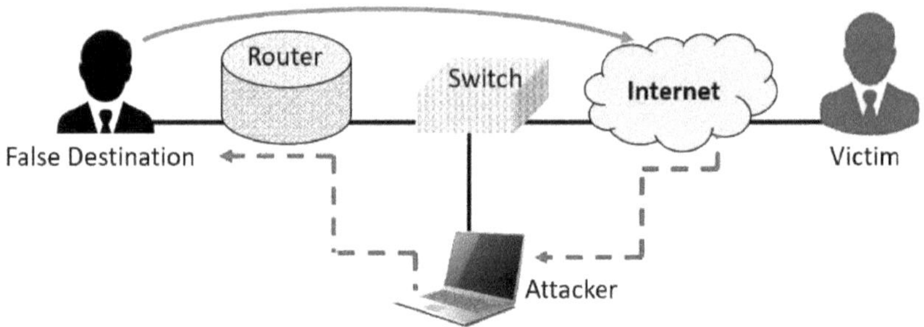

Figure 1.24: Routing Attacks on Blockchain Network

Sybil Attacks

A Sybil attack is a technique used by hackers to create multiple fake network identities, which they use to flood the network and crash the system. The name "Sybil" comes from a famous book character who was diagnosed with a multiple personality disorder.

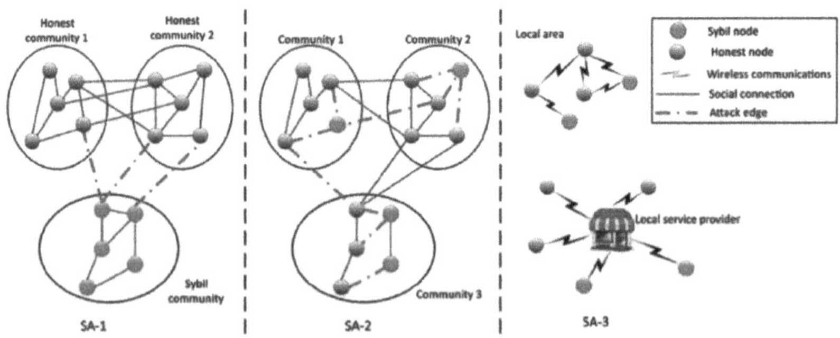

Figure 1.25: Sybil Attacks on Blockchain Network

51% Attacks

The process of mining in a blockchain network involves significant computing power, particularly for larger public blockchains. However, if a miner or a group of miners were to gather sufficient resources, they could obtain more than 50% of the network's mining power. This would give them control over the ledger and allow them to exploit it for malicious purposes.

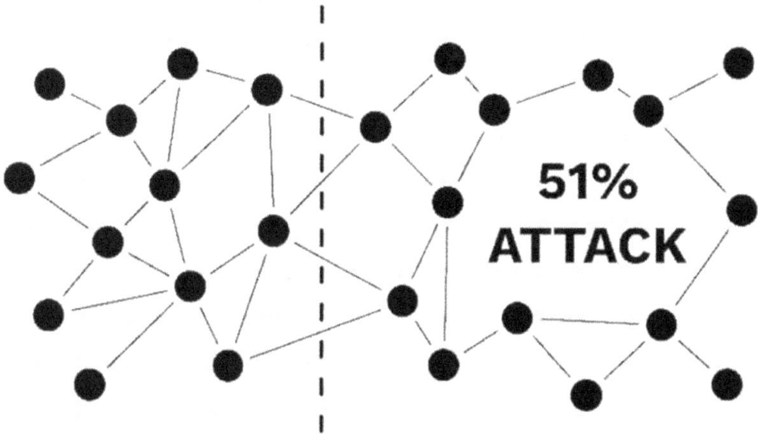

Figure 1.26: 51% Attacks on Blockchain Network

Blockchain-based Security for Business

When building an enterprise-based blockchain application, it is essential to prioritize security across all layers of the technology stack. This includes considering governance and permissions for the network. For an enterprise blockchain solution, a comprehensive security strategy is critical, which should include both traditional security controls and technology-unique controls.

For blockchain solutions in industries like mineral mining, which require extensive computing power for mining, additional security considerations must be considered. In addition to the security controls mentioned earlier, it is crucial to ensure the security of the mining hardware itself, as well as the storage and transfer of any mined cryptocurrency.

Some of the security controls specific to enterprise blockchain solutions include identity and access management, key management, data privacy, secure communication, smart contract security, and transaction endorsement. It is also recommended to employ experts who can help design a secure solution to achieve specific business goals, such as creating a production-grade platform for blockchain solutions that can be deployed on-premises or on a preferred cloud vendor.

Overall, in industries like mineral mining, where blockchain technology can be utilized to increase transparency and efficiency, it is crucial to prioritize security and take all necessary measures to prevent cyberattacks and other security threats.

Blockchain Security Recommendations & Best Practices

When developing a solution based on blockchain technology, it is recommended to consider the following important inquiries:

1. What governance model will be participating organizations or associates follow in the blockchain-based solution?
2. How can relevant regulatory provisions be fulfilled?
3. What specific data will be captured in each block?
4. What is the minimum-security posture required for blockchain clients to participate?
5. What logic will be used to resolve blockchain block collisions?
6. How are the identity details collected and managed, and are block payloads encrypted?
7. What is the disaster recovery plan for the blockchain participants?

To ensure the success of a blockchain-based solution, it is essential to consider several key questions in the design phase. These questions address various aspects such as governance, regulatory compliance, data capture, security posture, identity management, and disaster recovery planning.

In the case of private blockchains, it is critical to deploy them in a secure and resilient infrastructure that aligns with business needs and processes. Otherwise, they may be vulnerable to data security risks due to underlying technology vulnerabilities.

Furthermore, administrators must consider business and governance risks that emanate from the decentralized nature of blockchain solutions. To manage these risks, a blockchain security model must be developed based on a risk-based and threat-based approach.

The security controls in the model fall under three categories: unique blockchain security controls, conventional security controls, and business controls for blockchain. These controls aim to mitigate risks and threats to the blockchain solution comprehensively.

In the context of the mineral mining industry, blockchain technology can play a crucial role in ensuring transparency and accountability in the supply chain. By tracking the movement of minerals from the mining site to the end consumer, blockchain can help prevent fraud, exploitation, and human rights abuses. However, it is essential to consider the unique security challenges in the mining industry and design a blockchain solution that adequately addresses them. For example, the solution should incorporate robust identity and access management controls, secure communication protocols, and key management procedures to prevent unauthorized access and data breaches.

The security controls that mitigate the risks and threats of a blockchain solution can be categorized into three categories, which administrators must define:

- Enforce security controls that are specific to blockchain technology
- Apply conventional security controls
- Enforce business controls specific to the blockchain implementation.

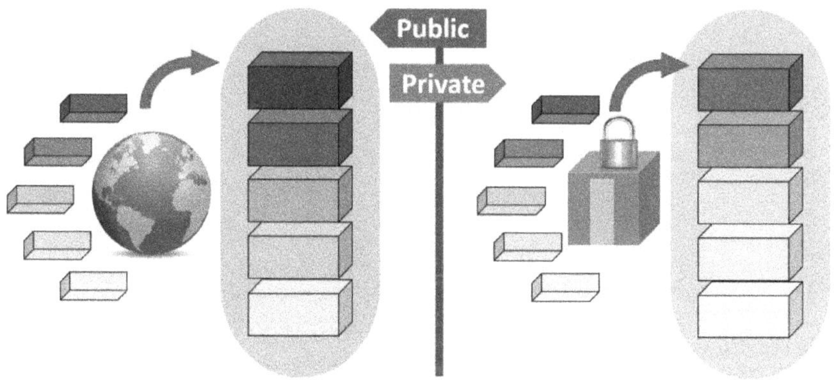

Figure 1.27: Public vs Private Blockchain

The Immutable Nature of Blockchain: Why Old Blocks Cannot be Manipulated or Deleted

To understand why old blocks in a blockchain network cannot be manipulated or deleted, it is important to have a basic understanding of the blockchain data structure and hashing. A cryptographic hash function, such as the SHA-256 hashing function, is used to calculate a hash value for a given data set. This hash value is a unique identifier that represents the data set and cannot be recalculated to reveal the original data.

The blockchain data structure consists of blocks that are linked to each other. Each block contains a hash value of the previous block, creating an unbroken chain of blocks. If someone tries to manipulate the data in a block, the hash value of that block will change, which will also change the hash value of the subsequent block. Therefore, any attempt to manipulate a block would break the link between that block and the subsequent blocks.

Since benign participants of a blockchain network always accept and extend the longest valid blockchain, an attacker would have to recalculate all the blocks created after the manipulated block. This is because the blockchain network always grows faster than the manipulated version on average, as long as the attacker does not have more than 50% of the total computing power. This makes a public blockchain with many participants highly secure.

To sum up, old data cannot be manipulated because of the immutable nature of blockchain's cryptographic hash function, which creates an unbreakable chain of blocks that cannot be tampered with. Any attempt to change the data in a block would break the link between that block and the subsequent blocks, making it highly secure.

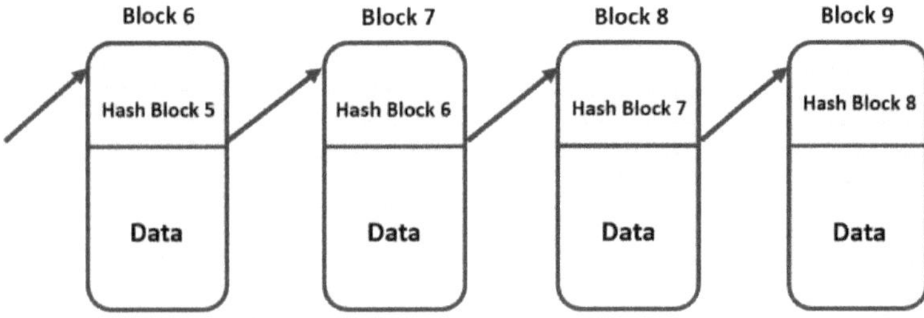

Figure 1.28: Cryptographic Hash used in a Blockchain Network

Understanding the Role of Developers, Miners, Investors, & Users in Determining the Rules & Security of Blockchain Networks

When it comes to the security of the blockchain, one of the key questions that arises is who determines the rules and ordinances. In the case of Bitcoin, the rules are set by the developers who work on the reference implementation of the Bitcoin software. These rules are then adopted by all the benign participants of the Bitcoin network. Similarly, Ethereum has a yellow paper that documents the rules, which are observed by all the network participants. Although a small group of people determines the rules, this does not pose a problem because the rule makers are observed by the miners. The miners operate the blockchain only with the rules that they agree with. If a blockchain has no miner, it is useless.

However, this does not mean that the miners have all the authority. They must choose the blockchain where they can profitably mine new blocks. The price of a cryptocurrency is driven by speculation, so investors have a significant influence on which blockchain the miners are operating on. The investors speculate on the further spread of the underlying cryptocurrency and thus, on more users who use Bitcoin, for example, for payment. Therefore, all users of a blockchain influence the investors and possess an indirect influence on the rules.

Furthermore, some argue that users do not care about the underlying cryptocurrency, as long as they can make payments. However, the services that make cryptocurrencies usable for everyone also have a significant influence. In addition to the factors mentioned earlier, there are also economic incentives that contribute to the security of a blockchain network. In a proof-of-work blockchain, miners compete to solve complex mathematical problems and are rewarded with newly minted cryptocurrency. This creates a self-regulating system in which miners are incentivized to behave honestly and follow the rules, as deviating from the rules could lead to loss of rewards and reputation. Furthermore, the transparency of blockchain technology allows for public scrutiny of the blockchain network. Any participant can view and audit the contents of the blockchain, which ensures that any malicious activity can be easily detected and traced back to its source. *(Ref. No.: 17- 20)*

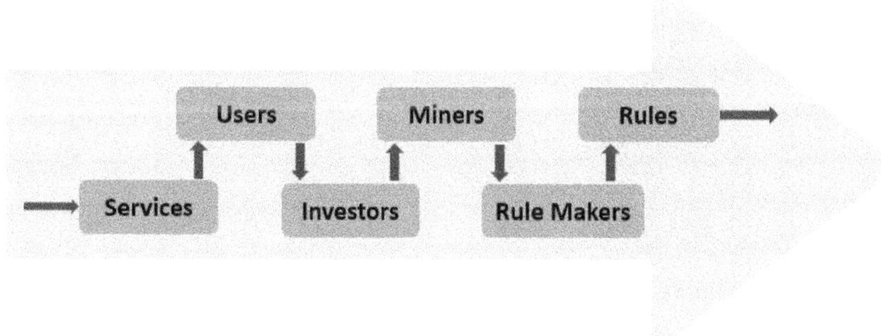

Figure 1.29: Authority in a Blockchain Network

Closing Thoughts

Distributed ledger technology, commonly known as blockchain technology, is still in its early stages of development. However, it has already demonstrated its potential to revolutionize the way governments, institutions, and corporations operate. Its ability to securely record and verify transactions offers a level of transparency and trust that was previously unattainable. In particular, blockchain technology has the potential to transform areas such as taxation, land registries, and voting procedures. By providing a tamper-proof, decentralized database of transactions, blockchain technology can help ensure that these important processes are conducted in a fair and transparent manner. For example, a blockchain-based voting system could greatly enhance the security and transparency of elections, by providing a verifiable, immutable record of each vote.

However, achieving a balance between decentralization and security is crucial to building a dependable and powerful blockchain network. By blending game theory and cryptography, blockchain networks can achieve a high degree of security as a distributed system. Game theory provides a framework for designing incentive mechanisms that encourage participants to act in the best interests of the network, while cryptography provides the underlying technology to secure the network. As private blockchains are developed for businesses and other applications, their security measures may need to adapt to meet the unique needs of each use case. For example, a private blockchain used for supply chain management may require different security measures than a public blockchain used for financial transactions. It is important to carefully balance the level of decentralization with the specific security needs of each application. Overall, the potential benefits of blockchain technology are vast, and we are only beginning to scratch the surface of what is possible. As the technology continues to evolve and mature, it is likely that we'll see even more innovative use cases emerge. Ultimately, blockchain technology has the potential to transform the way we conduct business, governance, and even our daily lives.

Chapter 2: Game Theory & Byzantine Fault Tolerance in Blockchain Consensus

"The blockchain symbolizes a shift in power from the centers to the edges of the networks."

— William Mougayar

The Importance of Game Theory in Cryptocurrency Networks

When considering blockchain technology, many people associate it with Bitcoin, decentralization, and security. Cryptocurrencies have managed to thrive despite numerous attempts to disrupt their networks, and one of the key reasons for this success is the application of game theory. Although the concept of game theory may be unfamiliar to some, it has proven to be instrumental in protecting the network of cryptocurrencies.

Game theory is a mathematical framework for analyzing strategic interactions between rational decision-makers. It is a way to model decision-making in situations where two or more individuals have competing interests, and the outcome of each individual's decision depends on the decisions made by others. Game theory has been widely applied in various fields, from economics to political science, and it has become increasingly relevant in the world of cryptocurrencies.

The application of game theory in cryptocurrencies can be seen in the design of the blockchain network itself. The network is designed to incentivize miners to act in the best interest of the network as a whole, rather than in their own self-interest. Miners are rewarded for contributing to the network's security, which requires them to use their computing power to solve complex mathematical problems. In order to be rewarded, miners must follow the rules of the network, including validating transactions and building on top of the longest valid chain.

However, there is always a possibility that some miners may try to act in their own self-interest, rather than in the interest of the network. For example, a miner may attempt to create a new block that includes invalid transactions, in the hope of receiving a reward without contributing to the network's security. If such behaviour were to become widespread, it could undermine the security and reliability of the network.

To prevent this from happening, the blockchain network is designed in such a way that miners have a strong incentive to act honestly. The game-theoretical approach used in the network assumes that miners will act rationally, and will seek to maximize their rewards. By making it more profitable for miners to act honestly than to cheat, the network ensures that the vast majority of miners will act in the interest of the network as a whole. This chapter aims to introduce readers to the concept of game theory and its application in cryptocurrency networks.

Game Theory: Strategic Decision-Making in Interactive Environments

Game theory is a branch of applied mathematics that examines strategic decision-making in interactive environments. In these environments, individuals or organizations engage in rational decision-making based on the rules of the game and the behaviour of other players. Initially developed to analyze the behaviours of businesses, markets, and consumers in economics, game theory has been adopted in other fields such as politics, sociology, psychology, and philosophy. Games are defined by the number of players involved, the actions and strategies of each player, the payoff associated with each strategy, and the best response or strategy of each player given the actions of others. They can be categorized into two types: zero-sum games and non-zero-sum games. In zero-sum games, one player's gain is equal to another player's loss, while in non-zero-sum games, this relationship does not hold true, and both players can gain or lose simultaneously. One way to represent games is through a payoff matrix, which is a tabular representation of all the possible outcomes of a game. There are two ways to define a game: normal form games and extensive form games. In this chapter, only normal form games will be considered. These games are represented in a payoff matrix form.

To understand the mechanics of a game and how it works, it is essential to clarify a few queries:

- Who are the players involved?
- What are the actions available to each player?
- What is the payoff associated with each action profile?
- What is the predicted outcome of the game?

To answer these queries, consider the following example:

- Who are the players? In this example, the players are Steve and Tracey, designated as Player 1 and Player 2, respectively.
- What are the actions available to each player? Each player can either choose to split the reward or steal the entire amount.
- What is the payoff associated with each action profile? The payoff matrix shows the outcome of each possible combination of actions. For example, if both players choose to split, they both receive a reward of 2. If one player chooses to steal and the other splits, the player who stole receives the entire reward of 3, and the other player receives nothing.

- What is the predicted outcome of the game? Based on the best response of each player, the predicted outcome of this game is (Steal, Split).

In summary, game theory is a powerful tool for understanding decision-making in interactive environments, and it has applications in a wide range of fields. By examining the behaviour of players, their actions, and the potential outcomes of the game, game theory provides a framework for predicting and analyzing strategic behaviour.

Eventually, the outcome of the game can be easily determined. If Player 1 chooses to Split, the optimal response for Player 2 would be to Steal. On the other hand, if Player 1 chooses to Steal, Player 2 would be indifferent between Stealing and Splitting. As the game is symmetrical, we can intuitively predict that the outcome of the game would be (Steal, Steal). It is important to note that not all outcomes are aimed at maximizing profits. In this particular game, both players could have benefitted by choosing to Split, but the expected result is (Steal, Steal). In the game played by Steve and Tracey, their action profile is (Steal, Split).

These strategic interactions, characterized by mutual awareness of optimal strategies and perfect knowledge of each player's available moves, fall under the category of complete information games. In such games, transparency reigns supreme, and participants are equipped with a comprehensive understanding of their own best response strategies as well as those of their fellow players. Furthermore, it is universally acknowledged that all participants share this level of information, creating a level playing field for strategic decision-making.

The principle of Prisoner's Dilemma

The classic example of the Prisoner's Dilemma encapsulates a fundamental concept in game theory. In this scenario, two apprehended criminals, A and B, face separate interrogations by a prosecutor. The prosecutor's objective is clear: persuade each criminal to testify against the other, ultimately leading to reduced charges for the testifying party. The consequences are stark: if A chooses to testify against B, A walks free while B faces a three-year prison sentence (and vice versa). Conversely, if both A and B betray each other by testifying, they both receive a two-year prison sentence. The twist comes when both criminals opt for silence. In this case, due to insufficient evidence, they are each sentenced to just one year in prison. The Prisoner's Dilemma serves as a captivating illustration of how individual rationality can lead to suboptimal outcomes when cooperation is forsaken.

#	B betrays (Defects)	B stays quiet (Cooperates)
A betrays (Defects)	Both jailed for 2 years.	A is free. B is jailed for 3 years.
A stay quiet (Cooperates)	B is free. A is jailed for 3 years.	Both jailed for 1 year.

Table 2.1: The principle of Prisoner's Dilemma

Therefore, there are several possible outcomes depending on each criminal's decision, but the optimal scenario for A (or B) is to betray and be set free. However, this would require the other criminal to stay silent, which cannot be predicted. Given the potential reward, many rational prisoners would likely choose to act in their self-interest and betray the other. However, if both A and B betray, they could end up serving two years in prison, which is not the best outcome. Thus, the best option for them as a pair would be to remain silent and receive a sentence of one year in prison instead of two.

The Prisoner's Dilemma has numerous variations, but this basic scenario illustrates the concept of using game theory models to analyze human behaviour and predict possible outcomes based on their rational decision-making process.

Nash Equilibrium

In Game Theory, a Nash Equilibrium refers to a solution in which no player has an incentive to change their strategy, given the strategies of the other players. In other words, it is a stable state where no player can benefit by changing their actions unilaterally. The concept is named after the Nobel laureate economist John Nash, who proved the existence of this equilibrium in his famous paper in 1950. The Battle of the Sexes game is a classic example of a game with multiple Nash Equilibria. In this game, both players want to spend time together, but they have different preferences for the activity they do. If the boy chooses to watch a cricket match, he gets a higher utility than going to a theatre, and similarly, the girl has a higher utility for going to a theatre than watching cricket. In this game, there are two Nash Equilibria in pure strategies: (Cricket, Cricket) and (Theatre, Theatre). In both equilibria, both players agree on the same activity, and neither has an incentive to change their strategy.

Matching Pennies is another game that demonstrates the concept of Nash Equilibrium. In this game, two players simultaneously choose between "heads" or "tails." If the choices match, Player 1 wins and takes Player 2's penny, and if they do not match, Player 2 wins and takes Player 1's penny. Interestingly, there is no pure strategy Nash Equilibrium in this game. Both players need to randomize their choices with equal probability (50/50) to achieve a mixed-strategy Nash Equilibrium. This equilibrium means that each player's strategy is an equal probability mix of "heads" and "tails." This type of equilibrium can arise when there is no dominant strategy for either player.

The concept of Nash Equilibrium is crucial in game theory and has important applications in many fields, including economics, political science, and computer science. For example, the security and decentralization of the blockchain technology rely on this concept. In the blockchain, different nodes work together to maintain the system's security and reach a consensus on new transactions. Each node has an incentive to follow the protocol and act honestly because deviation from the protocol can lead to punishment and a loss of rewards. The Nash Equilibrium ensures that no node has an incentive to deviate from the protocol, making the blockchain secure and decentralized.

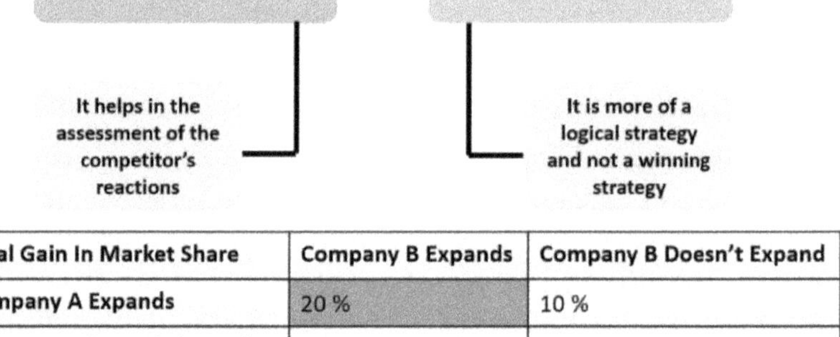

Total Gain In Market Share	Company B Expands	Company B Doesn't Expand
Company A Expands	20 %	10 %
Company A Doesn't Expand	10 %	0 %

Figure 2.1: Nash Equilibrium in Blockchain Network

Cryptocurrencies & Peer-to-Peer Networks

Cryptocurrency mining is the process where all nodes in a network compete to find the next block using consensus mechanisms like Proof of Work and Proof of Stake. Miners help in the mining process and earn cryptocurrencies as a reward. However, miners may try to cheat the network by not mining the correct block, and this is where game theory comes into play. In any cryptocurrency network, there are two players: users and miners. Users send and receive cryptocurrencies, while miners validate transactions and mine blocks.

The game theory behind cryptocurrencies works on the principle of incentives and punishments. Miners are incentivized to act honestly in the network because they earn rewards for mining new blocks. However, if a miner tries to cheat the network by not mining the correct block, they will not receive the reward, and their reputation in the network will be damaged. This can lead to a loss of potential rewards in the future.

On the other hand, users also have a role to play in maintaining the network's integrity. If a user tries to make a fraudulent transaction, they will not only lose their own funds but also face penalties like being banned from the network or losing their reputation. This incentivizes users to act honestly and maintain the network's integrity.

In contrast to the peer-to-peer network for cryptocurrencies, torrenting is an example of a peer-to-peer network that often suffers from inefficient functioning. While people download everything they need without any hesitation, many usually delete the files once they are done. People who download these files do not seed them as there is no incentive to do so. This leads to the inefficient functioning of the network. Lower seeds imply slower download speed and many other disadvantages. As participants are not punished for this action, the network can easily be broken down.

Overall, game theory plays a crucial role in ensuring the efficient functioning of peer-to-peer networks like cryptocurrencies. By incentivizing honest behaviour and punishing fraudulent behaviour, these networks can maintain their integrity and continue to provide value to their users.

Cheating & its Consequences.

In the world of cryptocurrencies, the potential for miners to engage in deceptive practices, such as accepting and adding invalid transactions to the blockchain, does exist. Double spending, a well-known form of cheating, poses a substantial threat as it can erode trust among network participants. The consequences of such behaviour can be dire, potentially leading to the network's collapse and a subsequent decline in the cryptocurrency's value. Surprisingly, despite this possibility for misconduct, many cryptocurrencies maintain stability. This isn't solely due to a lack of desire to cheat, but rather because the costs and penalties associated with dishonest actions often outweigh the potential gains. The delicate balance of incentives and disincentives that discourages cheating can be comprehensively explained through the concept of Nash Equilibrium, shedding light on how rational decision-making among participants contributes to the overall integrity of the cryptocurrency ecosystem.

Nash Equilibrium in Cryptocurrency Networks

Cryptocurrency mining can be thought of as a game that involves multiple players, specifically the nodes in the network, who compete to find the next block using consensus mechanisms such as Proof of Work or Proof of Stake. In this game, the miners' main objective is to maximize their profits and improve the security of the network. However, cheating can occur when miners accept invalid transactions to earn cryptocurrencies and add more invalid blocks, leading to a loss of trust in the network and ultimately the collapse of the cryptocurrency.

To prevent cheating, the game theory concept of Nash Equilibrium comes into play. Each node in the network must choose a strategy that maximizes their profits in the long run and enhances the network's security while considering the actions of other nodes in the network. The penalties for cheating are severe, making cheating a costlier option for miners than playing by the rules. This is especially true in the Proof of Work mechanism where mining is a very costly process requiring expensive computing hardware and increasing electricity costs.

Mining in the Proof of Stake mechanism also involves game theory, where miners stake cryptocurrencies in a smart contract before mining begins. Any wrongdoing or misdeeds by a node will result in penalties, including the deduction of cryptocurrencies from the staked amount. If a node's staked amount falls below the minimum requirement mentioned in the smart contract, they cannot stake at all, reducing their probability of validating transactions and earning rewards.

Moreover, cheating reduces people's faith in the cryptocurrency, leading to a loss of its value. Therefore, miners who cheat ultimately affect the security of the network, and users of cryptocurrencies would not value a network filled with cheating nodes. Hence, game theory plays a crucial role in producing a decentralized network like that of cryptocurrencies, where each node chooses a strategy that maximizes their profits and enhances the security of the network, leading to a fair and transparent game for all players involved.

Proof of Work Difficulty Adjustment:

The difficulty of finding the next block in a Proof of Work system is adjusted periodically to ensure that the rate of block creation remains relatively constant. This difficulty adjustment is based on the total computational power (hash rate) of the network, and is designed to maintain a target block creation time. The equation for calculating the difficulty adjustment in Bitcoin, for example, is:

*New Difficulty = Old Difficulty * (Actual Time Taken / Target Time Taken)*

where Actual Time Taken is the amount of time it took to mine the last 2016 blocks, and Target Time Taken is the ideal amount of time (in seconds) it should have taken to mine those blocks.

Mining Reward Calculation:

In a Proof of Work system, miners are rewarded with a certain amount of cryptocurrency for successfully mining a block. This reward is usually composed of two parts: the block reward and the transaction fees. The equation for calculating the total mining reward in Bitcoin, for example, is:

Total Mining Reward = Block Reward + Transaction Fees

where Block Reward is the fixed amount of new Bitcoin created with each block (currently 6.25 BTC), and Transaction Fees are the fees paid by users to have their transactions included in the block.

Expected Reward Calculation:

In a Proof of Work system, each miner's chance of successfully mining a block is proportional to their share of the total hash rate. The expected reward for a given miner can be calculated using the following equation:

*Expected Reward = Mining Reward * (Miner's Hash Rate / Network Hash Rate)*

where Mining Reward is the total reward for mining a block (as calculated in equation 2), Miner's Hash Rate is the miner's share of the total hash rate, and Network Hash Rate is the total hash rate of the entire network.

These equations demonstrate some of the ways that game theory and incentives are built into the design of cryptocurrency networks. By creating rewards for miners and penalties for cheating or misbehaviour, these networks are able to incentivize honest behaviour and maintain the integrity and security of the system.

Key Takeaways

1. Game theory suggests that the dominant strategy is the most suitable action for an individual, regardless of how other players behave.

2. Nash equilibrium refers to the optimal outcome of a game in which each player makes the best decision while considering the actions of their opponent.
3. One of the most well-known examples of the Nash equilibrium in game theory is the prisoner's dilemma.
4. Even though they are separate concepts, the dominant strategy may also be the Nash equilibrium.
5. Nash equilibrium can be achieved when all members of a group cooperate, or when none of them do.
6. Game theory and Nash equilibrium can also be applied to analyze situations beyond traditional two-player games, encompassing multi-player scenarios where individuals make choices that collectively impact outcomes. In such complex systems, understanding the dynamics of cooperation, competition, and strategic decision-making becomes essential for predicting and influencing group behaviours.

Game Theory & Cryptoeconomics in Decentralized Systems

Game theory models play a critical role in designing a stable and trustworthy financial system, particularly in the realm of cryptocurrencies like Bitcoin. The use of game theory in cryptocurrencies has led to the development of cryptoeconomics, which studies the economics of blockchain protocols and the potential outcomes resulting from participant behaviour. Cryptoeconomics also considers the behaviour of external agents who may not be part of the network but may try to disrupt it from within.

The Bitcoin blockchain network is designed as a distributed system with nodes distributed globally, making it necessary to rely on their agreement in the validation of transactions and blocks. However, these nodes cannot inherently trust each other. To address this issue, the Proof of Work consensus algorithm is one of the most crucial features of the Bitcoin network, protecting it from malicious activity.

The Proof of Work algorithm applies cryptographic techniques that make mining very expensive and demanding, creating an exceptionally competitive mining environment. This incentivizes mining nodes to behave truthfully so as not to risk losing the resources invested. In contrast, any malicious activity is discouraged and quickly punished. Mining nodes that engage in dishonest conduct risk losing a considerable amount of money and being expelled from the network. As a result, the most rational and likely decision made by a miner is to behave honestly and ensure the security of the blockchain network.

Overall, the use of game theory in cryptocurrencies and the development of cryptoeconomics have led to the creation of decentralized systems like Bitcoin that can prevent malicious activity and dishonest nodes from disrupting the network.

Decentralized Trust: The Role of Blockchain & Cryptocurrencies

Blockchain technology has become an integral part of the architecture of most cryptocurrencies. Decentralization, enabled by blockchain, has paved the way for trustless financial systems, allowing for transparent and reliable monetary transactions without the need for intermediaries. This development has made cryptocurrencies a viable alternative to traditional banking and payment systems that rely heavily on centralized trust.

However, achieving consensus on distributed networks in a secure and efficient way is a significant challenge. The participants in a cryptocurrency network must regularly agree on the current state of the blockchain through a consensus mechanism. This task becomes more complex when some of the nodes are malicious or behave dishonestly.

This challenge is known as the Byzantine Generals' problem, which gave birth to the principle of Byzantine Fault Tolerance. The Byzantine Fault Tolerance principle allows for a distributed network of computer nodes to agree on a decision, even in the presence of malicious nodes. This principle plays a crucial role in ensuring the security and reliability of blockchain-based systems.

Since the inception of Bitcoin in 2008, many different cryptocurrencies have been created, each with a unique mechanism. However, most cryptocurrencies share the blockchain as a core element of their architecture. The blockchain is deliberately designed to be decentralized, operating as a digital ledger maintained by a distributed network of computer nodes spread across the globe.

The Proof of Work consensus algorithm is one of the most widely used consensus mechanisms in blockchain-based systems. This algorithm applies cryptographic techniques that make the mining process very expensive and demanding, creating an extremely competitive mining environment. Therefore, this architecture incentivizes mining nodes to behave honestly, as any malicious activity is discouraged and quickly punished. Miners who exhibit dishonest behaviour risk losing the resources invested and may be kicked out of the network. Hence, the most rational decision a miner can make is to behave honestly and keep the blockchain network secure.

In summary, the decentralization enabled by blockchain technology has paved the way for trustless financial systems. However, ensuring the security and reliability of blockchain-based systems remains a significant challenge. The principle of Byzantine Fault Tolerance plays a crucial role in addressing this challenge, allowing for a distributed network of computer nodes to agree on a decision, even in the presence of malicious nodes.

Understanding the Byzantine Generals' Dilemma

The Byzantine Generals' Problem, first introduced in 1982, is a theoretical challenge that illustrates how a group of Byzantine generals may encounter communication difficulties while attempting to coordinate their actions. In this scenario, each general leads a personal army situated in distinct locations across the city they plan to attack. The generals must agree on a common decision, whether to attack or retreat, to execute the move in coordination. The problem arises due to the fact that the generals can only communicate with each other through messages forwarded by a courier, and these messages can get delayed, lost, or destroyed. Moreover, one or more generals may choose to behave maliciously and send fraudulent messages to mislead the others, resulting in a mission failure. This communication issue has a significant impact on distributed computing systems, such as blockchain networks, as nodes need to reach a consensus on the current state of the system. In a blockchain network, each general represents a network node, and the nodes must achieve consensus on the current state of the system. This implies that the majority of participants/nodes in a network within a distributed network must agree and execute the same action to avoid complete failure. Therefore, to achieve consensus in these types of distributed systems, at least two-thirds or more of the network nodes must be dependable and honest.

This concept can be illustrated mathematically using the Byzantine Generals' Problem. Suppose there are n generals and m of them are faulty. To achieve consensus, a voting process is initiated where each general casts a vote for either attack or retreat. The decision is made by taking the majority of votes, and if the number of faulty generals is less than or equal to (n-1)/3, the decision will be correct. Mathematically, the formula for consensus in the Byzantine Generals' Problem is:

Let n be the total number of generals

Let m be the number of faulty generals Then, $(n-1)/3 < m$ for the network to be vulnerable to Byzantine failures.

The Byzantine Generals' Problem represents a fundamental challenge in achieving consensus in distributed computing systems, including blockchain networks. The solution lies in having a significant number of honest nodes to maintain the system's security and avoid the risk of failures and attacks.

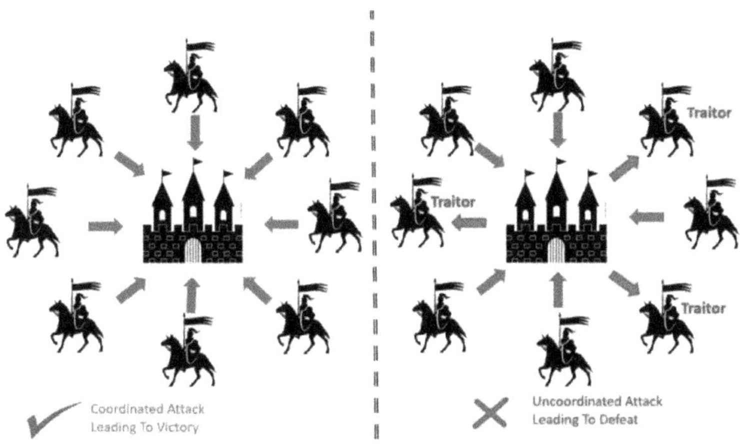

Figure 2.2: Byzantine Generals' Problem

Byzantine Fault Tolerance (BFT)

The principle of Byzantine Fault Tolerance (BFT) refers to the ability of a decentralized system to function properly despite internal network nodes experiencing failures or behaving maliciously. It is a critical aspect of ensuring the reliability and security of blockchain systems. BFT is based on the concept of consensus achievement among the network nodes, which means that they agree on the same decision regarding the current state of the system.

The Byzantine Generals' Problem is the foundation of the BFT principle, which presents the challenge of reaching consensus in a distributed network of computer nodes in a secure and efficient way, even when some nodes are malicious or fail. Therefore, a BFT system must be able to withstand arbitrary and unpredictable faults and failures in the network.

To achieve BFT in a blockchain network, various consensus algorithms have been developed. These algorithms define the rules for how nodes can communicate with each other and agree on the same decision. There are multiple approaches to the Byzantine Generals' Problem, which leads to different consensus algorithms that offer varying degrees of BFT.

For example, one of the most well-known BFT consensus algorithms is the Practical Byzantine Fault Tolerance (PBFT) algorithm. It requires at least two-thirds of the network nodes to be honest and agree on a decision for consensus to be achieved. In contrast, the Proof-of-Work (PoW) consensus algorithm used by Bitcoin and many other cryptocurrencies does not require Byzantine Fault Tolerance explicitly. Instead, it relies on a large number of nodes performing computational work to validate transactions and secure the network against malicious actors. *(Ref. No.: 15- 23, 24- 28)*

In summary, Byzantine Fault Tolerance is a critical property of decentralized systems, particularly in the context of blockchain networks. Various consensus algorithms have been developed to achieve BFT, and each has its own strengths and weaknesses. Ultimately, the choice of consensus algorithm depends on the specific requirements and goals of the blockchain network in question.

Benefits of Practical BFT

Byzantine Fault Tolerance (BFT) is a consensus algorithm that offers several benefits compared to the Proof-of-Work (PoW) consensus algorithm used by Bitcoin and other cryptocurrencies.

1. **Flexibility & Speed of Transactions:** One of the primary benefits of BFT is its ability to provide transaction finality without the need for confirmation, which is required in the PoW mechanism. With BFT, once a transaction is included in a block and added to the blockchain, it is considered final and cannot be altered. This provides more flexibility and speed in the transaction processing, as there is no need to wait for multiple confirmations.

2. **Low Energy Consumption:** BFT also offers a significant reduction in energy consumption compared to PoW. In PoW, the miners compete to solve complex mathematical puzzles to validate transactions and add new blocks to the blockchain. This process requires an enormous amount of computational power and electricity, leading to a high energy cost. However, BFT does not require such calculations and is more energy-efficient.

3. **Energy Efficiency:** BFT achieves distributed consensus without the need for complex mathematical computations, which is a significant advantage over PoW. In BFT, the network nodes communicate with each other to achieve consensus on the state of the system. They exchange messages and vote to determine the majority decision, and if

the majority agrees, the decision is considered final. This method is less resource-intensive and more energy-efficient.

4. **Transaction Finality:** Another significant advantage of BFT is that transactions do not require multiple confirmations, as in PoW. In PoW, each node individually verifies all transactions before adding a new block to the blockchain, which leads to a delay in transaction finality. In contrast, BFT offers transaction finality once a block is added to the blockchain, reducing the need for multiple confirmations, and providing faster transaction processing.

BFT offers several advantages over PoW, such as flexibility and speed of transactions, low energy consumption, energy efficiency, and transaction finality. These benefits make BFT a viable alternative to PoW for blockchain networks that prioritize efficiency, speed, and low energy consumption.

Proof of Work: A Genius Solution to Byzantine Faults

A consensus algorithm can be defined as the method by which a blockchain network achieves agreement among its participants. While Proof of Work (PoW) and Proof of Stake (PoS) are the most common implementations, the specific consensus algorithm used by a blockchain network depends on the rules of the system and the desired level of Byzantine Fault Tolerance (BFT). For instance, in the case of Bitcoin, the PoW consensus algorithm is the mechanism that enables the network to reach consensus through the verification and validation of transactions. PoW was chosen by Satoshi Nakamoto as the intelligent algorithm that would enable the creation of a BFT system for Bitcoin. Although the concept of PoW predates cryptocurrencies, PoW has been shown to be one of the most stable and reliable implementations for blockchain networks due to its cost-intensive mining process and the underlying cryptographic techniques. *(Ref. No.: 29- 39)*

While PoW is not 100% tolerant to Byzantine faults, it has proven to be a robust and dependable solution for blockchain networks. Thus, the PoW consensus algorithm, designed by Satoshi Nakamoto, is considered to be one of the most ingenious solutions to the Byzantine Faults.

Closing Thoughts

Expanding on the transformative potential of distributed ledger technology (DLT) in various industries, including mineral mining, is essential. DLT, often exemplified by blockchain networks, can fundamentally alter the way transactions are recorded, verified, and managed. This innovation extends beyond traditional applications such as taxation and land registries, reaching into the heart of industries like mineral extraction and supply chain management. One of the key strengths of blockchain technology is its ability to introduce unparalleled levels of transparency and accountability. In mineral mining, this translates to the tracking of mineral extraction processes from the source to the end user. Every step in the supply chain can be documented on an immutable and transparent blockchain, ensuring that each transaction is verifiable and tamper-proof. This transparency not only instills trust among stakeholders but also addresses critical issues like illegal mining and conflict minerals. To build a reliable and robust blockchain network, it's crucial to find the right balance between decentralization and security. This equilibrium is achieved by applying concepts from game theory, which helps design consensus mechanisms that prevent malicious actors from undermining the network's integrity. Moreover, cryptography plays a vital role in securing data and ensuring that transactions are only accessible to authorized participants.

For instance, consider the Proof of Work (PoW) consensus algorithm, famously used in Bitcoin. PoW introduced a decentralized financial system that has garnered popularity and trust due to its security measures. Similarly, Proof of Stake (PoS) blockchains, which differ in their approach to transaction validation and block creation, employ principles from game theory to fortify their security as distributed systems. The safety and reliability of a blockchain network hinge on its protocol and the diversity of participants. Larger and more distributed networks are generally more resistant to attacks and failures. In the context of the Byzantine Generals' Problem, Byzantine Fault Tolerant (BFT) systems have emerged. These are now being adopted in industries where security and reliability are paramount, such as aviation, space exploration, and nuclear energy. In the realm of cryptocurrencies, achieving consensus and addressing sustainability concerns are pivotal for the success of any blockchain ecosystem. While PoW and PoS have gained attention as BFT systems, ongoing efforts are directed toward enhancing their security, scalability, and environmental friendliness.

Despite their limitations, these innovative blockchain technologies hold great promise. They offer the potential to usher in new standards of transparency and accountability across a wide range of industries, including mineral mining. As the technology continues to evolve and mature, it is likely to play an increasingly central role in reshaping business processes and ensuring the responsible and efficient management of resources.

Chapter 3: Understanding Private, Public & Consortium Networks in Blockchains

"Blockchain should be used to address opportunities and problems that lack easier answers."

— Julie Sweet

Introduction

The world of blockchain technology is a vast and ever-evolving landscape, offering diverse opportunities and possibilities for businesses and individuals alike. As the adoption of blockchain continues to expand, it becomes crucial to understand the different types of networks that exist within this ecosystem. In this chapter, we delve into the intricacies of private, public, and consortium networks in blockchains, unravelling their distinct characteristics and exploring their applications. When Bitcoin, the pioneering cryptocurrency, was introduced in 2008, it brought with it a decentralized approach to digital transactions. The underlying technology, blockchain, has since become the backbone of numerous industries, revolutionizing the way we transact, verify, and secure digital assets. As blockchain technology gained traction, innovators began to recognize that its potential extends far beyond cryptocurrencies.

One of the fundamental distinctions between various blockchain networks lies in their governance and accessibility. Public networks, such as Bitcoin and Ethereum, operate in an open and permissionless manner, allowing anyone to participate and contribute to the network. On the other hand, private networks restrict access and limit participation to specific entities or individuals. These private networks often prioritize privacy, scalability, and efficiency, catering to the specific needs of organizations and enterprises. In addition to public and private networks, there exists a hybrid model known as consortium networks. Consortium networks combine elements of both public and private networks, involving a group of organizations collaborating to maintain and govern the blockchain network. These networks offer enhanced scalability and transaction throughput while maintaining a degree of control and confidentiality among the consortium members.

Understanding the nuances and trade-offs between private, public, and consortium networks is essential for navigating the blockchain landscape effectively. Each network type possesses unique strengths and considerations that can greatly impact the suitability and feasibility of blockchain solutions in different scenarios. By delving into the intricacies of these network models, we can gain valuable insights into how blockchain can be applied in various industries, including finance, supply chain management, mining industry, and more. There are three types of blockchains: private, public, and consortium chains. In this chapter, we will take a closer look at these different types of blockchains and their specific applications.

Before that, let us reiterate some key features that all three have in common:

1. **An Append-Only Ledger:** To qualify as a blockchain, a system needs to comply with the chain of blocks structure in which every block is connected to the last. This structure is also known as an append-only ledger.

2. **A Network of Peers:** A blockchain operates within a network of peers, where each participant (or node) maintains a complete copy of the blockchain's entire history. Unlike centralized systems where a central authority holds the data and controls access, blockchain participants are equal peers who communicate directly with each other. This decentralized nature eliminates the need for intermediaries and central control, increasing transparency and reducing the risk of a single point of failure.

3. **A Consensus Mechanism:** There must be a consensus mechanism in place for nodes to agree upon the correctness of transactions propagated across the network. This mechanism ensures that there is no bogus data being written to the chain.

The table below sums up some of the major differences.

#	Blockchain type		
	Public	Private	Consortium
Permissionless?	Yes	No	No
Who Can Read?	Anyone	Invited users only	Depends
Who Can Write?	Anyone	Approved participants	Approved participants
Ownership	Nobody	Single entity	Multiple entities
Participants Known?	No	Yes	Yes
Transaction Speed	Slow	Fast	Fast

Table 3.1: Types of Blockchain

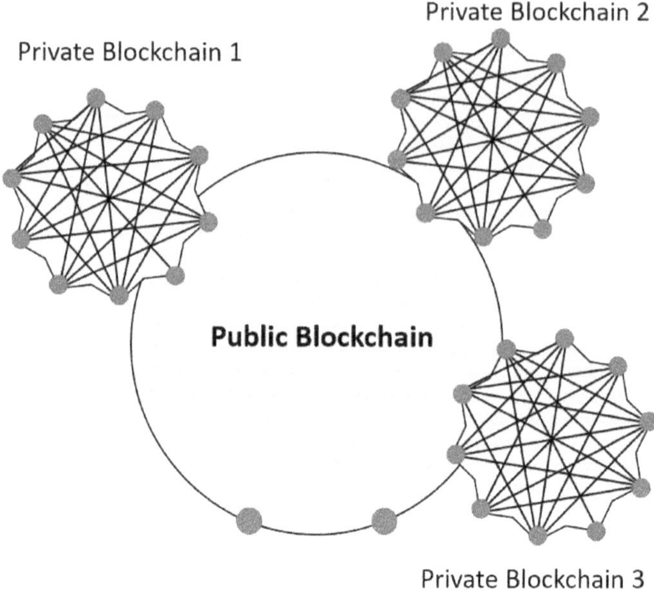

Figure 3.1: Private & Public Blockchain Network

Public Blockchains

Public blockchains are the most common type of distributed ledger, and they are widely used by cryptocurrency users. The term "public" refers to the fact that anyone can view transactions and join the network by downloading the necessary software. This type of blockchain is also referred to as "permissionless," meaning that there are no barriers to entry, and anyone can participate in the consensus mechanism, such as by mining or staking. Due to this open participation, public blockchains tend to have a decentralized topology, with no central authority controlling the network.

Public blockchains are often considered more censorship-resistant than private or semi-private blockchains. However, to prevent malicious actors from gaining an advantage, the protocol must include security mechanisms. Although this approach provides better security, it comes at the cost of performance, with lower throughput and scaling issues. Making changes to the network without causing a split can also be a challenge since all participants must agree on any proposed changes.

In summary, public blockchains are open and decentralized networks that allow anyone to join and participate in consensus mechanisms. However, ensuring security and achieving consensus can be challenging due to the lack of central control, which may result in lower performance and scalability issues.

Merits:
1. Complete trustability and transparency
2. No intermediaries involved in transactions
3. High level of security and resistance to censorship

Demerits:
1. Scalability issues due to the limited processing power of the network
2. Lack of transaction speed compared to centralized systems
3. High energy consumption required to maintain the network and process transactions.

Private Blockchains

When compared to public blockchains, private blockchains operate in a permissioned environment, meaning that access to the chain is restricted to authorized parties who are selected based on established rules. Unlike public blockchains, private blockchains are not decentralized since there is often a hierarchy in place that dictates control. Nevertheless, they are distributed, and many nodes store a copy of the chain on their machines. Private blockchains are suitable for enterprise environments where businesses want to reap the benefits of blockchain technology without exposing their network to the outside world.

Proof of Work (PoW), which is known to be energy-intensive, is a security measure that is essential for open environments such as public blockchains. However, in a private blockchain, where identities are known, governance is hands-on, and the risks that PoW is intended to deter are not as significant, an algorithm that is more efficient can be employed. Appointed validators, which are nodes selected to perform specific functions for transaction validation, are the basis of a more efficient algorithm. These validators usually form a group of nodes that must approve every block. In case any of the nodes act maliciously, they can be quickly identified and removed from the network. Since there is top-down control of the blockchain, coordinating a reversal is relatively straightforward.

In summary, private blockchains offer businesses a way to enjoy the advantages of blockchain technology while maintaining their network's security and privacy. By restricting access to the chain, a private blockchain offers a higher level of control, which allows for more efficient algorithms that can validate transactions faster and use less energy. The downside is that private blockchains are not as transparent and open as public blockchains and may require a trusted central authority to coordinate a reversal in case of a malicious attack or error.

Merits:

1. Higher transaction per second (TPS) due to less complicated consensus mechanisms and fewer participants.
2. Highly scalable as the private blockchains are built to cater to a specific group, thus minimizing the network's overall load.

Demerits:

1. Less secured compared to public blockchains because it is easier to compromise a network with fewer nodes and less rigorous consensus mechanisms.
2. Less decentralized because it is usually operated by a group of companies or organizations, meaning it lacks the broad participation of public blockchains.
3. Achieving trust is difficult because there is no public consensus mechanism, and users must trust the central authority governing the network.

Consortium Blockchains

The consortium blockchain is a hybrid of public and private blockchains, combining the best of both worlds. The most significant difference between the two systems is at the consensus level. Instead of an open system, where anyone can validate blocks, or a closed system where a single entity appoints block producers, a consortium chain sees a few equally-effective parties operating as validators. *(Ref. No.: 40- 44)*

In a consortium chain, the rules of the system are flexible and can be customized to meet the specific needs of the participating organizations. The visibility of the chain can be secured to validators, viewable to authorized individuals, or by all. Provided the validators can reach a fair consensus, modifications can be easily rolled out. Furthermore, if a certain threshold of these parties is acting honestly, the system won't run into any issues.

A consortium blockchain is most beneficial in a setting where multiple organizations operate in the same industry and require a common ground to carry out transactions or relay information. Joining a consortium of this kind could be beneficial to an organization, as it would allow them to share insights into their enterprise with other players.

The merits and demerits of consortium blockchain are as follows:

Merits:

1. Offers a more flexible system that can be customized to meet the specific needs of participating organizations
2. Provides a common ground for multiple organizations operating in the same industry to carry out transactions and share information

Demerits:

1. The security level is not as high as public blockchains
2. The level of decentralization is not as high as public blockchains
3. Achieving trust among consortium members may be difficult.

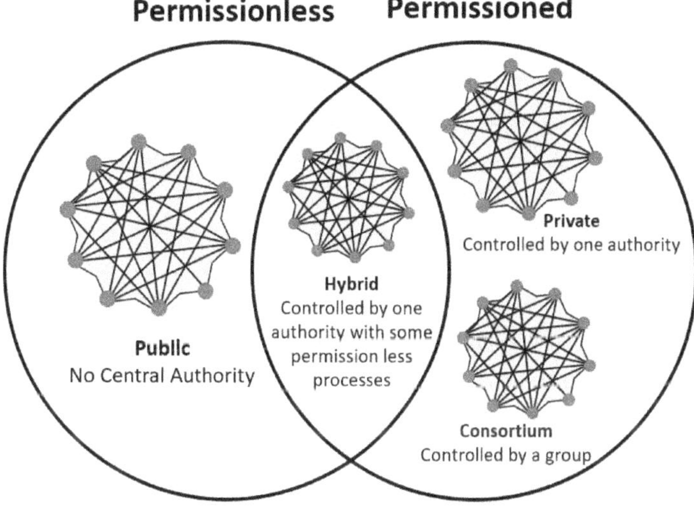

Figure 3.2: Permissioned & Permissionless Blockchains

Understanding the Differences & Use Cases of Public Private & Consortium Blockchains

In essence, public, private, and consortium blockchains are not competing technologies; instead, they offer different solutions to different problems. Public blockchains are best suited for applications where censorship resistance and security are critical, but speed and throughput may be slower due to the nature of consensus mechanisms such as Proof of Work. On the other hand, private blockchains prioritize speed and efficiency, as they do not need to worry about the critical factors of failure that public blockchains do. Private blockchains are often deployed in situations where an individual or organization must maintain control and keep data private.

Consortium blockchains, on the other hand, offer a middle ground between public and private blockchains. They leverage the benefits of both by allowing a group of equally-powerful parties to act as validators, eliminating the need for centralized control. As a result, consortium blockchains can be more efficient than public blockchains due to their smaller node count. They also mitigate some of the counterparty risks associated with private blockchains. Consortium blockchains are most useful for applications where multiple organizations operate in the same industry and require a common ground to carry out transactions or share information.

In summary, choosing the right type of blockchain technology depends on the specific needs of the application. Public blockchains offer a high level of security and censorship resistance, but can be slower and less efficient. Private blockchains prioritize speed and efficiency but have less security and decentralization. Consortium blockchains provide a middle ground that leverages the benefits of both public and private blockchains while mitigating some of the risks associated with each.

Bitcoin Nodes: Definition & Types

In the context of computer or telecommunication networks, a node can serve as a redistribution point or a communication endpoint. It typically refers to a physical network device, but virtual nodes are also used in some cases. A network node is a component that can create, receive, or transmit a message. In the case of Bitcoin, there are several types of nodes: complete nodes, supernodes, miner nodes, and SPV clients.

Complete nodes are full Bitcoin network nodes that store the entire blockchain and validate all transactions. They act as a trustless peer in the network, allowing users to verify their own transactions without relying on a third party. Supernodes are complete nodes that have high connectivity and processing power, making them important in maintaining the health of the network. Miner nodes, on the other hand, are specialized nodes that contribute computing power to solve complex mathematical equations and earn new Bitcoin as a reward. Lastly, SPV (Simplified Payment Verification) clients are lightweight nodes that do not store the entire blockchain but instead rely on other nodes for verification. They are useful for mobile or low-power devices that cannot handle the storage and processing requirements of complete nodes.

Bitcoin Nodes: The Backbone of the Bitcoin Network

Exploring the intricacies of blockchain technology, which often operates as a distributed system, it becomes evident that the network of computer nodes is a critical component that enables Bitcoin to function as a decentralized peer-to-peer digital currency. This design ensures that Bitcoin is resistant to censorship and eliminates the need for a middleman or centralized entity to facilitate transactions, regardless of their location in the world. Each computer or device that connects to the Bitcoin interface is considered a node, as they communicate with one another to facilitate the transmission of data about transactions and blocks across the distributed network of computer systems using the Bitcoin peer-to-peer protocol. There are various types of Bitcoin nodes, each defined according to its specific functions. Full nodes are the ones that provide guidance and security to the Bitcoin network and are therefore critical to its functioning. These nodes are also referred to as fully validating nodes as they are responsible for verifying transactions and blocks against the system's consensus rules. Additionally, full nodes are capable of relaying new transactions and blocks to the blockchain. A complete node typically downloads a replica of the Bitcoin blockchain network with each block and transaction, although it is not strictly necessary to do so. A complete Bitcoin node can be installed using a specialized software implementation, with the most popular one being Bitcoin Core. To run a Bitcoin Core complete node, a desktop or PC with a current version of Windows, Mac OS X, or Linux, 200GB of free disk space, 2GB of RAM, and a high-speed internet connection with an upload speed of at least 50kB/s is required. An unmetered connection or a reference to high upload limits is also necessary. The upload usage of online complete nodes may reach or exceed 200GB/month, with a download usage of 20GB/month. It is important to note that complete nodes must run for at least 6 hours a day, ideally continuously (24/7).

Numerous volunteer groups and users run complete Bitcoin nodes to support the Bitcoin ecosystem, with approximately 9700 public nodes running on the Bitcoin network as of 2018. It is important to note that this number only includes public nodes, which are visible and accessible, and does not account for hidden nodes operating behind firewalls, using hidden protocols like Tor, or those configured not to listen for connections. *(Ref. No.: 44- 52)*

Listening Nodes (Supernodes)

Listening Nodes, also known as Supernodes, play a crucial role in the functioning of decentralized networks like the blockchain. Essentially, these nodes are publicly visible and function as a virtual redistribution point that communicates with every other node that establishes a connection with it. This means that a supernode serves as both a data source and communication bridge, facilitating the transfer of blockchain history and transaction data to multiple nodes across the globe. Supernodes are highly dependable and usually operate 24/7, with numerous designated connections to ensure the efficient flow of information within the network. As a result, they require a higher computational power and better internet connection than hidden nodes. Their high availability and reliability make them essential for maintaining the network's performance and ensuring the security of the transactions.

Miners' Nodes

To mine Bitcoin and participate in the network's transaction verification process, specialized hardware and software programs are necessary in the current competitive environment. These software programs, known as mining software, are separate from the Bitcoin Core software and operate in parallel to mine new Bitcoin blocks. A miner can choose to work alone, referred to as a solo miner, or join a mining pool, known as a pool miner. When working alone, a miner operates their own full node that contains a complete copy of the blockchain. In contrast, pool miners work together, each contributing their own computational resources, known as hash power, towards the same goal of mining new blocks. In a mining pool, only the pool administrator needs to run a complete node, referred to as a pool miner's complete node. By joining a mining pool, miners increase their chances of successfully mining new blocks and earning rewards, as they are able to combine their hash power and share in the rewards earned by the pool. However, they also have to share the rewards with other members of the pool, and the pool operator typically charges a fee for their services. Working alone as a solo miner can be more challenging, as the chances of mining a block and earning a reward are lower, but the entire reward goes to the individual miner. Ultimately, the decision to mine alone or in a pool depends on a miner's resources, preferences, and goals.

Lightweight or SPV Clients

SPV, or Simplified Payment Verification, clients are a type of lightweight client that uses the Bitcoin network without acting as a full node. Unlike full nodes,

SPV clients do not contribute to the network's security as they do not have a copy of the blockchain nor participate in transaction verification and validation. Instead, they rely on other complete nodes, or supernodes, to provide them with the necessary data.

SPV clients are particularly useful for those who do not want to download the entire blockchain or do not have enough storage space on their device to do so. With SPV, users can check whether a transaction has been included in a block without having to download the entire block data. This approach saves time and resources, making it a popular choice for many cryptocurrency wallets. However, it is important to note that relying solely on SPV clients can come with risks. Because these clients do not contribute to the network's security, they are more vulnerable to certain types of attacks. Therefore, it is recommended to use SPV clients in combination with other security measures, such as running a full node or using multi-signature wallets. In contrast to SPV clients, full mining nodes require specialized mining hardware and software to mine Bitcoin blocks. These nodes collect pending transactions and create a candidate block that is then mined. If a miner successfully mines a valid block, it is broadcast to the network, and other full nodes can validate it. In this way, consensus rules are maintained and secured by the distributed network of validating nodes, rather than by the miners alone.

Client vs. Mining Nodes

It is important to understand that running a complete node is not the same as running a mining node. While mining requires expensive hardware and software, anyone can run a fully validating node without any special equipment. Before mining a block, a miner must collect a set of valid transactions that have been accepted by the network nodes. They then create a candidate block that includes these transactions and attempt to mine that block by solving a complex mathematical problem. If they are successful, they broadcast the solution to the network for other full nodes to verify the validity of the block. It is important to note that the consensus rules, which determine the validity of blocks and transactions, are established, and maintained by the network of validating nodes and not the miners. Mining nodes play a critical role in securing the network by processing transactions and creating new blocks, but they do not have the final say in the decision-making process. This responsibility lies with the network of validating nodes that work together to establish consensus and maintain the integrity of the blockchain. *(Ref. No.: 52- 54)*

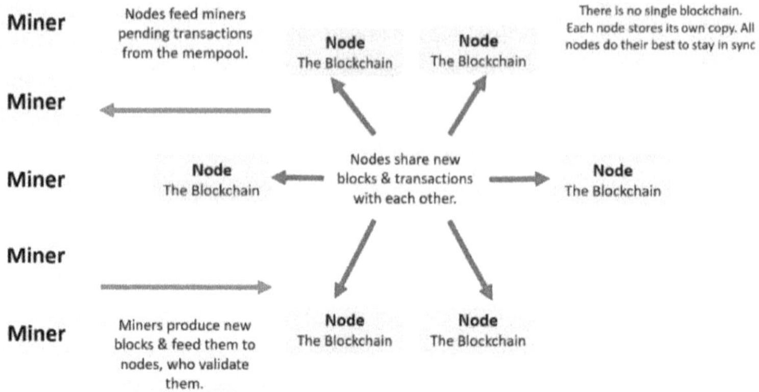

Figure 3.3: Client vs. Mining Nodes

Closing Thoughts

Blockchain technology has revolutionized various industries, providing numerous benefits through public, private, and consortium blockchains. Just like in the coal mining industry where miners have to carefully choose the right tool for the job, users of blockchain technology have to select the most appropriate option that suits their needs. In the case of Bitcoin, the P2P network protocol ensures the network's integrity and trust by allowing nodes to communicate and detect dishonest nodes. Running a full validating node is essential in guaranteeing security, privacy, and trust in the network. This protects users against cyber-attacks, fraud, and double-spending, while granting them complete control over their funds. Although full validating nodes do not offer financial rewards, they offer significant benefits, making them highly recommended. Choosing the right blockchain and employing full validating nodes are crucial for the success of any blockchain ecosystem. This is like how selecting the right coal mining tool and equipment is crucial for the success of coal mining operations. To ensure continued growth and success of blockchain technology, there is a need for the continued development and implementation of new technologies and protocols. This will help increase widespread adoption and success of blockchain technology across various industries, just like how the development and implementation of new coal mining technologies and protocols have improved the productivity and safety of coal mining operations. With its potential to transform how transactions are recorded and verified, blockchain technology has the capacity to shape the future of the digital economy, just like how coal mining shaped the industrial revolution.

Chapter 4: Cryptocurrency, Blockchain & Peer-to-Peer Networks

"I'm passionate about blockchain as a whole, decentralized finance as a whole."

— Spencer Dinwiddie

Cryptocurrency: A Decentralized Digital Currency with Freedom & Autonomy

Cryptocurrency can be compared to digital cash that allows individuals to purchase various items and services, such as books, flights, and resorts, just like they would with physical money. The key difference is that cryptocurrency is entirely digital, which makes it possible to send funds to anyone, anywhere in the world.

However, it is essential to note that cryptocurrency differs from conventional online payment gateways like PayPal or bank transfers. With these traditional payment methods, centralized organizations control the flow of funds, holding onto customers' money and facilitating transfers to other entities on their behalf. In contrast, cryptocurrencies operate through a decentralized network of thousands of entities and individuals who act as their banks. They use open-source software to connect with each other directly without the need for any intermediaries.

This decentralized system eliminates the need for traditional sign-ups, registrations, or the use of third-party payment processors. Instead, users can download decentralized apps onto their devices and start sending and receiving cryptocurrency within minutes. This level of autonomy and freedom from centralized control is one of the most impressive aspects of cryptocurrencies.

Why is it termed Cryptocurrency?

The term "cryptocurrency" is a combination of "cryptography" and "currency." Cryptography is a complex mathematical process that secures our funds and prevents unauthorized access or spending. However, with the help of autonomous decentralized applications, the intricacies of this process are hidden from users, and they can use cryptocurrency without even realizing what is happening behind the scenes. Unlike traditional currency, which is subject to centralized control by banks or other financial institutions, cryptocurrencies are not owned or controlled by anyone.

The use of advanced cryptography ensures the security of the system, making it virtually impossible for anyone to tamper with or steal funds. Although there are already numerous secure payment applications available, cryptocurrency offers several unique features that are not found in traditional payment systems. Therefore, it is important to explore the advantages of cryptocurrency to fully understand its potential benefits.

Cryptocurrency Permissionless

Cryptocurrencies are permissionless, which means that no one can prevent anyone from using them. In contrast, centralized payment systems can halt transactions or freeze accounts at their discretion, limiting users' control over their funds.

Censorship-Resistant

Cryptocurrencies are also censorship-resistant due to their decentralized nature. Unlike centralized payment systems that rely on a single point of control, cryptocurrencies operate on a distributed blockchain network where no single entity has complete control. This means that even if a hacker or other malicious actor were to compromise a single node on the network, the rest of the network would remain secure and operational. In other words, the decentralized nature of cryptocurrencies makes them more resilient to cyber-attacks, censorship, and other types of interference.

A Reasonably Priced & Fast Payment Method

Cryptocurrencies are also a fast and affordable method of payment. Transactions can be completed within seconds, regardless of the sender or recipient's location. Moreover, the transaction fees for cryptocurrency transactions are significantly lower than those for traditional international wire transfers. This means that users can send or receive funds with minimal fees, making cryptocurrencies an attractive option for international payments. Additionally, the speed of cryptocurrency transactions is unparalleled, as there are no intermediaries involved, and funds can be transferred directly from one user to another.

Satoshi Nakamoto & Blockchain Technology

Bitcoin, the world's first cryptocurrency, was invented by an individual or group of people operating under the pseudonym Satoshi Nakamoto. To this day, their true identity remains unknown. In 2008, Satoshi Nakamoto published a 9-page paper detailing the workings of the Bitcoin blockchain network. The software was released anonymously in 2009, and since then, it has inspired the development of many other cryptocurrencies, each with their own unique features. Blockchain is the underlying technology that powers Bitcoin and other cryptocurrencies. It is a decentralized database that is append-only, meaning that once information is added, it cannot be deleted or modified. Each entry, or block, in the database is cryptographically synchronized with the previous one.

This creates a chain of blocks, hence the name "blockchain." The immutability of the blockchain is maintained through the use of cryptography, game theory, and consensus algorithms. One of the unique features of blockchain is its decentralization, which means that there is no central authority controlling it. This makes the network censorship-resistant, as it is virtually impossible for hackers or other cyber attackers to bring down the network. Transactions on the blockchain network are also fast and inexpensive, with funds being transferred within seconds and at a fraction of the cost of traditional international wire transfers. In the cryptocurrency space, there is a common saying - "Do Your Own Research" (DYOR). This means that investors should not rely solely on information from a single source before making financial investments in a specific project. Due diligence is always recommended.

Understanding Cryptocurrency Mining & its Role in Blockchain Security

Cryptocurrency mining is a crucial aspect of the decentralized nature of blockchain technology, as it enables the verification and validation of transactions without the need for a central authority. In addition to ensuring network security, mining also creates new units of cryptocurrency and incentivizes honest participation in the network through transaction fees.

To delve further into the mining process, when a new transaction is made on the blockchain, it is first added to the network's memory pool. This pool contains all unconfirmed transactions waiting to be validated by miners.

The role of a miner is to collect transactions from the memory pool and organize them into a candidate block. Each block has a specific size limit, and the miner must ensure that the block they create is within this limit. They then add a special code known as a "nonce" to the block, which is essentially a random number that, when combined with the block's data, creates a hash that meets a specific target value.

The hash target is an essential aspect of the mining process. It acts as a measure of the computational effort required to mine a block. Miners must use their computational resources to find a nonce that will create a hash that meets the target value. This process requires a significant amount of computational power and is often referred to as a proof-of-work algorithm.

Once a miner successfully finds a hash that meets the target value, they broadcast their candidate block to the network. The other nodes on the network then verify the hash and validate the transactions contained in the block. If the block is valid, it is added to the blockchain, and the miner is rewarded with newly created cryptocurrency and any transaction fees associated with the transactions in the block.

As the number of miners on a network increases, the difficulty of the mining process also increases to maintain a consistent rate of new blocks added to the blockchain. This ensures that the network remains secure and that the block time (the time it takes to add a new block to the blockchain) remains relatively constant. *(Ref. No.: 55- 57)*

Overall, cryptocurrency mining plays a vital role in ensuring the security and integrity of blockchain networks. By incentivizing honest participation in the network, it enables the creation of a decentralized, trustless system that can be used for a variety of applications beyond cryptocurrency.

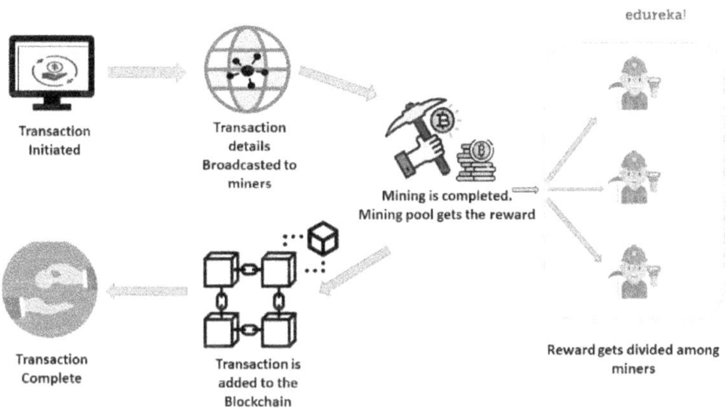

Figure 4.1: Mining In Blockchain Network

Taking a more in-depth look at the mining process.

Step 1 - Hashing Transactions

The first step in the mining process is to take all the pending transactions from the memory pool and hash them one by one. Hashing is a process where a specific input is transformed into a fixed-length output using a hash function. The resulting output, known as a hash, acts as a unique identifier for the input data. In the context of mining, the hash of each transaction includes a string of alphanumeric characters that represent the transaction's information. The miner will then take each transaction's hash and list them individually in a candidate block. Along with these transaction hashes, the miner also includes a custom transaction known as the Coinbase transaction. This transaction is where the miner rewards themselves with newly created cryptocurrency and is usually the first transaction in a new block. After the Coinbase transaction is included, the miner will add all the pending transactions they want to validate in the block. Once all the transactions are included, the miner will move on to the next step of the mining process, where they will try to solve a complex mathematical problem to validate the block. It is important to note that each block has a specific size limit, and miners need to ensure that the block they create does not exceed this limit. This is why miners prioritize including transactions with the highest transaction fees as they will receive a larger reward for validating them.

Step 2: Creating a Merkle Tree

Once all the transactions are hashed and listed in the candidate block, they are then organized into a data structure called a Merkle Tree (also known as a hash tree). This structure is created by pairing up transaction hashes and hashing them together. The resulting hash outputs are then paired up and hashed again, and this process is repeated until there is only one hash left. This final hash is known as the root hash, or Merkle root, and represents all the previous hashes that were used to generate it. The Merkle Tree is a critical component of the mining process as it helps ensure the security and integrity of the blockchain network. By organizing the transaction hashes in a tree-like structure, the Merkle Tree enables miners to efficiently and securely verify the authenticity of each transaction without having to hash every single transaction in the block. Instead, miners only need to verify the Merkle root, which acts as a summary of all the transactions in the block. In addition to enhancing security and efficiency, the Merkle Tree also enables nodes on the blockchain network to quickly and easily verify if a specific transaction is included in a block. By referencing the Merkle root, nodes can quickly traverse the tree to find the relevant transaction hash and confirm its inclusion in the block.

Importance of the Merkle Tree in Blockchain

1. **Security & Integrity:** It ensures the security and integrity of blockchain data by making it nearly impossible to tamper with transactions within a block.

2. **Efficiency:** It improves efficiency by allowing quick verification of the authenticity of transactions within a block through the Merkle root.

3. **Quick Transaction Verification:** Nodes can rapidly confirm whether a specific transaction is part of a block by referencing the Merkle root.

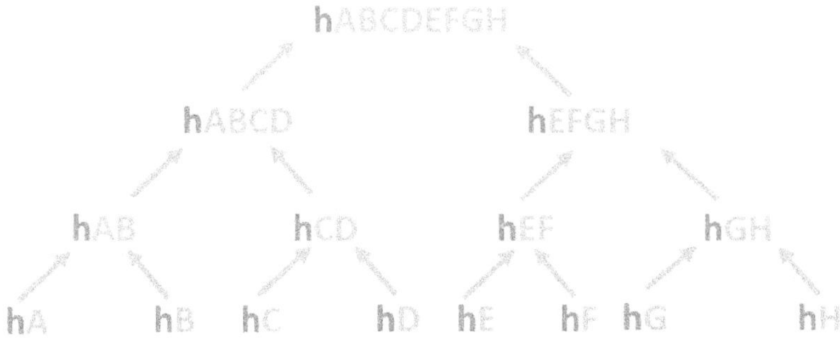

Figure 4.2: Merkle Tree, also known as a Hash Tree

Step 3: Finding a Legitimate Block Header (Block Hash)

To identify each block uniquely, a block header is created by the miners. This block header is made by combining the hash of the previous block with the root hash of the current candidate block. Additionally, a nonce, which is a random value, is added to the mix. A nonce is a one-time-use value that is used in various cryptographic hash functions and authentication protocols. During the process of mining, the miner combines the root hash, the hash of the previous block, and the nonce value, and applies a hash function to find a legitimate hash. A legitimate hash is one that is less than a predetermined target value, as set by the protocol. In Bitcoin mining, the block hash must start with a specific number of zeros, which is known as the mining difficulty. Miners have to repeatedly change the nonce value until a valid hash is found.

Once the Merkle Tree is created, the next step in the mining process is to create a block header, which is a unique identifier for each block. The block header is made up of several pieces of information, including the hash of the previous block, the root hash of the current block (created from the Merkle Tree), and a nonce value.

A nonce is a random value that is added to the block header to make it unique. The miner's goal is to find a valid hash that is less than a specific target value set by the protocol. To do this, the miner changes the nonce value and repeatedly applies a hash function until a valid hash is found.

Step 4 - Broadcasting the Mined Block

The process of mining a block does not stop after finding a legitimate block hash. The next step is to broadcast the mined block to the network. Once a miner has found a legitimate block hash, they will broadcast the newly mined block to the network. The rest of the nodes on the network will receive this block and will perform several checks to ensure its validity. If the block passes these checks, it will be added to their copy of the blockchain.

Once the block has been confirmed and added to the blockchain, all the miners will start working on mining the next block. This is because the blockchain is a chronological sequence of blocks, and each block includes a reference to the preceding block. Therefore, the miners have to continue their work to maintain the continuity of the blockchain.

However, not all miners may have been able to find a legitimate block hash. In this case, they discard their candidate block and start over with a new set of pending transactions. This is because the mining race is an ongoing process that requires a constant amount of computational power to remain secure and decentralized.

Overall, the process of mining a block involves several intricate steps, starting from hashing transactions and creating a Merkle tree, to finding a legitimate block header (block hash) and broadcasting the mined block to the network. All of these steps work together to maintain the integrity of the blockchain network and ensure that transactions are validated and recorded in a decentralized and secure manner.

Mining & Difficulty in Cryptocurrency: How Blocks are Mined & Difficulty is Adjusted

In cryptocurrency, the term "difficulty" refers to the level of effort required to mine a block. The difficulty of mining a block is regularly adjusted by the protocol to ensure that the rate at which new blocks are created remains consistent. This is what makes the issuance of new coins consistent and predictable. The difficulty adjusts in response to the amount of computational power, or hash rate, committed to the network.

The Proof of Work (PoW) consensus algorithm in a blockchain network enforces certain guidelines that cause the difficulty to rise or fall depending on the amount of hashing power on the network. This is done to ensure that blocks are not produced too quickly, which could compromise the security of the network, and to ensure the ongoing security of the network. For example, Bitcoin sets the block time at approximately ten minutes, which is the average time it takes to find a new block. If blocks are consistently taking longer to find, the difficulty will be increased. If blocks are being discovered too quickly, the difficulty will be decreased.

As new miners join the network and competition increases, the hashing difficulty will rise, preventing the average block time from decreasing. Conversely, if many miners decide to leave the network, the hashing difficulty will go down, making it easier to mine a new block. These adjustments help to maintain the block time constant, regardless of the overall hashing power of the network.

However, sometimes it happens that miners broadcast a legitimate block at the same time, causing the network to end up with competing blocks. In this case, miners begin to mine the next block based on the block they received first, leading to the network splitting temporarily into different versions of the blockchain. The competition among these network blocks will continue until a subsequent block is mined on top of either one of the competing blocks. When a new block is mined, whichever block that came before it will be considered by the winner. The block that gets abandoned is referred to as an orphan block or a stale block, leading all miners that picked this block to switch back to mining the chain of the winner block. An orphan block is a block whose parent block is unknown or non-existent. These types of blocks have been formed in older versions of the Bitcoin Core software, where network nodes could receive blocks regardless of the lack of data about their ancestry. However, since the release of Bitcoin Core v.0.10 in early 2015, Bitcoin orphan blocks (in the literal sense) are no longer possible. It is worth noting that not all cryptocurrencies are mineable. While Bitcoin is the most widespread and well-established example of a mineable cryptocurrency, other cryptocurrencies use different consensus algorithms that do not require mining, such as Proof of Stake (PoS).

Overview of Proof of Work & Cryptocurrency Mining Techniques

Proof of Work (PoW) is the original and most well-known consensus mechanism used by blockchain networks. Satoshi Nakamoto, the anonymous creator of Bitcoin, introduced PoW in the Bitcoin whitepaper in 2008. The mechanism aims to establish consensus among distributed network participants without the need for a third-party intermediary. PoW achieves this by requiring miners to solve complex cryptographic puzzles using specialized mining hardware that consumes vast amounts of computational power. This computational work disincentivizes bad actors, as the cost of attempting to modify the blockchain becomes prohibitively expensive.

In a PoW network, transactions are confirmed by miners who compete to be the first to solve the cryptographic puzzle and add a new block to the blockchain. The first miner to do so receives a block reward in the form of newly minted cryptocurrency, as well as any transaction fees associated with the transactions in the block. The block reward varies across different blockchains, but for example, as of December 2021, a Bitcoin miner can receive 6.25 BTC for successfully adding a block to the blockchain.

It is important to note that the amount of BTC in a block reward decreases by half every 210,000 blocks, which occurs approximately every four years, due to Bitcoin's halving mechanism. This is an intentional feature designed to limit the supply of Bitcoin and ensure its scarcity.

Cryptocurrency Mining Techniques: Exploring Different Methods

Cryptocurrency mining is an essential process to verify and validate transactions, create new coins, and maintain the blockchain network. To achieve this, miners use specialized computing units to solve complex cryptographic equations. Here, we explore different mining methods and how they work.

CPU Mining: CPU mining is the earliest and simplest way to mine cryptocurrencies. It involves using a PC's CPU to carry out the hash functions required by Proof of Work. Initially, the mining difficulty was low, and anyone with an ordinary CPU could attempt to mine Bitcoin and other cryptocurrencies. However, as more people started to mine, and the network's hash rate increased, profitable CPU mining became increasingly difficult. Moreover, specialized mining hardware with greater computational strength made CPU mining almost impossible. Today, CPU mining is no longer a feasible option, as all miners use specialized hardware.

GPU Mining: Graphics Processing Units (GPU) are designed for processing a wide variety of programs in parallel, including cryptocurrency mining. GPUs are relatively cheap and more flexible than ASIC mining hardware. Some altcoins can be mined with GPU's; however, the efficiency depends on the mining difficulty and algorithm.

ASIC Mining: An Application-Specific Integrated Circuit (ASIC) is designed to serve a single unique purpose. In crypto, it refers to specialized hardware developed for mining. ASIC mining is remarkably efficient but expensive. As ASIC miners are at the cutting-edge of mining technology, the price of a unit is much higher than CPUs or GPUs. Moreover, continuous improvements in ASIC technology quickly render older ASIC models unprofitable, which means that they frequently need to be replaced. This makes ASIC mining one of the most expensive approaches to mine, even without including energy costs.

In summary, each mining technique has its pros and cons. CPU mining is no longer viable, while GPU mining is suitable for some altcoins. However, ASIC mining is the most efficient but also the most expensive option. Miners need to consider several factors, including mining difficulty, algorithm, and energy consumption, before choosing a mining method.

Mining Pools: A Strategy for Cryptocurrency Mining

In cryptocurrency mining, the block reward is granted to the first miner who successfully finds the appropriate hash. However, with the enormous computing power required for mining, the possibility of finding the right hash is exceedingly small. This means that miners with a small percentage of the mining energy stand a very small chance of coming across the subsequent block on their own. To solve this problem, miners have turned to mining pools. Mining pools are groups of miners who pool their resources (hash energy) to increase the probability of successfully mining the next block. By working together, mining pools can increase their computing power, thereby increasing their chances of winning block rewards. When a mining pool successfully finds a block, the miners split the reward equally amongst all participants in the pool, based on the amount of work contributed. This allows individual miners to benefit from the resources of the pool in terms of computing hardware and electricity expenses. However, the dominance of mining pools raises concerns about a potential 51% attack on the network. If a single mining pool controls more than 50% of the network's total hash power, it could potentially manipulate the blockchain by withholding transactions or double-spending. Therefore, it is important to ensure that no single pool dominates the network.

Overall, mining pools are a popular strategy for cryptocurrency mining that allows individual miners to increase their chances of earning block rewards while minimizing their expenses. However, it is important to be aware of the risks associated with mining pool dominance and take appropriate measures to prevent them.

Peer-to-Peer Networks Explained

A Peer-to-Peer (P2P) network, a concept in computer science, involves a group of interconnected devices that collectively store and share documents. Each node in the network is an individual peer, and they have identical abilities and perform the same tasks. In financial technology, P2P refers to the trading and exchange of digital assets or cryptocurrencies through a decentralized network. P2P platforms allow traders and consumers to execute trades without the need for central intermediaries. Additionally, some P2P websites may connect borrowers and lenders. Although P2P networks have various use cases, they gained popularity in the 1990s with the advent of document-sharing applications. Today, P2P networks are a vital part of many cryptocurrencies and play a significant role in the blockchain industry. P2P networks are also leveraged in other distributed computing applications such as web search engines, online marketplaces, streaming platforms, and the Inter Planetary File System (IPFS) web protocol.

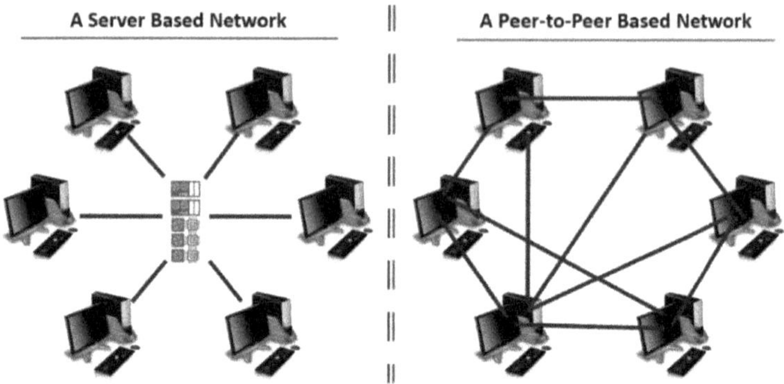

Figure 4.3: Difference between a P2P & Server Based Network

Understanding Peer-to-Peer Networks:

To put it simply, a Peer-to-Peer (P2P) network is a decentralized system in which multiple nodes, or users, are connected to each other and share files. Unlike traditional client-server systems, P2P networks do not have a centralized administrator or server. Each node acts both as a client and a server, making the sharing of documents more efficient and faster. P2P networks are maintained through a distributed network of users who hold a duplicate of the documents, and every node can upload and download files from other nodes. This architecture makes P2P networks resistant to cyberattacks since they do not have a single point of failure.

P2P systems are used for various purposes, but they are mainly known for their use in the exchange/trading of cryptocurrencies or digital assets. They enable consumers and traders to execute trades without the need for centralized intermediaries, allowing for a more decentralized financial system. There are three main types of P2P networks: unstructured, structured, and hybrid. Unstructured P2P networks are the simplest form, where nodes connect to each other randomly. In contrast, structured P2P networks use a specific algorithm to connect nodes in a more structured way. Hybrid P2P networks combine features of both unstructured and structured P2P networks. Overall, P2P networks have become increasingly important in the blockchain industry, powering most cryptocurrencies' networks. But they are also leveraged in other distributed computing applications, such as web search engines, streaming platforms, online marketplaces, and the Inter Planetary File System (IPFS) web protocol.

Types of the peer-to-peer network

P2P networks can be classified into three types based on their architecture, namely:

1. Unstructured P2P Networks
2. Structured P2P Networks
3. Hybrid P2P Networks

Let us discuss each type in detail below.

Unstructured P2P Networks: Unstructured P2P networks are characterized by a lack of organization among the nodes. In other words, the participants communicate randomly with one another. These types of networks are considered robust against high levels of churn activity, which refers to numerous nodes frequently joining and leaving the network. However, unstructured P2P networks may require higher CPU and memory utilization because search queries are dispatched out to the highest number of peers possible. This can flood the network with queries, especially if only a small number of nodes are supplying the desired content.

Structured P2P Networks: In contrast to unstructured P2P networks, structured P2P networks present an organized structure, allowing nodes to efficiently search for files even if the content is not widely available. This is often achieved through the use of hash functions that facilitate database lookups. While structured networks can be more efficient, they tend to offer higher levels of centralization and typically require higher setup and maintenance costs. Structured networks are also less robust when faced with high rates of churn.

Hybrid P2P Networks: Hybrid P2P networks combine the traditional client-server model with some aspects of the Peer-to-Peer structure. For example, a central server can be used to facilitate the connection between peers. When compared to the other types, hybrid models tend to offer improved overall performance. They typically integrate some of the primary benefits of both approaches, achieving significant levels of efficiency and decentralization simultaneously. It is worth noting that the type of P2P network used depends on the specific use case and requirements of the system. Each type has its own advantages and disadvantages, and the decision to use one over the other will depend on factors such as network size, data availability, and desired levels of centralization.

Distributed vs. Decentralized

It is important to note that although the P2P structure is designed to be decentralized, not all P2P networks are completely decentralized. Some P2P networks rely on a central authority to manage and guide network activities, making them highly centralized. For example, some P2P file-sharing systems allow users to download and share files with other users, but they cannot participate in other activities, such as managing search queries. Moreover, the degree of decentralization in P2P networks can vary widely. Even small networks controlled by a limited user base with shared goals can be considered to have a higher degree of centralization, despite the lack of a centralized network infrastructure. Therefore, it is essential to understand the level of centralization in a P2P network before using it for any purpose. In highly centralized networks, the central authority can exercise significant control over the activities of the users. On the other hand, completely decentralized networks provide more freedom to users, but may be more difficult to manage and coordinate.

The Role of P2P Architecture in Blockchain Network

During the early stages of Bitcoin in 2008, Satoshi Nakamoto defined it as a "Peer-to-Peer Electronic Cash System." Its purpose was to provide a digital form of currency that can be transferred through a P2P network, which manages a decentralized distributed ledger known as a blockchain network. The P2P architecture that is fundamental to blockchain technology is what allows Bitcoin and other cryptocurrencies to be transferred globally without the need for intermediaries or any central server. Anyone can set up a Bitcoin node and take part in verifying and validating blocks, without the need for centralized banking institutions to process or record transactions. In a Bitcoin network, the blockchain acts as a decentralized digital ledger that publicly records all financial activity. Every node in the network holds a copy of the blockchain and compares it to other nodes to ensure the accuracy of the information. Any malicious activity or inaccuracies are quickly rejected by the network. If someone tries to modify the data, the network will discard any inaccurate data, making it very secure. P2P architecture has a significant impact on the blockchain network in terms of how nodes participate in network activities.

Nodes can take on various roles, and in the context of cryptocurrency blockchains, full nodes are those that offer security to the network by verifying transactions against the system's consensus rules. Full nodes verify transactions using the consensus algorithm set by the network and hold a complete, updated copy of the blockchain, enabling them to participate in the collective work of verifying the actual state of the distributed ledger. It is important to note that not all full validating nodes are miners, and full validating nodes help make the network more secure and are responsible for having a complete and updated copy of the blockchain's ledger. The decentralized nature of P2P architecture and blockchain technology means that anyone can participate in the Bitcoin network, helping to validate and verify blocks, just like an open P2P network. It is important to understand that there is no need for a central authority to record or process transactions on a blockchain, including the Bitcoin network. Every public activity is recorded in a digital ledger, making it accessible almost everywhere in the world in an instant. The decentralized and open nature of P2P architecture and blockchain technology makes them truly amazing and similar!

Advantages

The Peer-to-Peer (P2P) architecture of blockchains provides numerous advantages, with the most critical being the extra security it offers in comparison to traditional client-server setups. The distribution of blockchain data across a vast network of nodes makes it almost impervious to Denial-of-Service (DoS) attacks, which are a common problem for many systems. Additionally, the consensus requirement of a majority of nodes to validate new information makes it almost impossible for malicious actors to manipulate the data on the blockchain, especially for larger networks like Bitcoin. The distributed P2P network, coupled with the majority consensus requirement, provides blockchains with a high degree of resistance to malicious cyber activities, which is why Bitcoin (and other blockchains) can achieve the Byzantine Fault Tolerance. This resistance also means that cryptocurrency blockchains are immune to censorship by central authorities, unlike traditional financial institutions where governments can freeze or exhaust accounts with the help of financial organizations. This also extends to private payment processing and content platforms that may attempt to censor transactions or content. Another advantage of using P2P architecture in blockchains is data immutability. Once data is written, it cannot be altered, and the larger the network, the less likely it is for data to be tampered with. A 51% attack would require an entity to control the majority of nodes, making it difficult to carry out. Blockchains with P2P architecture can run independently without censorship from central authorities, while traditional banks require complete control over customer information, and they can also restrict transactions at their discretion.

Limitations

Although the P2P architecture of blockchain networks provides many advantages, there are also some limitations to consider. One of the significant limitations is the requirement for a massive amount of computing power to add transactions to the network. As each node must have an up-to-date copy of the distributed ledger, the computing power required can reduce the network's performance and scalability, hindering its widespread adoption. However, cryptographers and blockchain developers are exploring solutions that can be used as scaling options, such as the Lightning Network, Ethereum Plasma, and the Mimblewimble protocol. Another potential problem is related to attacks that can occur during hard fork events. As most blockchains are decentralized and open-source, groups of nodes can duplicate and update the code to form a new, parallel network. While hard forks are normal and not a threat to the network, if certain security strategies are not followed, each chain could become vulnerable to replay attacks. Additionally, the distributed nature of P2P networks makes it exceptionally challenging to manipulate and regulate the strategy, creating potential security issues in the blockchain niche. In summary, while P2P architecture has significant advantages for blockchain networks, it also has some limitations that need to be addressed. The high computing power requirement for adding transactions and the potential for replay attacks during hard fork events are among the primary concerns. However, ongoing research and development in the blockchain field are exploring solutions to address these limitations, allowing for greater scalability and wider adoption of blockchain technology. *(Ref. No.: 58- 62)*

Advantages of Peer-to-Peer Networks: Resilience, Scalability & Efficient File Sharing

Peer-to-peer networks have become very common in recent times, but there are several other advantages to using this type of network that gives it an edge over others. Here are a few reasons why:

1. **Resilience:** P2P networks are very difficult to bring down. Even if one unit goes down, other nodes continue to operate and communicate, ensuring that the network remains functional. This makes P2P networks more resistant to failures, cyber-attacks, and censorship compared to centralized networks.

2. **Scalability:** P2P networks are highly scalable because adding new peers is a quick and easy process. Unlike centralized networks, there is no need for any central configuration on the server, making it easier to accommodate more peers.

3. **Efficient File Sharing:** P2P networks excel in the realm of file sharing, offering an ideal platform for distributing digital content swiftly and efficiently. This efficiency arises from the unique way P2P networks handle file storage and retrieval. In a P2P environment, a file is not stored on a single central server but is rather distributed across multiple connected peers. Each peer holds a portion of the file, which collectively forms the complete file. This decentralized file storage approach enables concurrent downloading from multiple locations, without imposing undue strain on a central server. As a result, users seeking to download a file can access and retrieve different parts of it simultaneously from various peers. This parallelism significantly accelerates the download process, making file sharing on P2P networks notably faster and more efficient than traditional centralized networks.

Closing Thoughts

Cryptocurrency mining and P2P architecture are both crucial elements of the blockchain technology that underpins cryptocurrencies. While mining helps to keep the network stable and the issuance of new coins steady, it also has its potential drawbacks such as fluctuations in profitability due to market prices and electricity expenses. Therefore, it is important to research and evaluate all possible risks before engaging in cryptocurrency mining. P2P architecture is equally important in blockchain technology, as it offers superior security, decentralization, immutability, and censorship resistance by distributing transaction ledgers across vast networks of nodes. Its applicability is not limited to blockchain technology alone but can also be employed in other distributed computing applications like file-sharing networks and energy trading platforms. As blockchain technology continues to evolve, both cryptocurrency mining and P2P architecture will continue to play significant roles in its success and development. It is highly likely that we will see new applications and use cases emerge for both mining and P2P architecture. The success of blockchain technology will depend on continued innovation and development in these areas, as well as on the ability of the industry to adapt to new challenges and changing market conditions.

Chapter 5: Consensus Algorithms in Cryptocurrency: An Overview & Types

"The right way to think about the blockchain is that it's going to replace the entire Internet."

— Brock Pierce

Introduction

The concept of a peer-to-peer (P2P) network is not a new one, as it has been around for several decades. However, with the advent of blockchain technology, P2P networks have gained even more attention as they offer numerous advantages over traditional centralized networks. One of the critical components of a P2P network, particularly in the context of cryptocurrencies and digital ledgers, is the consensus algorithm. A consensus algorithm allows nodes or users to coordinate with each other in a distributed setting to agree on a single source of truth. This is particularly important in a decentralized setup where no single entity has control over the system. In this context, consensus algorithms are essential for ensuring fault tolerance and maintaining the integrity of the system. In this chapter, we will explore the importance of consensus algorithms in cryptocurrencies and distributed digital ledgers and how they are used to ensure a single source of truth in a decentralized setup. We will also delve into the common properties of various consensus algorithms, including the requirement for validators to provide a stake, the role of transparency, and the importance of dissuading malicious nodes from acting dishonestly.

Consensus Algorithms & Cryptocurrency

A consensus algorithm is a mechanism that allows machines or users to coordinate in a distributed setting to agree on a single source of truth, even in the presence of failures. In a centralized setup, a single entity has power over the system, and changes can be made at will without a complex governance system for reaching consensus among multiple administrators. However, in a decentralized setup, consensus is paramount. In a distributed database, how does an entity reach agreement on which entries should be added?

Consensus algorithms are essential to the functioning of cryptocurrencies and distributed digital ledgers. In cryptocurrencies, users' financial balances are recorded in a blockchain, and it is crucial that every node maintains an identical duplicate copy of the database to prevent conflicting information. Public-key cryptography ensures that users cannot spend each other's coins, but there still needs to be a single source of truth that network contributors can rely on to determine whether funds have already been spent.

To coordinate with network contributors, Satoshi Nakamoto, the anonymous founder of Bitcoin, proposed a Proof of Work (PoW) approach. However, there are other consensus algorithms in existence that share common properties. Firstly, block validators must provide a stake, which is a form of value that dissuades malicious nodes from acting dishonestly. If they cheat, they lose their stake. This stake can be in the form of computing power, cryptocurrency, or even reputation. There is also a reward available, typically the protocol's native cryptocurrency, which is made up of fees paid by other users, freshly-generated cryptocurrency units, or both. Finally, transparency is required to detect cheating. The goal of the consensus algorithm is to make sure that every representative in the system can agree on a single source of truth, even in the presence of failures. The system must be fault-tolerant, and in a decentralized setup, the consensus algorithm is essential for the network to function properly.

Types of Consensus Algorithms:

Proof of Work (PoW) & Proof of Stake (PoS) is the most-discussed consensus algorithms. But there is an extensive variety of other consensus algorithms also, and each comes with its very own advantages & disadvantages.

Other Consensus algorithms are as follows:

1. Delayed Proof of Work (DPoW)

2. Leased Proof of Stake (LPoS)

3. Proof of Authority (PoA)

4. Proof of Burn (PoB)

5. Delegated Proof of Stake (DPoS)

6. Hybrid Proof of Work /Proof of Stake

Understanding the Proof of Work (PoW) Consensus Algorithm

The Proof of Work (PoW) consensus algorithm is a crucial component of many blockchain networks, including Bitcoin. It works by requiring validators, also known as miners, to hash data until they arrive at a specific solution that meets certain requirements set by the protocol. A hash is a seemingly random string of letters and numbers generated by running data through a hash function. Even if a single detail is changed, the output hash would be completely different. This characteristic makes it impossible for a user to predict the input data based on the output hash, making it ideal for proving that a user had knowledge of a particular piece of data before a certain time. To create a valid block, the protocol sets specific conditions that must be met, such as requiring the block's hash to begin with a certain sequence of characters. The only way for a miner to create a block that meets these requirements is to use automation and scripts to brute-force inputs. With major blockchains, the bar is set very high, requiring miners to have powerful hashing hardware and significant computational resources.

The miner's stake in the network is the expense of the hardware and electricity required to run it. ASICs, specialized hardware used for mining cryptocurrencies, are built for this specific purpose, and have no other practical use. A miner's only way to recoup their investment is by successfully adding a new block to the blockchain network, which yields a significant reward. Despite the computational complexity of the process, the network can easily verify that a miner has created a valid block. If a miner produces a hash that meets the protocol's requirements, the block will be accepted, and the miner will receive a reward. If the hash is not valid, the network will reject the block, and the miner will have wasted time and electricity for nothing.

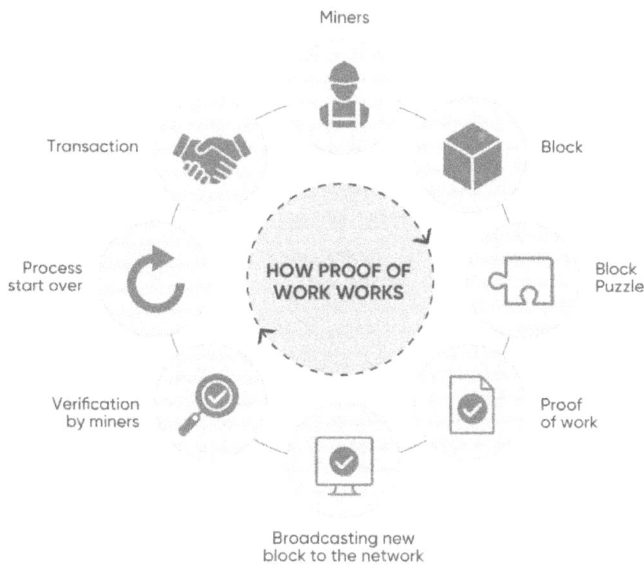

Figure 5.1: Proof of Work Consensus Algorithm

Pure Proof of Work (PoW) Consensus Algorithm & the Role of Miners in Network Governance

The Pure Proof of Work (PoW) consensus algorithm is a key feature of many blockchain networks, including the Bitcoin network. In PoW, miners play a critical role in introducing new blocks to the chain by correctly guessing the solution to a complex mathematical problem. Each time a miner makes a legitimate guess, they are able to assemble a block that the network accepts.

While miners are free to mine any chain, they are incentivized to mine at the longest chain. This is because the longest chain is the one with the maximum gathered Proof of Work, or in other words, the maximum number of hashes or guesses. Once a miner sees a legitimate new block, they will attempt to discover the answer for the subsequent block that permits them to construct on the pinnacle of the preceding one. This is what allows the blockchain to function as a ledger for economic transactions.

When a transaction takes place in a block that sends money to a wallet, and if numerous blocks are constructed on the pinnacle of that block (confirmations), it becomes doubtful that the block (with all the transactions) will be re-written. This is because if an entity controls sufficient hashing electricity to surpass the "honest chain," it may re-write (or reorganize) the blockchain through mining on an "old" block in preference to the modern-day block.

This type of attack is also called a 51% attack. The attacker spends in block X via sending to an exchange, then begins mining a parallel chain in private (blocks, are not broadcasted, to the blockchain network). Once the desired number of confirmations has passed, the attacker trades the coins for something else and withdraws that from the exchange. When the withdrawal clears, they launch the parallel chain, and if it has greater PoW (blocks) than the authentic chain, the blockchain network accepts it as a valid chain and the version of history characterized by the original chain (including the attacker's stake) will disappear. The assailant is then free to spend these coins again.

As miners are the only entities who can upload blocks to the chain in natural PoW cryptocurrencies, this presents them with a robust function in governance. For an alternative to the network consensus policies to be adopted, it needs to have the help of a majority of hash electricity. "Soft forks" require sufficient miners to understand a brand-new rule set so customers can transact and count on their transactions to be nicely processed and included in blocks. A "hard fork" could diverge the blockchain network into components, and via the commonly shared rule of "the chain with maximum PoW is the proper chain to follow," miners could determine which one is accepted as legitimate.

Overall, PoW consensus algorithm and the role of miners in network governance are crucial aspects of blockchain technology that enable secure and trustworthy transactions on decentralized networks. While PoW has its limitations, it remains a fundamental building block of the blockchain ecosystem and continues to play an important role in the development of new blockchain technologies.

Proof of Work & the Benefits of Cryptocurrency Blockchains

Cryptocurrency systems rely on a distributed ledger called a blockchain, which is maintained using a consensus algorithm known as Proof of Work (PoW). The PoW mechanism is a fundamental part of the Bitcoin protocol, which was created as an alternative to the centralized and inefficient traditional international monetary system. By implementing PoW, a decentralized peer-to-peer financial network can provide real-time payment settlements without the need for intermediaries, reducing transaction costs.

PoW is implemented by a network of mining nodes that use specialized hardware (ASICs) to solve complex cryptographic problems. These miners compete to mine a new block approximately every ten minutes by solving a cryptographic puzzle. Once a miner solves the puzzle, they are allowed to add a new block to the blockchain, and they are rewarded with newly created coins and all the transaction fees for that block. However, mining comes at a cost, as it requires a considerable amount of energy, failed attempts, and expensive ASIC hardware.

Despite its limitations, PoW is still considered the most secure and dependable consensus algorithm for blockchain networks. It is a mechanism for preventing double-spending, which is a crucial security feature of cryptocurrencies. PoW was first introduced by Satoshi Nakamoto in 2008 and was designed to secure the decentralized ledger used by cryptocurrencies. However, the technology itself was not new, and an early instance of a PoW algorithm was Adam Back's HashCash, which was developed to mitigate spam by requiring senders to perform a small amount of computing before sending an email.

PoW is a computationally intensive process that involves solving a cryptographic puzzle. The difficulty of the puzzle is adjustable, making it possible to adjust the rate at which new blocks are added to the blockchain. This feature allows the network to maintain a steady flow of new blocks while preventing malicious actors from creating blocks at a faster rate than the rest of the network. PoW has been successful in creating a trustless decentralized network, but it has limitations, such as scalability issues, which restrict the number of transactions that can be processed per second. Nevertheless, PoW remains the standard for creating a fault-tolerant solution for blockchain networks.

What is a Double-Spend?

Cryptocurrencies operate on a digital ledger known as a blockchain, and one of the major concerns in digital cash systems is preventing double-spending. Double-spending occurs when the same funds are used more than once, which can lead to the collapse of the currency. This is not a problem with physical cash since it is not possible to spend the same bill twice.

However, in digital cash systems, it is possible to duplicate a digital document and send it to multiple recipients, which can result in multiple uses of the same funds. To prevent this, a consensus algorithm like Proof of Work (PoW) is used.

The PoW algorithm involves miners using specialized hardware (ASICs) to solve complex cryptographic problems to mine a new block every ten minutes. To add a new block to the blockchain, a miner must find a solution to that block, completing a proof of work. The miner is rewarded with newly created coins and transaction charges of that particular block.

The PoW algorithm plays a crucial role in preventing double-spending. For instance, when a user wants to send some amount of cryptocurrency to another user, they broadcast their transaction to the network. The miners on the network then validate the transaction and check whether the user has enough funds to complete the transaction. Once the transaction is verified, it is included in the blockchain, and the user's account is debited with the amount they sent.

Since the miners are incentivized to mine a new block by being rewarded with newly created coins and transaction charges, it is in their best interest to prevent double-spending. If a user tries to spend the same funds twice, the miners will recognize that the funds have already been spent, and the second transaction will be rejected. Mathematically, the probability of successfully double-spending can be calculated using the following formula:

$$P(Double\text{-}spending) = q^n$$

Where P(Double-spending) is the probability of double-spending, q is the proportion of computational power controlled by the attacker, and n is the number of confirmations required before the transaction is considered final. Therefore, the longer a transaction has been confirmed (i.e., the larger the value of n), the lower the probability of double-spending becomes. This is because the attacker would need to have control over the majority of the computational power on the network, which becomes increasingly difficult as more confirmations are added to the transaction.

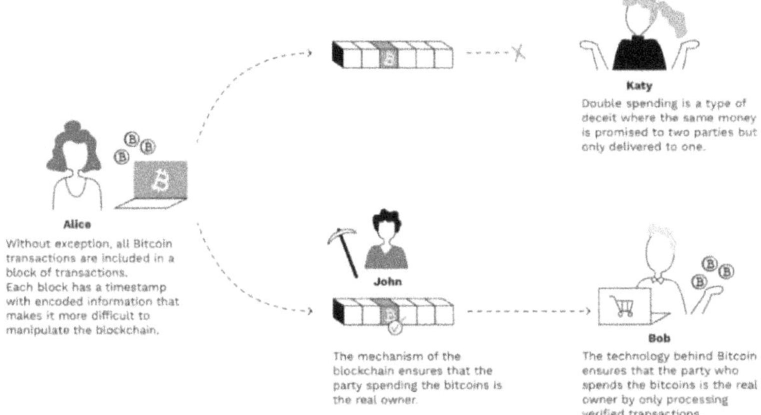

Figure 5.2: Double Spending in Blockchain Network

How Proof of Work (PoW) Ensure the Legitimacy & Security of Blockchain Transactions?

To ensure that transactions are legitimate, a consensus mechanism is required for decentralized networks. In the case of blockchain, this is achieved through Proof of Work (PoW). PoW ensures that every user who wants to update the blockchain must prove that they have done some computational work in the form of solving a complex cryptographic puzzle. This computational work is referred to as mining.

In a blockchain network, users broadcast their transactions to the network, but these transactions are not immediately considered legitimate. To become legitimate, the network must reach a consensus that the transaction is valid. The consensus process involves a group of users, called miners, solving a cryptographic puzzle to create a new block of transactions. The first miner to solve the puzzle adds the new block to the blockchain and receives a reward in the form of newly minted cryptocurrency.

The process of mining involves using specialized hardware, called ASICs, to solve the cryptographic puzzle. The puzzle is designed to be difficult to solve, but easy to verify once a solution is found. This is where the computational work comes in, as miners must perform numerous calculations to solve the puzzle.

PoW provides security to the blockchain by making it prohibitively expensive for any one miner to launch an attack on the network. As a result, the network becomes more secure with each additional miner that participates in the consensus process.

The use of PoW ensures that transactions are legitimate and that users are not spending cash that they do not have the right to spend. By using PoW, users can trust that the blockchain network is secure and that transactions are being added to the blockchain in a fair and transparent manner.

Understanding the Proof of Work (PoW) Consensus Algorithm

To gain a better understanding of the Proof of Work (PoW) concept, let us consider a blockchain network that utilizes the same basic principles. In this network, transactions are grouped into blocks, and then announced to the network so that users creating a block will include them in the candidate block. The transactions are only considered legitimate once their candidate block becomes a confirmed block, meaning it has been added to the blockchain. However, appending a block is not a cheap process. The PoW algorithm requires a miner to use some of their computational resources for the privilege of appending a block. The miner must hash the block's data until a solution to a complex mathematical puzzle is discovered.

In the PoW algorithm, a user must provide data whose hash fits certain conditions. The most effective way to do this is to pass the entire data through a particular hash function and test if it satisfies the conditions. If it does not, the user has to change the data slightly to get a unique hash. Changing even one character in the data will result in a completely unique outcome, making it a guessing game for the user trying to create a block. The variable data used for each attempt is known as a nonce.

Mining is the process of gathering blockchain data and hashing it together with a nonce until a specific hash is discovered. If a hash is found that satisfies the conditions set out by the protocol, the miner receives the right to broadcast the new block to the network, and the other participants of the network update their blockchains to include the new block. The conditions to discover a legitimate hash are exceptionally tough for major cryptocurrencies today, as the higher the hash rate on the network, the harder it is to find a legitimate hash. This is achieved to ensure that the blocks aren't found too quickly, which can be expensive in terms of computational cycles and electricity.

However, the protocol rewards the miner with cryptocurrency in case a legitimate hash is discovered. It is expensive for anyone to mine, but the miner would be rewarded if they produce a legitimate block. Non-mining users can confirm that a block is legitimate without expending much computational power by checking the hash of a given input. In case a user attempts to cheat by adding fraudulent transactions into the block and generating a legitimate hash, Public-Key cryptography comes into play.

Cryptographic tricks permit any consumer to verify whether or not someone has a right to move the funds they're trying to spend. When a transaction is created, it is signed, and anyone on the network can examine a particular signature with a specific public key to check whether or not they match. They can also check if a user could actually spend a fund and whether the sum of the inputs is higher than the sum of the outputs. Any block that contains an invalid transaction will be automatically rejected by the network, making it highly priced for anyone to attempt to cheat. A malicious participant would waste their computational resources with no reward. Proof of Work makes it expensive to cheat, but profitable to act honestly, ensuring that any rational miner seeks a Return on Investment (ROI) and behaves in a way that guarantees revenue.

Here are some mathematical functions and equations related to Proof of Work:

Hash Function: A hash function is used to generate a unique identifier for a block, known as a block hash. The most commonly used hash function in cryptocurrencies is SHA-256 (Secure Hash Algorithm 256-bit), which generates a 256-bit (32-byte) hash value.

Example: *Hash = SHA256(Data)*

Nonce: A nonce is a random number that a miner includes in the block header when attempting to solve the Proof of Work puzzle. The miner keeps changing the nonce until the block hash meets the difficulty target set by the network.

Example: *Block-Header = Data + Nonce Block-Hash = SHA256(Block-Header)*

- **Block-Header:** This is a portion of data that includes various details about a block, such as the timestamp, transaction data, and a nonce.
- **Data:** This is the actual information or transactions that are being added to the blockchain in that specific block.
- **Nonce:** A nonce is a number that miners change when trying to create a new block so that the resulting block hash meets certain criteria.

- **Block-Hash:** This is the unique identifier or fingerprint of a block, calculated by applying the SHA-256 cryptographic hash function to the block header.

Difficulty Target: The difficulty target is a parameter that adjusts the difficulty of the Proof of Work puzzle to maintain a constant block creation rate. The higher the difficulty target, the harder it is to find a valid block hash.

Example: *Difficulty-Target = 2^256 / (Target-Block Creation Time * Network Hash Rate)*

- **Difficulty Target:** This is a value that determines how difficult it is to find a valid block in a blockchain's proof-of-work system. It is represented as a target value that a block's hash must be below to be considered valid.
- **2^256:** This is a very large number, specifically 2 raised to the power of 256. In the world of cryptography and blockchain, it is used as a constant representing the maximum possible value in a 256-bit number system.
- **Target Block Creation Time:** This is the desired average time it should take to mine a new block in the blockchain. For example, in Bitcoin, the target block creation time is approximately 10 minutes.
- **Network Hash Rate:** This is a measure of the total computational power (hashing power) of all miners in the blockchain network. It indicates how many hash calculations the entire network can perform per second.

Network Hash Rate: The network hash rate is the total computational power of all miners in the network. It is measured in hashes per second (H/s) or kilohashes per second (KH/s) or megahashes per second (MH/s) or gigahashes per second (GH/s).

Example: *Network Hash Rate = Total Hash Rate Of All Miners In The Network*

Block Reward: A block reward is the incentive given to a miner who successfully mines a block. In Bitcoin, the current block reward is 6.25 BTC (as of May 2023). However, the block reward is reduced by half every 210,000 blocks (known as halving).

Example: *Block Reward = 6.25 * 10^8 Satoshis*

- Block Reward refers to the amount of cryptocurrency given to miners as a reward for successfully mining a new block in a blockchain. In this example, the block reward is specified as "6.25 * 10^8 Satoshis."
- Satoshis are the smallest unit of Bitcoin. One Bitcoin (BTC) is equal to 100,000,000 Satoshis (10^8). So, the block reward in this case is 625,000,000 Satoshis, which is 6.25 BTC. This is the amount that a miner receives when they successfully add a new block to the Bitcoin blockchain.

Mining Profitability: Mining profitability is the measure of how much revenue a miner can generate by mining a block, considering the cost of mining (electricity, hardware, etc.) and the block reward.

Example: *Mining Profitability = Block Reward – Mining Cost*

- Mining Profitability represents the potential profit a miner can earn from mining cryptocurrency. It is calculated by subtracting the Mining Cost from the Block Reward.
- Block Reward is the amount of cryptocurrency that a miner receives as a reward for successfully mining a new block in a blockchain.
- Mining Cost refers to the expenses incurred by a miner to operate their mining hardware, such as electricity costs, hardware depreciation, and maintenance expenses.

Proof of Work vs. Proof of Stake

Proof of Work (PoW) and Proof of Stake (PoS) are two popular consensus algorithms used in blockchain technology. While PoW requires miners to solve complex mathematical problems to validate transactions and create new blocks, PoS replaces miners with validators who are chosen based on the amount of cryptocurrency they hold and have locked up as a "stake" in the network.

In PoW, the first miner to solve the problem and validate the transaction is rewarded with newly created cryptocurrency and transaction fees. However, in PoS, validators are randomly selected and given the opportunity to create a new block. If the block is legitimate, the validator receives a reward made up of transaction fees from the block. Validators are required to lock up a stake in the network as a form of collateral. If they act dishonestly, their stake (or a portion of it) can be taken as a penalty.

One of the biggest advantages of PoS is that it consumes far less energy compared to PoW, which requires a lot of computational power to solve complex problems. PoS, on the other hand, only requires validators to hold cryptocurrency as a stake in the network. However, PoW is a proven consensus algorithm that has been used successfully for over a decade and has ensured trillions of dollars' worth of transactions. PoS is a newer technology and has not been tested at the same scale, so its security and reliability are still being evaluated.

Cryptocurrencies using PoW:

Several cryptocurrencies rely on the Proof of Work (PoW) consensus algorithm, including

1. Litecoin
2. Ethereum
3. Monero
4. Dogecoin

In these systems, miners compete to solve complex mathematical problems to validate transactions and add new blocks to the blockchain. Successful miners receive rewards in the form of the cryptocurrency they are mining.

Common Cryptographic protocols used in Proof of Work Systems:

One of the most commonly used consensus algorithms based on proof-of-work is the SHA-256, which was originally introduced with Bitcoin. There are also several other proof-of-work algorithms that are used in different cryptocurrencies, such as Scrypt, SHA-3, scrypt-jane, and scrypt-n. These algorithms are designed to be computationally intensive and require significant processing power to solve cryptographic puzzles in order to validate transactions and earn rewards. The use of different proof-of-work algorithms can also affect the level of security, decentralization, and energy consumption of a particular cryptocurrency network.

Features of Proof of Work System:

The popularity of the Proof of Work consensus protocol can be attributed to two primary factors:

1. The difficulty in finding a solution to the mathematical problem is high, which ensures that the protocol is secure and resistant to attacks. This high level of difficulty is achieved through the use of complex

mathematical puzzles, which require significant computational power and energy to solve.

2. Once a solution to the puzzle is found, it is easy to verify its correctness. This means that other nodes in the network can quickly check and confirm that the solution is valid, without needing to redo the work. This feature allows for efficient processing of transactions and helps to maintain the integrity of the blockchain.

Proof of Stake (PoS): An Alternate Consensus Algorithm to Proof of Work (PoW)

Proof of Stake (PoS) is an alternative consensus algorithm to Proof of Work (PoW) that was proposed in the early days of Bitcoin. Unlike PoW, PoS does not require specialized computing hardware or massive energy consumption. In PoS, participants lock up their cryptocurrency as an internal resource in a hardware or software wallet, which makes an ordinary PC sufficient for staking. To be eligible for staking, a participant must preserve a minimum amount of funds, and rules may differ with each protocol.

In PoS, participants bet on blocks, and the protocol chooses one. If a participant's block is selected, they receive a proportion of the transaction fees depending on their stake. The more funds a participant has locked up, the more they stand to benefit. However, if a participant tries to cheat by featuring invalid transactions, they will lose a portion or all of their stake. This mechanism is similar to PoW, where acting honestly is more profitable than performing dishonestly.

In contrast to PoW, there is generally no freshly-created cash as a part of the reward for validators in PoS Instead, the blockchain's native currency must be issued in some other way, such as an ICO or IEO. Alternatively, the protocol can launch with PoW before later transitioning to PoS

Pure PoS has only been deployed in smaller cryptocurrencies so far. It remains to be seen if it can function as a viable alternative to PoW. Although it appears theoretically sound, it could be very exclusive in practice. Once PoS is rolled out on a network with a massive amount of value, the system becomes a playing field of game theory and financial incentives. Anyone with the expertise to "hack" a PoS system would likely only do so if they could benefit from it, so the only way to discover if it is feasible is on a live network. The Casper upgrade will be executed as a part of a sequence of upgrades to the Ethereum network (collectively referred to as Ethereum 2.0), and we will soon see PoS tested on a large scale.

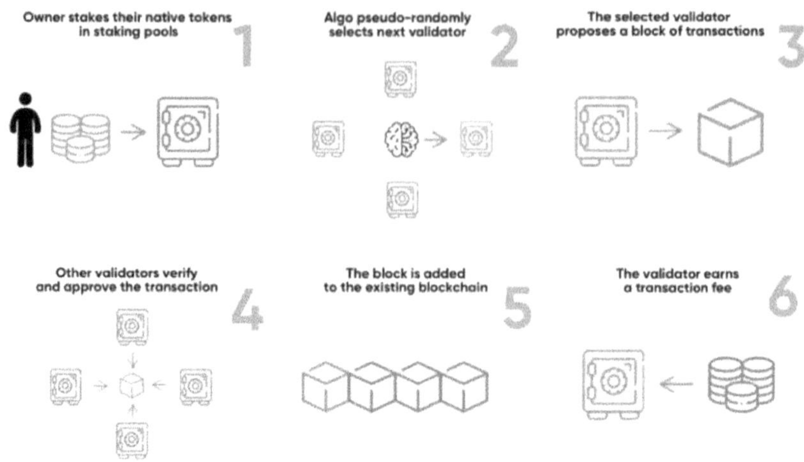

Figure 5.3: Mechanism of Staking in the PoS System

Proof of Stake (PoS) Consensus Mechanism & Its Challenges

Proof of Stake (PoS) consensus is an alternative method for determining who can add new blocks and confirm the current state of the blockchain. Unlike Proof of Work, where miners compete to solve complex mathematical problems using their computing power, PoS selects the next block miner through a process based on the number of coins held in wallets (or "staked"). This process assumes that those with the highest stake will be responsible for important decision-making in the blockchain network.

While PoS eliminates the need for energy-intensive mining, it introduces a new problem known as "nothing at stake." This is because, in the case of a forked chain, PoS forgers are incentivized to validate blocks on both chains, as it costs them little to no work on an additional chain, and they can collect rewards on both chains. This can lead to disagreements within the community, as there should only be one chain, and agreeing on the state of that single chain is the entire purpose of the consensus mechanism.

PoS also has challenges related to the distribution of tokens. PoW miners have significant expenses (hardware, electricity) and must often sell a large portion of their mined coins to cover their mining expenses. This means that many mined coins are available for purchase on the decentralized market, rather than being stockpiled by miners. In contrast, PoS forgers have very low operating costs, so they do not have the same pressure to sell the coins they receive for maintaining the network. As a result, large holders/miners who engage in Proof of Stake tend to increase their share of the circulating coins as they manage block rewards and transaction fees from users of the blockchain network.

This can create a scenario similar to feudalism, where the network is effectively owned and operated by coin holders, and users pay them a kind of rent for using it. There is generally a cutoff below which it is not possible to participate directly in PoS, which can further exacerbate this concentration of power. Despite these challenges, PoS is still a promising alternative to PoW, and its potential will become clearer as it is deployed on larger networks, such as the upcoming Casper upgrade to the Ethereum network.

Here are some mathematical functions and equations used in Proof of Stake (PoS):

1. **Validator Selection Probability:** This is the probability that a validator is chosen to create the next block. It is calculated as follows:

$$P = (W\_i / T) * (1 + r)$$

where P is the validator selection probability, W_i is the weight of the ith validator, T is the total weight of all validators, and r is a random number between 0 and 1.

2. **Block Validation Probability:** This is the probability that a validator correctly validates a block. It is calculated as follows:

$$P = (S / T)$$

where P is the block validation probability, S is the stake of the validator, and T is the total stake of all validators.

3. **Slashing Condition:** This is the condition under which a validator's stake is slashed (i.e., reduced) due to misbehaviour. It is defined as:

$$f(S, t) >= c$$

where $f(S, t)$ is a function of the validator's stake S and the time t, and c is a constant threshold value.

Finality Condition: This is the condition under which a block is considered final (i.e., irreversible). It is defined as:

$$P\_f = 1 - (1 - P\_v)^k$$

where P_f is the finality probability, P_v is the block validation probability, and k is the number of confirmations required for finality.

Proof of Stake (PoS) Consensus Algorithm: A Viable Alternative to Proof of Work (PoW)

The Proof of Stake (PoS) consensus algorithm is an increasingly popular alternative to the Proof of Work (PoW) algorithm used in many blockchains. PoS was developed to overcome the shortcomings and problems associated with PoW-based systems. One of the most significant problems that PoS addresses is the high costs involved in PoW mining, such as electricity consumption and hardware expenses. PoS blockchains eliminate the need for mining and instead validate new blocks based on the number of coins that are staked by participants. The more coins a participant stakes, the greater the chances of being selected as a block validator, also known as a minter or forger.

Compared to PoW systems, which require external investments such as power consumption and hardware, a PoS blockchain is secured through an internal investment, which is the cryptocurrency itself. Additionally, PoS systems make attacking a blockchain more expensive, as a successful attack would require ownership of at least 51% of the existing coins. This means that failed attacks could result in significant monetary losses. Despite the benefits and compelling arguments in favor of PoS, these systems are still in their early stages and have yet to be tested on a large scale.

The PoS consensus algorithm was initially proposed on the Bitcoin talk discussion forum in 2011 as a solution to problems with the then-popular PoW algorithm. While both algorithms aim to achieve consensus in the blockchain network, their approaches are quite different. PoW relies on mining, which involves using computing power to solve complex mathematical problems. In contrast, PoS relies on staking, which involves holding a specific amount of cryptocurrency to participate in block validation. Despite its relative novelty, PoS has gained popularity due to its efficiency and security benefits.

Understanding the Functioning of the Proof of Stake (PoS) Algorithm

The Proof of Stake (PoS) algorithm is a consensus mechanism used in some blockchain networks. Unlike the Proof of Work (PoW) algorithm which involves mining, PoS relies on a process known as forging. In a PoS network, the selection of the next validator to forge the next block is done through a pseudo-random election process that considers various factors such as staking age, randomization, and the node's wealth. The length of time a stake has been held in the network determines the probability of a node being selected as the validator.

To participate in the forging process, users must lock a certain portion of coins in the network as their stake. The more coins a user stakes, the higher the probability of being chosen as the validator. However, to prevent the network from being dominated by nodes with the highest stake, other selection strategies such as Randomized Block Selection and Coin Age Selection are implemented. In the Randomized Block Selection approach, validators are chosen based on a combination of the lowest hash fee and the highest stake. On the other hand, the Coin Age Selection method selects nodes based on the length of time their tokens have been staked. Once a validator has been selected, they are responsible for checking the validity of transactions in the block, signing the block, and adding it to the blockchain. In return, the validator receives transaction fees as their reward. However, unlike PoW-based systems where new coins are created as a reward for miners, PoS networks generally use transaction fees as the reward.

To prevent malicious actors from introducing fraudulent blocks to the network, validators who are found to have done so have their stake and rewards withheld for a certain period of time. This period allows the network to confirm that there are no fraudulent blocks before releasing the stake and rewards back to the validator.

It is important to note that each cryptocurrency using the PoS algorithm has its own set of regulations and strategies that are designed to ensure the best possible combination for their network and users. Overall, the PoS algorithm offers several benefits such as lower energy consumption, reduced hardware requirements, and increased network security.

PoS relies on validators who have staked their cryptocurrency as collateral to participate in block creation.

1. **Staking:** Users who want to participate in the PoS consensus mechanism are required to stake a certain amount of cryptocurrency. This is done by locking their funds into the network as collateral.
2. **Validator Selection:** The selection of the validator is determined by several factors such as the staking age, randomization, and the node's wealth. The more cryptocurrency a validator has staked, the higher the chances of being selected as the validator.

---

3. **Block Creation:** Once a validator has been selected, they create a new block by validating transactions on the network. The validator then signs the block, and adds it to the blockchain.
4. **Reward:** As a reward for their work, the validator receives transaction fees associated with the transactions in the block they have created.
5. **Release of Staked Funds:** If a validator wants to stop participating, their staked funds along with the earned rewards are released after a certain period of time, giving the network time to affirm that there are no fraudulent blocks introduced to the blockchain by the node.

The PoS algorithm is more complex than the PoW algorithm and there are several different methods used to determine the validator selection process. The most commonly used techniques are "Randomized Block Selection" and "Coin Age Selection".

1. **Randomized Block Selection:** Validators are selected based on a combination of the lowest hash fee and the highest stake. The probability of being chosen as the validator is proportional to the amount of cryptocurrency staked.
2. **Coin Age Selection:** Validators are selected based on how long their tokens have been staked. The longer a validator has held their tokens as stake, the higher the probability of being chosen as the validator. The coin age is calculated by multiplying the number of days the coins have been held as stake by the number of coins that are staked.

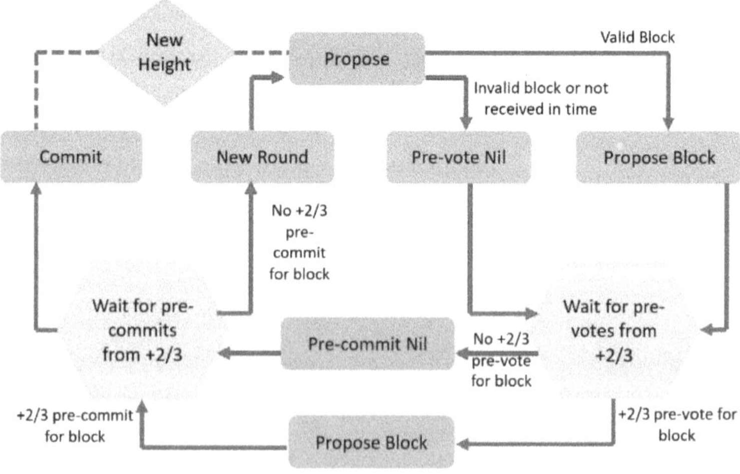

Figure 5.4: Mechanism of PoS Consensus Algorithm

Security

The security of the Proof of Stake algorithm is ensured by the financial incentive provided to the validator node. The validator puts their own cryptocurrency stake at risk if they validate or create fraudulent transactions. If a fraudulent transaction is detected, the validator will lose a portion of their stake and their right to participate as a validator in the future. In this way, the validator is motivated to behave honestly and validate only legitimate transactions. The cost of validating fraudulent transactions is higher than the benefit of the reward for a validator as long as their stake is worth more than the reward. This helps prevent malicious behaviour and ensures the security of the network.

To control the network and approve fraudulent transactions, a validator would need to own a majority of the cryptocurrency stake in the network, known as a 51% attack. However, this would be very impractical and expensive, especially for a valuable cryptocurrency. The Proof of Stake algorithm provides numerous advantages over other algorithms, such as energy efficiency and increased security. It is easy and affordable for a greater number of users to run nodes, and the randomization process makes the network more decentralized, reducing the need for mining pools to mine blocks. This stability also facilitates the price of a particular coin to stay more stable as there is less of a need to launch many new coins for a reward.

It is essential to note that the cryptocurrency industry is rapidly evolving and developing, with several other algorithms and strategies being experimented with. Therefore, the Proof of Stake algorithm may not be the only method used in the future to secure blockchain networks.

Criteria	Proof of Work (PoW)	Proof of Stake (PoS)
Validation Method	Requires miners to solve complex mathematical problems to validate transactions and create new blocks	Validators (or "forgers") are chosen based on the number of coins they hold and are willing to "stake" (lock up as collateral)

Security Mechanism	External investment (electricity consumption and mining hardware)	Internal investment (cryptocurrency held and staked by participants)
Network Scalability	Limited due to energy-intensive mining requirements	More scalable due to less energy consumption
Attack Vulnerability	Susceptible to 51% attacks where an attacker controls more than half of the network's mining power	Susceptible to "nothing-at-stake" attacks where forgers validate blocks on multiple forks of the blockchain
Decentralization	May lead to centralization due to economies of scale in mining	More decentralized as participants can stake any amount of cryptocurrency
Monetary Policy	Block rewards and transaction fees go to miners	Block rewards and transaction fees go to forgers and stakers
Environmental Impact	High energy consumption and carbon emissions	Lower energy consumption and carbon emissions

Table 5.1 Difference PoW Vs PoS

Delayed Proof of Work Explained

Delayed Proof of Work (dPoW) is a security mechanism developed by the Komodo project, which serves as an updated version of the traditional Proof of Work (PoW) consensus algorithm. The primary goal of dPoW is to enhance network security by utilizing the hash power of the Bitcoin blockchain. This mechanism allows Komodo developers to not only secure their own network but also any third-party chain that may become part of the Komodo ecosystem in the future. This means that dPoW can be implemented for any project that develops an independent blockchain using a UTXO model.

UTXO, which stands for Unspent Transaction Output, refers to the digital currency that remains in a user's account after completing a cryptocurrency transaction, such as bitcoin. With dPoW, once a block is validated on the Komodo blockchain, it is not immediately added to the main blockchain but instead is first submitted to the Bitcoin network for an additional layer of security. This process ensures that the Komodo blockchain remains secure and robust, even against potential attacks. Overall, dPoW is a powerful security mechanism that provides added protection for blockchain networks and has the potential to significantly improve the overall security of the blockchain ecosystem.

Benefits of Delayed Proof of Work (dPoW) for Blockchain Security

Delayed Proof of Work (dPoW) is a security mechanism that aims to enhance network security for blockchains that use a UTXO model. The Komodo project developed this mechanism as an updated model of the Proof of Work (PoW) consensus algorithm. One of the benefits of dPoW is that it can be used to secure not only the Komodo network but also any third-party chain that becomes a member of the Komodo ecosystem in the future.

To apply dPoW, the Komodo system takes a snapshot of its own blockchain at intervals of ten minutes. This snapshot is then written into a block on the Bitcoin network, a process known as notarization. The notary nodes in Komodo's network execute a transaction on the Komodo chain to write a block hash from every dPoW-protected blockchain onto the Komodo ledger. Using the OP_RETURN command, the notary nodes store a single block hash onto the Komodo chain.

The reason why the notary nodes select an old block hash that is about ten minutes old is to confirm that the entire network consents to the blocks being valid. Each blockchain network still implicates consensus for every block. The notary nodes merely document a block hash from a previously mined block, and further, write a block hash from the Komodo chain onto the Bitcoin ledger. This technique is also completed by executing a BTC transaction & using OP_RETURN to write the data into a block on the Bitcoin chain.

Once this notarization to Bitcoin occurs, Komodo's notary nodes write that block data from the BTC chain back onto the chain of every other shielded chain. The network will no longer accept any re-organizations that attempt to change a notarized block or any blocks that were created prior to the most recently notarized block. This process creates a backup of the entire Komodo system, which is saved within the Bitcoin blockchain.

Currently, dPoW is being used with Bitcoin, but the dPoW algorithm has the potential to be used as a tool for leveraging both the security and features of any other blockchain that makes use of a UTXO model. By leveraging the hash power of a secure blockchain like Bitcoin, the dPoW mechanism offers a way to enhance the security of smaller chains without requiring the significant computing resources that are necessary for PoW mining. As such, dPoW offers a practical and efficient solution for enhancing the security of blockchain networks.

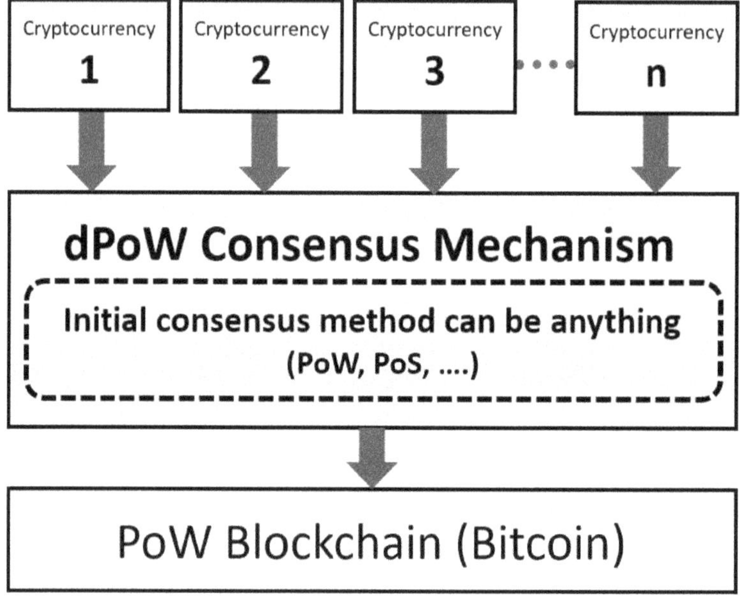

Figure 5.5: Delayed Proof of Work Consensus Algorithm

Proof of Work (PoW) vs Delayed Proof of Work (dPoW)

To enhance the security of blockchain networks, consensus algorithms like Proof of Work (PoW) and Delayed Proof of Work (dPoW) have been developed. The PoW algorithm uses the mining process to maintain network security and prevent cyber-attacks such as DDoS. The mining process involves solving complex cryptographic puzzles that require significant computational power and financial investment. This process also verifies the legitimacy of transactions and creates new cryptocurrency units as a reward for the miner who solves the puzzle. PoW is directly associated with the amount of computational power being devoted to the network, which makes smaller networks less secure than larger ones.

In contrast to PoW, the dPoW algorithm is not considered a consensus algorithm but is rather a security mechanism that is carried out in addition to ordinary PoW consensus regulations. It is designed to make blockchains more secure and immune to the 51% attack, which is when a group of miners control more than 50% of the network's hash rate and can manipulate the network. The dPoW mechanism achieves this by notarizing blocks at regular intervals and writing them onto the Bitcoin blockchain, which creates a backup of the entire blockchain system.

Each time a block is notarized, dPoW "re-sets" the blockchain's consensus policies, making it impossible for previously notarized blocks to be reorganized. The network will no longer accept a sequence that starts at block "XXX, XX0" or prior, even if it is the longest one. For example, when a blockchain's network receives confirmation that block XXX, XX1 has been notarized, the longest chain rule initiates over at block XXX, XX2. This process makes blockchains more secure and protects them from 51% attacks. Though currently used with Bitcoin, dPoW has the potential to be used with any blockchain that uses the UTXO model, making them more secure and reliable. *(Ref. No.: 63- 86)*

Leased Proof of Stake Consensus Explained

The Waves network employs a consensus algorithm known as Leased Proof-of-Stake (LPoS), which is a modified version of the Proof-of-Stake (PoS) consensus algorithm. In LPoS, the nodes that validate transactions and create new blocks are known as "forgers" and they are selected based on the number of WAVES tokens that they hold or lease from other users. The more WAVES tokens a node has, the more likely it is to be selected as a forger and earn block rewards. However, not all users have the required number of WAVES tokens to become a forger. In such cases, they can lease their tokens to a forger, and in return, they receive a share of the forger's block rewards. This system incentivizes users to hold or lease more WAVES tokens, thereby increasing the network's security and decentralization.

In addition to LPoS, the Waves network also uses the Waves-NG protocol, which is designed to improve transaction throughput and reduce network congestion. This protocol allows for the creation of micro-blocks, which are smaller and faster to validate than traditional blocks. This, in turn, enables the network to process a greater number of transactions per second, making it more scalable and efficient.

The Waves network offers a comprehensive blockchain ecosystem that includes a variety of tools and services for businesses and developers. The platform allows for easy creation and customization of new tokens, and offers a secure and user-friendly approach for interacting with decentralized applications (dApps) and web services. The integrated decentralized exchange (DEX) enables peer-to-peer trading of cryptocurrencies, while the Waves Keeper browser plug-in provides a secure way for users to manage their digital assets and interact with the blockchain. Overall, the LPoS consensus algorithm and Waves-NG protocol make the Waves network a robust and scalable platform for businesses and developers seeking to leverage blockchain technology.

Scalability

Waves has been focused on addressing the scalability issues of existing blockchains from the beginning, recognizing that scalability is essential for widespread blockchain adoption. For instance, although Bitcoin is highly secure, it is incredibly slow and can process only about 7 transactions per second (TPS). Moreover, the Bitcoin network has certain limitations that make it impractical to be used as a global currency. This necessitates second-tier solutions like the Lightning Network for Bitcoin to function as a currency on a larger scale.

In contrast, Waves has taken a different approach by prioritizing high on-chain scalability before considering second-tier applications. The Waves network employs a Leased Proof-of-Stake (LPoS) consensus algorithm combined with the Waves-NG protocol to enable a high degree of scalability and transaction throughput. By implementing these protocols, Waves can process up to 100 transactions per second, which is significantly more than Bitcoin. Additionally, Waves' approach is focused on creating a blockchain ecosystem for business processes that offers comprehensive toolkits to cater to various requirements, including easy creation of custom cryptocurrency tokens, trustworthy smart contracts, and a decentralized exchange.

Furthermore, the Waves Keeper browser plugin offers a safe and convenient way to interact with dApps and web offerings. By prioritizing high on-chain scalability, Waves has made significant strides towards creating a blockchain ecosystem that can handle the transaction volume required for mass adoption. This approach makes it possible to process a high volume of transactions quickly and efficiently, making Waves an attractive option for businesses and individuals looking to use blockchain technology on a large scale.

Balance Leasing

Waves blockchain network uses a unique consensus algorithm, known as Leased Proof of Stake (LPoS), to secure its transactions. The LPoS is a modified version of the Proof of Stake (PoS) algorithm that was initially used by Waves. Unlike PoS, where validators (miners) are required to stake a specific amount of their cryptocurrency as collateral to validate transactions and add new blocks to the blockchain, LPoS allows anyone to lease their WAVES tokens to the mining nodes.

Once a user leases their tokens, they are locked in their account and cannot be traded or transferred, but the user retains full control of them. The leased WAVES tokens increase the staked weight of the miner, which improves their chances of discovering the subsequent block. This approach helps to improve network security since it becomes more difficult for attackers to carry out a 51% attack as more WAVES are used to secure the network.

Furthermore, users can lease WAVES tokens from their cold storage address to a mining node, which drastically reduces the risk of tokens being hacked from computers that are online since the leased funds are not transferred to the miner. This approach ensures that the network is secure while allowing users who do not run full nodes to participate in the network's consensus.

Moreover, Waves' approach to scalability is also unique. They prioritize on-chain scalability before regarding second-tier solutions, which has allowed them to increase transaction throughput and accommodate mass blockchain adoption. The Waves-NG protocol, combined with LPoS, allows the network to process up to 1,000 transactions per second, making it more suitable for use as a standard currency worldwide than Bitcoin's 7 transactions per second.

A New Strategy to Consensus: Waves-NG

In December 2017, Waves-NG was launched as a new protocol for the Waves network. It is based on the Bitcoin-NG proposal, which was introduced in 2015 by Emin Gün Sirer, a professor of Information Technology at Cornell University. While the original Bitcoin protocol selects each miner retrospectively, Waves-NG selects the subsequent miner in advance, allowing for quicker processing of transactions.

In the Waves-NG protocol, the selected miner creates an empty 'key block', which is the block that will eventually be added to the blockchain. Small blocks called 'microblocks' are then added in near-real-time to this key block, allowing transactions to be processed in just a few seconds. This approach has allowed Waves to update the concept of Bitcoin-NG for a proof-of-stake network, creating the first deployment of Bitcoin-NG for an open, public blockchain.

Another feature of the Waves network is Mass Transfers, which allows up to 100 transfers to be made within a single transaction with reduced fees. This feature enables users to conduct huge airdrops or perform weekly pay-outs to individuals/participants who lease their WAVES to mining nodes. The Mass Transfer feature, combined with Waves-NG, allows for a very high rate of throughput on the network.

In October 2018, stress tests were conducted on the Waves network to test the throughput potential of the new protocol. The results showed that the public, open blockchain protocol could support over 6.1 million transactions within a 24-hour period, with an average of 4,200 transactions per minute and peak throughput achieving hundreds of transactions per second. These results confirm the scalability and efficiency of the Waves network.

here are some mathematical equations and functions that support the concept of Leased Proof of Stake (LPoS):

Staked weight: The staked weight of a miner is the total number of WAVES tokens that are being used to secure the network. It is calculated by summing up the number of WAVES that the miner owns plus the number of WAVES that are leased to the miner. The staked weight is used to determine the probability of a miner discovering the subsequent block. The function for staked weight can be expressed as:

$$Staked\ Weight = Miner's\ WAVES\ Balance + Leased\ WAVES$$

Mining probability: The probability of a miner discovering the subsequent block is proportional to their staked weight. The higher the staked weight, the higher the mining probability. The function for mining probability can be expressed as:

$$Mining\ Probability = Staked\ Weight\ /\ Total\ Staked\ Weight$$

51% Attack: A 51% attack is a situation where an attacker gains control of 51% or more of the network's staked weight. This allows the attacker to manipulate the blockchain and carry out fraudulent transactions. The probability of a successful 51% attack is inversely proportional to the amount of staked weight controlled by the attacker. The function for the probability of a successful 51% attack can be expressed as:

probability of successful 51% attack = (attacker's staked weight) / (total staked weight - attacker's staked weight)

These mathematical concepts help to illustrate the importance of staked weight and how it affects the security and integrity of the Waves network. By leasing WAVES tokens to mining nodes, users can contribute to the network's staked weight and help to make the network more secure.

Proof of Authority Explained

The Proof of Authority is a consensus mechanism that has gained popularity in the cryptocurrency space due to its high scalability and efficiency in handling a large number of transactions per second (TPS). Unlike the Proof of Work and Proof of Stake algorithms, which rely on resource-intensive computations and large amounts of computing power to secure the network, Proof of Authority leverages a different approach.

Proof of Authority is based on the concept of identity verification, where a predetermined set of nodes or validators are given the authority to validate transactions and add them to the blockchain. This set of validators is typically composed of known and trusted entities, such as individuals or institutions with a proven track record of reliability and trustworthiness.

Compared to the PoW and PoS algorithms, which rely on the participation of a large number of nodes to achieve consensus, Proof of Authority networks require a smaller number of validators to validate transactions. This results in faster transaction confirmations and higher TPS, making it a more efficient alternative for certain use cases.

Bitcoin and other PoW-based blockchains have limited overall performance in terms of TPS due to the distributed nature of their network, which requires consensus from a large number of nodes before transactions can be confirmed. PoS networks have presented a better overall performance, but their scalability is still limited. In this context, Proof of Authority presents itself as a more efficient alternative to handle a larger number of transactions per second.

What is Proof of Authority?

Proof of Authority (PoA) is a consensus mechanism that provides a practical and efficient solution for blockchain networks, particularly in private settings. Proposed by Ethereum co-founder Gavin Wood in 2017, PoA leverages identity verification to secure the network instead of the traditional staking of coins.

The PoA consensus algorithm is based on a limited number of validators who are considered trustworthy entities. These validators are not required to stake coins, but their reputation serves as the basis of their authority. In a PoA blockchain, blocks and transactions are confirmed by pre-authorized participants who act as moderators of the system.

One of the key benefits of the PoA model is its scalability. Since it relies on a limited number of validators, it can handle a larger number of transactions per second compared to other consensus mechanisms like Proof of Work or Proof of Stake. This makes PoA an ideal alternative for logistical applications like supply chain management, where high transaction volumes are expected.

The PoA model also allows companies to maintain their privacy while enjoying the benefits of blockchain technology. For example, Microsoft Azure has implemented PoA in its platform to offer solutions for private networks. The system does not require a native currency like Ether gas, as there is no need for mining.

Proof of Authority (PoA) is a consensus mechanism that operates differently than other consensus mechanisms like Proof of Work and Proof of Stake. PoA leverages identity verification to secure the network, making it an ideal solution for private blockchain networks.

In PoA, blocks and transactions are confirmed by a limited number of validators, called nodes. These nodes are selected based on their identity, reputation, or some other criteria, instead of staking coins or performing computational work. To become a node, an entity must prove their identity and be approved by existing nodes.

The consensus process in PoA involves the validators taking turns proposing and validating blocks. Each block proposal contains a list of transactions, and the validator who proposes the block gets to include a transaction fee. The other validators then validate the block and add it to the blockchain if it meets the consensus rules.

To prevent Sybil attacks, where an entity creates multiple fake identities to control the network, PoA uses a reputation-based system. Each node is assigned a reputation score based on their behaviour, such as the number of valid blocks proposed or the number of times, they fail to validate a block. The reputation score is used to determine the order in which nodes get to propose blocks.

The mathematical functions used in PoA include hashing and signing. Hashing is used to create a unique, fixed-length digital fingerprint of a block or transaction. This fingerprint is then signed by the node who proposes the block or transaction using their private key. The other nodes can then verify the signature using the public key of the proposing node.

The PoA consensus algorithm can be expressed using the following equation:

Let n be the number of validators in the network, and let f be the maximum number of malicious or faulty validators that the network can tolerate without compromising security.

The PoA consensus algorithm is secure if:

$$n > 2f$$

This means that the network must have at least twice the number of validators as the maximum number of malicious or faulty validators. For example, if the network can tolerate up to 10% of malicious or faulty validators, then it must have at least 21 validators to ensure security.

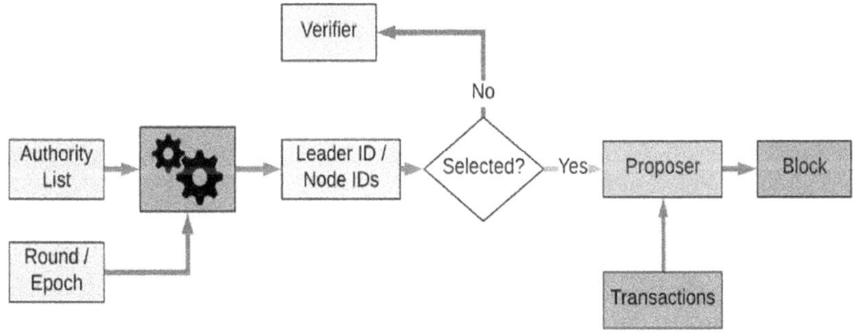

Figure 5.6: Proof of Authority Consensus Algorithm

Proof of Authority vs Proof of Stake

Proof of Authority (PoA) and Proof of Stake (PoS) are two consensus algorithms used in blockchain networks. While both of them aim to achieve consensus and maintain the security of the blockchain, they have some differences in their approach.

Proof of Authority is a consensus algorithm that relies on a limited number of validators, which are pre-selected and considered trustworthy. These validators are identified by their reputation or identity, and they do not stake any coins to participate in the validation process. The PoA algorithm is efficient and scalable, making it an ideal choice for private blockchain networks. In PoA, the validators are responsible for creating and validating blocks. They are rewarded for their services and are incentivized to maintain the security and reliability of the network. On the other hand, Proof of Stake is a consensus algorithm that involves validators staking a certain number of coins to participate in the validation process. The validators are selected randomly based on the number of coins they have staked. In PoS, the validators are incentivized to validate transactions correctly and maintain the security of the network as they risk losing their stake if they validate invalid transactions.

While PoA is sometimes considered a modified version of PoS, the two algorithms differ significantly. PoS is more suitable for public blockchains, whereas PoA is ideal for private blockchain networks. PoS is more decentralized than PoA, as it involves a larger number of validators. However, PoA is more efficient and scalable, as it involves a limited number of validators who are pre-selected based on their reputation or identity.

Conditions for Proof of Authority Consensus

Here is an elaboration on the conditions that the Proof of Authority (PoA) consensus algorithm typically relies upon:

1. **Legitimate & Trustworthy Identities:** PoA requires validators to verify their actual identities. This is to ensure that only legitimate and trustworthy entities are allowed to participate in the consensus process. Validators must have a reputation to uphold, which incentivizes them to act in the best interest of the network.
2. **Difficulty to Become a Validator:** PoA also requires a candidate to make investments, such as money and reputation, at stake. This process reduces the risks of selecting questionable validators and incentivizes a

long-term commitment to the network. By requiring a tough process to become a validator, PoA ensures that only dedicated and reliable entities participate in the consensus process.

3. **A Standard for Validator Approval:** PoA requires a standardized process for selecting validators. The recognition mechanism behind PoA is based on the knowledge of a validator's identity, and a standardized process ensures that all validators go through the same rigorous technique. This helps weed out bad players and ensures the integrity and reliability of the network.

Overall, these conditions ensure that only legitimate and trustworthy entities with a reputation to uphold are allowed to participate in the consensus process. Validators must have a vested interest in the network's success and be committed to its long-term growth. By requiring a standardized process for selecting validators and verifying their identities, PoA ensures the integrity and reliability of the network.

Limitations of the Proof of Authority Consensus Algorithm

Although Proof of Authority (PoA) consensus algorithm has some limitations, it is still considered a valuable blockchain solution for specific applications. Some of the limitations include:

1. **Lack of Decentralization:** PoA foregoes decentralization, making it a centralized system, which could be concerning for the cryptocurrency space. However, it provides an efficient solution for huge corporations with logistical needs.
2. **Immutability:** PoA systems may raise concerns about censorship and blacklisting, leading to questions about their immutability.
3. **Identity Visibility:** PoA validators' identities are visible to anyone, making them susceptible to third-party manipulation. Competitors could try to influence publicly recognized validators to compromise the system from within.

Despite these limitations, PoA still has unique advantages, including:

1. **High Throughput & Scalability:** PoA systems have high throughput, making them ideal for private blockchain applications.
2. **Investment of Reputation:** In PoA-based blockchain networks, validators play a pivotal role in proposing and validating blocks or transactions. Unlike some other consensus mechanisms, PoA relies heavily on the reputation and identity of these validators. Validators are

typically well-established entities or individuals within the network, often with a known history of ethical behaviour and adherence to network rules. Their reputation is built over time based on their consistent validation of transactions and blocks in a truthful and honest manner. This reputation serves as a critical incentive mechanism. Validators have a strong incentive to maintain their reputation because it directly impacts their status within the network. If a validator is found to be acting maliciously, dishonestly, or in a manner that jeopardizes the network's integrity, their reputation suffers. This can result in the removal of their validation rights, or in some PoA networks, they may even face financial penalties.

3. **Trustworthiness:** Proof of Authority is renowned for its ability to maintain the legitimacy of blockchain transactions and blocks by imposing strict criteria on those who serve as validators. These validators are entrusted with the critical task of proposing and validating new blocks, making it essential to ensure their reliability and trustworthiness.

Proof of Burn Explained

Proof of Burn (PoB) is an emerging consensus algorithm being tested as an alternative to the traditional Proof of Work (PoW) and Proof of Stake (PoS) algorithms utilized by most blockchain networks. The consensus algorithm is crucial for the overall stability of the network and for validating financial transactions. In PoW, miners compete to solve a complex cryptographic problem, and the first miner to find a legitimate solution broadcast their proof of work (the block hash) to the network. The network then verifies the legitimacy of the proof, and if it is valid, the miner earns the right to permanently add the block to the blockchain network and receives a reward of newly generated bitcoins. On the other hand, PoS algorithms use digital signatures to prove the ownership of coins, and the confirmation of new blocks is carried out by selected forgers or minters. The more coins a forger has at stake, the higher the probability of being selected as a block validator. Unlike PoW, the majority of PoS systems do not offer block rewards, and forgers only receive transaction fees. The Proof of Burn algorithm, however, has its unique way of achieving consensus and validating blocks. Instead of solving complex cryptographic problems or using digital signatures, the PoB algorithm involves burning or destroying coins. The idea is that by destroying a certain number of coins, the user demonstrates their commitment to the network, and in return, earns the right to validate transactions and add blocks to the blockchain.

Although Proof of Burn has some similarities with PoW and PoS, its distinct approach offers some potential benefits, such as reducing the energy consumption associated with PoW mining and preventing centralized control of the network, which can occur in PoS systems. However, it is still an experimental algorithm, and more research is needed to determine its effectiveness and suitability for widespread use in blockchain networks.

Proof of Burn (PoB): A Sustainable Alternative to Traditional Blockchain Consensus Algorithms

Proof of Burn (PoB) is an innovative consensus algorithm that provides a sustainable alternative to the resource-intensive Proof of Work (PoW) used in many blockchain networks, particularly Bitcoin. PoB was initially proposed by Iain Stewart and has garnered attention in the cryptocurrency space as a way to address environmental concerns associated with PoW. In PoB-based networks, participants deliberately "burn" or destroy a certain amount of their cryptocurrency holdings. This process serves as a form of investment in the blockchain network and demonstrates their dedication to its success. Unlike PoW, which demands powerful mining hardware and consumes significant electricity, PoB eliminates the need for such resources.

The validation process in PoB doesn't rely on solving complex mathematical puzzles or using computational power. Instead, it operates on the principle that the more coins a participant burns, the greater their digital mining power becomes. This increased mining power enhances their chances of being selected as the next block validator. Essentially, PoB equates the amount of burned coins with mining strength. While PoB shares similarities with PoW and PoS in terms of validating transactions and adding blocks to the blockchain, it offers distinct advantages. These advantages include reduced energy consumption compared to PoW mining and a mechanism to mitigate centralized control, which can be a concern in some PoS systems.

However, it is important to note that PoB is still considered experimental in the blockchain community. Further research and testing are required to determine its effectiveness and suitability for widespread adoption. Nevertheless, the concept of "burning" coins as a symbol of commitment to the blockchain network presents an intriguing possibility for more sustainable and inclusive participation in blockchain ecosystems. As environmental concerns continue to grow in the cryptocurrency space, PoB represents a novel approach to addressing these issues while maintaining the integrity of blockchain networks.

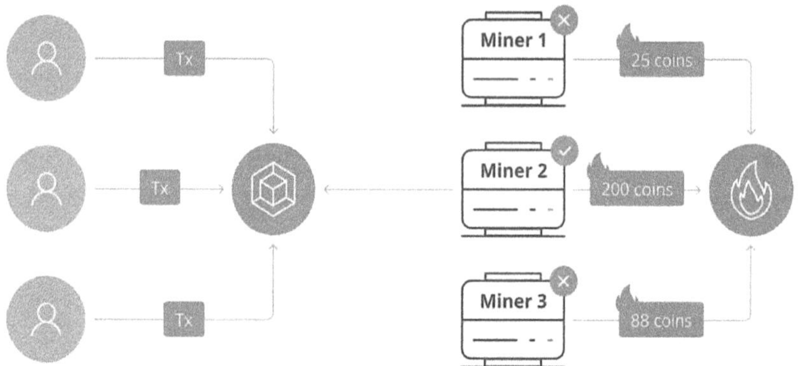

Figure 5.7: Mechanism of Proof of Burn Algorithm

How does the Proof of Burn Algorithm Secure Blockchain Networks?

The Proof of Burn (PoB) algorithm utilizes a unique approach to secure blockchain networks by burning cryptocurrency coins. This involves sending the coins to a public address, known as an eater address, which renders them unusable and inaccessible. The burning of coins reduces market availability, causing a potential increase in their value. However, it is also a way of investing in the security of the network, as it requires a significant commitment of resources to perform.

In contrast to Proof of Work (PoW) algorithms that require miners to invest in expensive hardware and electricity, PoB blockchains do not rely on computational power. Instead, the security of the network is based on the investment made through the burning of coins. This approach creates an incentive for miners to behave honestly and assist the network, in order to prevent the initial investment from being wasted.

Similar to PoW blockchains, PoB systems provide block rewards to miners. Over time, these rewards are expected to cover the initial investment made through the burned coins. The implementation of PoB can vary among projects, with some burning Bitcoins while others use their native coin for consensus. Regardless of the implementation, the Proof of Burn algorithm represents a sustainable and innovative alternative to traditional blockchain consensus algorithms. *(Ref. No.: 63- 86)*

There are different ways to mathematically model the Proof of Burn consensus algorithm. Here are a few examples:

1. **Probability of being selected as a block validator:**

In Proof of Burn systems, the probability of being selected as a block validator is proportional to the number of coins burned. We can represent this mathematically with the following equation:

$$P = C / (B + C)$$

Where:

- P is the probability of being selected as a block validator
- C is the number of coins burned by the miner
- B is the total number of coins in circulation

This equation assumes that all burned coins are irretrievable and cannot be used for other purposes.

2. **Expected return on investment:**

Miners in Proof of Burn systems expect to receive block rewards that compensate for their initial investment in burned coins. We can model the expected return on investment (ROI) with the following equation:

$$ROI = R / C$$

Where:

- ROI is the expected return on investment, expressed as a percentage
- R is the total block rewards received by the miner during a certain period of time
- C is the initial investment in burned coins

Assuming that the block rewards are proportional to the number of coins burned, we can write:

$$R = (B / (B + C)) * T$$

Where:

T is the total transaction fees collected during a certain period of time

Substituting this equation into the ROI equation, we get:

$$ROI = (B / (B + C)) * T / C$$

This equation shows that the ROI is proportional to the ratio of burned coins to total coins in circulation, and the total transaction fees collected.

3. Security of the network:

The security of Proof of Burn networks relies on the assumption that burning coins is a costly and irreversible process. We can express this mathematically with the following equation:

$$B = f(P, C)$$

Where:

B is the cost of burning coins, expressed as a function of the probability of being selected as a block validator and the amount of coins burned

P & C are as defined in the first equation above

This equation implies that the cost of burning coins increases as the probability of being selected as a block validator increases, and as the amount of coins burned increases. Therefore, a miner who wants to control the network would need to burn a significant number of coins, which would be financially costly and risky.

Proof of Burn (PoB) vs Proof of Stake (PoS)

PoB and PoS are two different consensus algorithms used in blockchain networks, with some similarities and differences. Both require block validators to invest their coins in order to participate in the consensus mechanism, but there are some key differences between the two. In PoS blockchains, validators must stake their coins by locking them up for a certain period of time. If they decide to leave the network, they can take back their coins and sell them on the market, which does not create a permanent market shortage. This means that coins are only taken out of circulation temporarily, and there is no permanent scarcity created.

On the other hand, in PoB blockchains, validators have to destroy their coins permanently, which creates a permanent economic scarcity. This means that the more coins a validator burns, the more valuable the remaining coins become. In other words, PoB creates a deflationary economic model, which is different from the inflationary economic model used in PoS. Another difference between PoB and PoS is the way in which they select validators. In PoS, validators are chosen based on the number of coins they have staked. The more coins a validator has, the higher their chances of being chosen to validate a block. In PoB, validators are chosen based on the number of coins they have burned. The more coins a validator burns, the higher their chances of being chosen to validate a block.

Overall, both PoB and PoS have their own advantages and disadvantages, and the choice of which consensus algorithm to use depends on the specific needs and goals of a blockchain network.

Advantages & Disadvantages of Proof of Burn

Proof of Burn (PoB) is a consensus algorithm that has been proposed as a more sustainable alternative to Proof of Work (PoW) and Proof of Stake (PoS). Here are some advantages and disadvantages of using PoB in blockchain-based systems:

Possible Advantages:

1. More environmentally sustainable compared to Proof of Work (PoW) consensus algorithm.
2. Lower electricity consumption, as it does not rely on powerful mining hardware like ASICs.
3. No need for physical mining equipment, as coin burns represent digital mining rigs.
4. Coin burns create a virtual investment in the blockchain and reduce the circulating supply, which can lead to market scarcity and potentially increase the coin's value.
5. Encourages long-term dedication from miners, as burning coins demonstrates their commitment to the network.
6. Coin distribution and mining tends to be less centralized than in some other consensus algorithms.

Possible Disadvantages:

1. PoB's eco-friendliness is debatable as the Bitcoins being burned might be generated through PoW mining, which requires significant computing resources.
2. Not yet tested on a larger scale, so its effectiveness cannot be fully verified.
3. More testing is needed to confirm its security and efficiency.
4. Verification of the work completed through miners can be slower than in other consensus algorithms.

5. PoB is not always as fast as PoW blockchains.
6. The technique of burning coins is not always transparent or easy for an average consumer to verify.

Delegated Proof of Stake Explained

Delegated Proof of Stake (DPoS) is a consensus algorithm that is gaining popularity as an alternative to Proof of Work (PoW) and Proof of Stake (PoS) mechanisms. DPoS is widely recognized as a more efficient and democratic model than PoS, as it requires fewer resources and is more sustainable and eco-friendlier.

The DPoS algorithm was developed by Daniel Larimer in 2014, and it is used by a number of cryptocurrency projects, including Bitshares, Steem, Ark, and Lisk. DPoS-based blockchains rely on a voting system where stakeholders outsource their work to a third party, who may be referred to as delegates or witnesses. These delegates are responsible for achieving consensus during the generation and validation of new blocks.

The voting strength in DPoS is proportional to the number of coins each user holds. The voting system varies from project to project, but in general, each delegate presents an individual concept when seeking votes. Delegates earn rewards for their work, which are proportionally shared with their electors. As a result, DPoS creates a voting system that is directly dependent on the delegates' reputation. If an elected delegate misbehaves or does not work efficiently, they will likely be removed and replaced by another one.

DPoS blockchains are more scalable than PoW and PoS, with the ability to process more Transactions Per Second (TPS). This makes them ideal for applications that require high throughput and low latency, such as payment systems and decentralized exchanges.

While the core concept of DPoS is based on a voting system, it can also be explained mathematically using equations and functions. Here is an explanation of DPoS with some mathematical equations and functions:

1. **Voting System:** DPoS relies on a voting system where stakeholders can vote for delegates to validate blocks. Each stakeholder can cast a number of votes that is proportional to the number of coins they hold. For example, if a stakeholder holds 100 coins and the total number of

coins in the network is 1,000, they will have 10% of the total voting power.

*Voting Power (VP) = (Number of Coins held by the Stakeholder / Total Number of Coins in the Network) * 100*

2. **Block Production:** Delegates are responsible for validating blocks in the network. The number of delegates can vary from one blockchain to another. The delegate with the most votes will be elected to validate blocks. For example, if there are 21 delegates in the network, the delegate with the top 21 votes will be elected to validate blocks.

3. **Block Reward:** Delegates receive rewards for validating blocks. The reward is a combination of the block reward and transaction fees. The block reward is fixed and decreases over time, while the transaction fees vary based on the number and size of transactions in the block.

Block Reward = Fixed Reward + Transaction Fees

4. **Delegate Performance:** Delegates need to maintain a good reputation to be elected and receive rewards. If a delegate fails to validate blocks or acts maliciously, they may lose their position and rewards. This creates an incentive for delegates to maintain good performance.

Delegate Performance = Block Validation Success Rate + Good Behaviour Rate

DPoS vs PoS

Proof of Stake (PoS) & Delegated Proof of Stake (DPoS) are both consensus algorithms that require participants to hold a certain amount of cryptocurrency in order to validate transactions and generate new blocks. However, DPoS offers an additional democratic voting system that allows block producers to be elected by stakeholders. This means that instead of a random selection process like in PoS, stakeholders can vote for delegates they believe are best suited to maintain the network's integrity and security.

The democratic voting system in DPoS incentivizes elected delegates to be accountable, efficient, and transparent, as their performance will determine whether they get re-elected in the next round of voting. This is in contrast to PoS, where validators may not be as motivated to maintain the network's security, as they are not elected by the stakeholders and are randomly selected.

Moreover, DPoS blockchains have been observed to be faster in terms of processing transactions per second compared to PoS blockchains. This is due to the fact that DPoS blockchains rely on a smaller set of elected delegates to validate transactions, as opposed to a larger number of validators in PoS. This reduces the time needed for reaching consensus on transactions, making DPoS more scalable and efficient. Overall, DPoS offers a more democratic and efficient consensus mechanism than PoS, making it a popular choice for many blockchain projects.

DPoS vs PoW

DPoS & PoW are two different consensus algorithms that are used in blockchain systems. PoW relies on a massive amount of computational work to secure an immutable, decentralized, and transparent distributed ledger, while DPoS streamlines the block production process, allowing for quick and efficient processing of large quantities of blockchain transactions.

Although DPoS and PoS are similar in the sense of stake holding, DPoS offers a novel democratic voting system, by which block producers are elected. In a DPoS system, delegates are motivated to be sincere and efficient, or they will be voted out. DPoS systems limit the use of staking to the election of block producers, with the actual block production being predetermined, in contrast to the competition-based system of PoW. Every witness gets a turn at block production, which makes DPoS quite different from PoW and even PoS.

DPoS's incorporation of stakeholder voting serves as a means for identifying and motivating honest and efficient delegates, also known as witnesses. The voting system is directly dependent on the delegates' reputation, and if an elected node misbehaves or does not work efficiently, it will likely be expelled and replaced by another one. This creates a more sustainable and eco-friendly system than PoW, which requires a lot of computational power and external assets.

While PoW is still considered the most secure consensus algorithm and is used where maximum cash transmittance occurs, PoS is quicker than PoW and has more use cases. DPoS offers an even more efficient and democratic model than PoS, and its incorporation of stakeholder voting and predetermined block production make it a unique consensus algorithm. In most cases, DPoS presents a better overall performance in terms of transactions per second compared to PoW and PoS.

Hybrid PoW/PoS Consensus Explained

A hybrid PoW/PoS consensus algorithm combines the features of both PoW and PoS, with the aim of achieving the strengths of both mechanisms while mitigating their weaknesses. In a hybrid system, both PoW and PoS are used to validate new blocks, and the algorithm switches between the two mechanisms depending on the specific block being validated. In a PoW/PoS hybrid consensus algorithm, PoW is used to create new blocks, while PoS is used to confirm them. The mining process of PoW allows for a secure initial creation of the block, while PoS allows for a quicker validation of the block. The PoS aspect allows validators to place a stake or collateral in the network, which provides them with a proportional probability of being chosen to validate new blocks. The PoS consensus mechanism allows validators to gain rewards for validating new blocks, while penalizing them for malicious behaviour. By combining the two mechanisms, hybrid PoW/PoS consensus aims to achieve a more secure, efficient, and scalable blockchain network. The hybrid approach can improve security by making it more difficult for malicious actors to compromise the network. Furthermore, the hybrid approach can also make the network more energy-efficient and reduce the overall cost of running the network.

One example of a blockchain that employs a hybrid PoW/PoS consensus algorithm is the Dash cryptocurrency. In the Dash network, PoW is used to create new blocks, while PoS is used to confirm them. This allows for a more secure network than pure PoS, while also maintaining faster transaction confirmation times than pure PoW. Overall, the hybrid PoW/PoS consensus algorithm provides a unique and potentially effective solution for improving the security, efficiency, and scalability of blockchain networks.

The hybrid PoW/PoS consensus algorithm represents an innovative approach that blends the key attributes of both Proof of Work (PoW) and Proof of Stake (PoS) algorithms, seeking to create a more robust and well-rounded blockchain network. In this hybrid system, PoW and PoS collaboratively participate in the block creation and validation process.

To delve into the workflow of this consensus algorithm, it commences with PoW miners, individuals, or entities within the network, who embark on a computational race to solve intricate mathematical puzzles. This PoW element serves two fundamental purposes: firstly, it ensures the validity of transactions by demanding miners to expend computational resources, thereby making malicious attacks like 51% attacks substantially more challenging. Secondly, it spearheads the creation of new blocks by selecting a miner who successfully solves the puzzle to propose the next block.

Once a PoW miner successfully creates a block, the focus shifts to the PoS phase. PoS validators, who are participants in the network, enter the scene. These validators assume the role of block validators, and their primary responsibility is to scrutinize and validate the newly created block's contents. To reinforce their commitment to honest validation, PoS validators are required to stake a certain amount of cryptocurrency as collateral. This collateral acts as a financial incentive for validators to carry out their role faithfully since they risk losing their staked assets if they engage in dishonest or erroneous validation.

By combining the computational and resource-intensive PoW component with the economic incentive and commitment to accuracy inherent in PoS, the hybrid PoW/PoS consensus algorithm strives to achieve a harmonious balance between the strengths and weaknesses of these two prominent consensus mechanisms. This fusion ultimately aims to enhance the security, efficiency, and sustainability of blockchain networks, providing a solid foundation for a wide range of applications and use cases across the blockchain ecosystem.

The mathematical equations used in hybrid PoW/PoS consensus algorithm are as follows:

1. Proof of Work (PoW) equation:

To validate a block using PoW, miners must solve a mathematical problem, which is usually a cryptographic puzzle. The mathematical problem must be hard enough to solve but easy enough to verify. The equation for PoW is:

$$Hash\ (block + nonce) < target$$

where block is the block data, nonce is a random number, and target is the difficulty level.

2. Proof of Stake (PoS) equation:

To validate a block using PoS, validators must stake their coins as collateral and be selected to validate a block. The equation for PoS is:

$$Chance\ Of\ Being\ Selected = (Number\ Of\ Coins\ Staked\ /\ Total\ Number\ Of\ Coins\ In\ The\ Network)$$

Once the validators are selected, they validate the block and add it to the blockchain.

3. Hybrid PoW/PoS equation:

In hybrid PoW/PoS, the equations for PoW and PoS are combined to create a consensus algorithm that leverages the strengths of both algorithms. The equation is as follows:

*Hash (Block + Nonce) < Target * (Number Of Coins Staked / Total Number Of Coins In The Network)*

In a hybrid consensus model that combines Proof of Work (PoW) and Proof of Stake (PoS), blocks are validated by both PoW miners and PoS validators. This dual-validation approach enhances network security by protecting against 51% attacks through PoW and incentivizing honest behaviour among validators through PoS. This combination of consensus mechanisms results in a more secure and decentralized blockchain network.

Hybrid Proof of Work/ Proof of Stake (PoW/PoS) Consensus Algorithm

Hybrid PoW/PoS is a consensus algorithm used by some cryptocurrencies, such as Decred, to combine the benefits of Proof of Work and Proof of Stake algorithms. In this algorithm, miners (who participate in PoW) and stakeholders (who participate in PoS) work together to secure the network and make decisions about the network's governance.

In Decred, participants in the PoS element of the hybrid algorithm must time-lock their DCR to purchase "tickets". The cost of an individual ticket is set by a market-like mechanism that aims for a fixed number of stay tickets (40,960). If there are more tickets than the target amount, the fee goes up, and if there are fewer tickets, the fee goes down. Once a ticket is purchased, the DCR used to buy it is locked and cannot be spent until the ticket is called to vote, or until it expires after approximately 142 days. This introduces a potential cost for PoS, intended to ensure that PoS voters have a stake in the network's success.

PoS participants have three distinct roles to play in the hybrid algorithm: block voting, voting on changes to the consensus rules, and voting on project-level management using the Politeia Proposal System. Block voting is the way in which PoS voters engage most directly in maintaining consensus. When a PoW miner reveals a legitimate block, they broadcast it to the network, but for the block to be considered legitimate, it must include votes from at least three of five randomly selected tickets. PoS voters keep their wallets open and ready to respond with votes when their tickets are called or engage Voting Service Providers to do this on their behalf.

Nodes on the network will not consider a new block legitimate until it includes at least three votes. If a majority of the tickets called to vote reject the preceding block's transactions, then they are returned to the mempool. This limits the power of PoW miners to veto changes to the network's consensus rules voted on by the stakeholders. Proof of Stake voters can reject any miner conduct they dislike by voting "no" if they detect malicious or inefficient behaviour, preventing bad PoW miners from writing transactions and receiving rewards. This PoS verification layer significantly boosts the network's security and resistance to most attacks.

The hybrid PoW/PoS design increases the cost of attacking the network because there are unique systems that must be circumvented by a cyber attacker. The PoS component is configured such that tickets can only be obtained quite slowly, and buying the maximum number causes the cost to increase sharply. Additionally, once these tickets have been purchased, the funds used to buy them can be time-locked, leaving an attacker exposed to any devaluation of their locked cash that occurs as a result of an attack. The requirement that each block is voted on by randomly selected stakeholders means that the blockchain must be shared with all network participants as it is mined, improving the network's security. Decred's hybrid system has been designed to give stakeholders power over the PoW miners, adding an additional layer of security to the network.

Decred's Governance & Funding Model: Consensus Change Voting and Politeia Platform

Consensus Change Voting is a key feature of Decred's governance model, which gives stakeholders a prominent role in decision-making. Decred's consensus policies include a rigorous upgrade ratification process that ensures any changes to the network's consensus guidelines are approved by a supermajority of 75% of voting tickets. This process is initiated when a certain percentage of miners and voters are running upgraded software with proposed changes to the rules.

During the four-week voting period, stakeholders can cast their votes to approve or reject the proposal. If the proposal receives the required 75% support, it is accepted, and the rule change is activated one month later. If it does not receive the necessary supermajority, the proposal is rejected, and a re-vote begins.

Decred's governance model also includes a project management component called Politeia. Through this platform, stakeholders can vote on how the block rewards are spent, which functions should be added, and policies that should be implemented. The block rewards are distributed among PoW miners, PoS voters, and the Treasury, which funds the development of open-source software that furthers the project's goals.

Overall, Decred's governance model emphasizes stakeholder participation and decision-making, ensuring that the network remains decentralized and resistant to centralization. The Consensus Change Voting and Politeia platforms provide stakeholders with a voice in the network's development, making Decred a community-driven project.

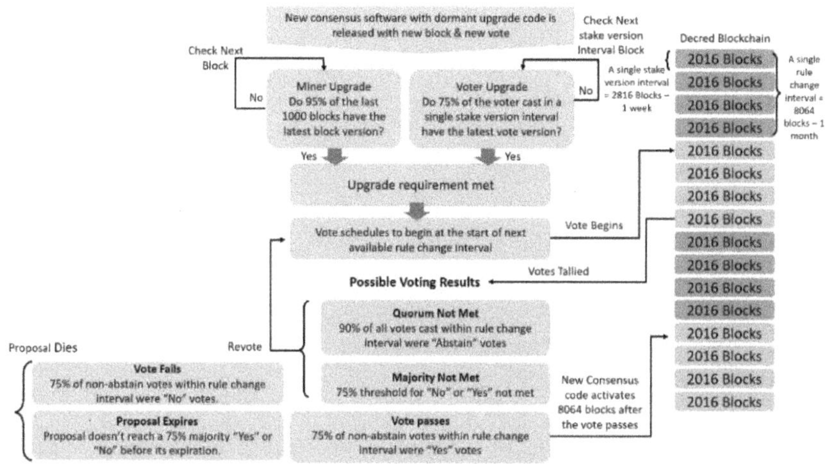

Figure 5.8: PoS in Blockchain Governance

Here is an example snippet of code that could be used to illustrate the Consensus Change Voting process in Decred:

```
# Example code for Decred consensus change voting
# First, we define the required percentage of miners and voters
running upgraded software
MINER_THRESHOLD = 0.95
VOTER_THRESHOLD = 0.75
# Next, we define the voting period length in blocks
VOTING_PERIOD_LENGTH = 2016
# When a proposed change is ready for a vote, we initiate the
process by checking the percentage of upgraded nodes

def initiate_vote():
```

```
num_upgraded_miners = count_upgraded_miners()
num_upgraded_voters = count_upgraded_voters()
if (num_upgraded_miners >= MINER_THRESHOLD) and
(num_upgraded_voters >= VOTER_THRESHOLD):
    start_voting_period()


# During the voting period, stakeholders can cast their votes
def cast_vote(vote):


# Logic for casting a vote goes here
# After the voting period, we check the vote results and
activate the rule change if it was approved


def finalize_vote():
  if vote_passed():
    activate_rule_change()
```

This is just a simplified example, but it gives an idea of how the consensus change voting process could be implemented in code. The actual implementation in Decred is likely to be more complex, considering factors such as ticket splitting and the mechanics of the PoS voting process.

This code provides a basic framework for implementing the Consensus Change Voting process in Decred. It includes functions for initiating a vote, casting votes, and finalizing the vote results. The code also considers the required percentage of upgraded nodes and the length of the voting period. This code provides a good starting point for understanding how the consensus change voting process could be implemented in practice. *(Ref. No.: 87- 154)*

Closing Thoughts

In conclusion, consensus algorithms serve as the bedrock of blockchain technology, providing the essential elements of trust, security, and stability to the network. Throughout this chapter, we have delved into various consensus algorithms that play vital roles in shaping the blockchain landscape. These algorithms include Proof of Work (PoW), Proof of Stake (PoS), Hybrid PoW/PoS, Delegated Proof of Stake (DPoS), Proof of Burn, Proof of Authority (PoA), Leased Proof of Stake (LPoS), and Delayed Proof of Work (dPoW).

The Hybrid PoW/PoS consensus algorithm skilfully combines the strengths of both PoW and PoS, delivering heightened security and energy efficiency. DPoS, on the other hand, empowers blockchain networks with swifter transaction processing times and scalability by delegating decision-making authority to selected participants. In the case of Proof of Burn, users demonstrate their dedication to the network by burning cryptocurrency, reducing the need for energy-intensive mining hardware, and improving sustainability.

PoA brings trust and reliability to private blockchain networks by utilizing trusted nodes for transaction validation and block generation. LPoS allows users to lease their stake to others, earning rewards in return. dPoW enhances security by utilizing a secondary blockchain network for added protection.

The renowned PoW consensus algorithm, famously used in Bitcoin, involves miners solving intricate mathematical puzzles to validate transactions and reap block rewards. PoS, conversely, requires validators to hold a specific amount of cryptocurrency as a stake, motivating them to validate transactions honestly for rewards.

It is essential to recognize that each consensus algorithm carries its unique strengths and limitations, and the choice of algorithm hinges on the particular requirements and goals of the blockchain network in question. As blockchain technology continues its evolution, we anticipate the emergence of innovative consensus algorithms with even more advanced features and capabilities, propelling the widespread adoption of blockchain technology across diverse industries and sectors. The dynamic and ever-expanding world of blockchain consensus algorithms promises a future of enhanced trust, security, and efficiency in the digital realm.

Chapter 6: Staking in Cryptocurrency: Benefits & Risks

"There is an opportunity to recreate the financial world as we know it in the parallel universe that is the blockchain. We are writing rules for this whole new universe."

— Patrick M. Byrne

Understanding Staking: A Less Resource-Intensive Alternative to Mining

Staking is a popular alternative to mining, which involves holding and locking up cryptocurrencies in a wallet to support the security and operations of a blockchain network. Staking enables participants to earn rewards for locking up their coins and is becoming an increasingly popular method of earning passive income in the crypto space. There are two ways to stake coins: directly from a software or hardware wallet or through an exchange's staking services. Many exchanges offer staking services to their users, making it easier for individuals to participate in staking without the technical know-how of setting up a node themselves.

To understand staking, it is essential to first comprehend the Proof of Stake (PoS) algorithm. PoS is a consensus mechanism that enables blockchains to operate more efficiently while maintaining a good degree of decentralization. In contrast to Proof of Work (PoW), which involves miners competing to solve a complex mathematical puzzle to add the next block to the blockchain, PoS assigns the right to validate the subsequent block to the nodes that have locked up their coins. The probability of being chosen to validate the next block is proportional to the number of coins locked up, meaning that participants who stake more coins have a higher chance of being selected. This approach eliminates the need for the energy-intensive computations required in PoW mining and promotes scalability, making it more attractive to many in the crypto space.

While Proof of Stake (PoS) brings with it numerous advantages, it is not without its unique set of challenges. One notable requirement of PoS is that participants must possess a minimum quantity of cryptocurrency to engage in the network. While this is intended to promote commitment and participation, it also carries the potential risk of centralization, as wealthier participants may wield more influence. Another concern is the susceptibility of PoS networks to a "nothing at stake" attack. In such scenarios, participants can cast their votes across multiple blockchain forks without putting any of their assets at risk. This situation can lead to a state of confusion regarding the true state of the network, potentially undermining its security and integrity. Despite these challenges, staking has gained significant popularity as a means of earning passive income within the cryptocurrency realm. The PoS consensus algorithm has undeniably contributed to more efficient and cost-effective blockchain operations. However, vigilance and ongoing evaluation of PoS networks' effectiveness and security are essential as the cryptocurrency industry continues to evolve. Ensuring the resilience and reliability of PoS networks is paramount to their sustained success in the ever-changing landscape of blockchain technology.

Evolution of Proof of Stake (PoS)

The concept of Proof of Stake (PoS) can be traced back to a 2012 paper by Sunny King and Scott Nadal for the cryptocurrency Peercoin. In the paper, they presented a design for a peer-to-peer cryptocurrency that was derived from Bitcoin, the cryptocurrency created by Satoshi Nakamoto. The Peercoin network was launched with a hybrid PoW/PoS mechanism where PoW was used to mint the initial supply of coins. However, over time, the significance of PoW was reduced, and the network's security became increasingly dependent on PoS. The hybrid PoW/PoS mechanism of Peercoin was an innovative approach that sought to address some of the issues associated with PoW, such as high energy consumption and centralization of mining power. By incorporating PoS, Peercoin was able to achieve a higher degree of decentralization, as anyone who held coins could participate in securing the network, rather than just those who had the resources to invest in expensive mining equipment.

Since then, the concept of PoS has been adopted by numerous cryptocurrencies, including Ethereum, which is currently in the process of transitioning from PoW to PoS through a series of technical enhancements called ETH 2.0. The PoS algorithm has proven to be an efficient and effective way to facilitate consensus in a decentralized manner while reducing energy consumption and promoting a higher degree of participation from network stakeholders.

Understanding Staking & DPoS in Blockchain Technology

Delegated Proof of Stake (DPoS) is an alternative consensus mechanism to Proof of Work and Proof of Stake, developed by Daniel Larimer in 2014. DPoS was first used as a part of the BitShares blockchain, but soon other networks adopted the model, including Steem and EOS, which were also created by Larimer.

DPoS works by allowing users to commit their coin balances as votes, where voting power is proportionate to the number of coins held. These votes are then used to elect a number of delegates who manage the blockchain on behalf of their voters, guaranteeing security and consensus. Typically, the staking rewards are distributed to these elected representatives, who then distribute part of the rewards to their electors proportionally to their individual contributions. DPoS allows for consensus to be accomplished with a lower number of validating nodes, enhancing network performance. However, it can also bring about a lower degree of decentralization as the network is based on a small, selected group of validating nodes who deal with the operations and overall governance of the blockchain network. They participate in the processes of reaching consensus and defining key governance parameters.

The DPoS model is designed to encourage users to participate in the network by delegating their votes to representatives who they believe will manage the network in their best interests. This system incentivizes representatives to act in the best interests of their electors, as their position is based on the number of votes they receive. It also allows users to signal their influence via other participants of the network, without having to run their own validating nodes.

In contrast to the Proof of Work-based blockchain that depends upon mining to feature new blocks to the blockchain network, Proof of Stake-based blockchains produce and validate new blocks through the technique of staking. Staking includes validators who lock up their coins so that they may be randomly selected through the protocol at precise intervals to create a block. Usually, participants that stake larger amounts have a higher chance of being selected as the succeeding block validator. This strategy permits blocks to be produced without relying on specialized mining hardware, which includes ASICs.

Staking rewards are a crucial incentive for participants to hold and stake coins on the network. These rewards can vary depending on the specific blockchain network and the number of staked coins. For example, in the case of Cardano's staking mechanism, participants who stake their ADA tokens can earn a yearly reward of around 5% on average. In addition to earning rewards, staking also helps to secure the network by requiring participants to have a vested interest in its success. This means that if a participant misbehaves or acts against the network's interests, they stand to lose their staked coins, thereby incentivizing them to act in the network's best interests.

On a practical level, staking involves holding coins in a suitable wallet that allows users to perform various network functions in return for staking rewards. It may also include adding funds to a staking pool, which aggregates the coins of multiple participants to increase their chances of being selected as a validator. In general, staking is a greener and more sustainable alternative to mining, as it requires less computational power and energy consumption.

Understanding Staking Rewards in PoS Blockchain Networks

Understanding the calculation of staking rewards is essential for individuals interested in participating in a Proof of Stake (PoS) blockchain network.

Different networks can have various ways of estimating staking rewards, which can be adjusted on a block-by-block basis, considering multiple factors. The primary factors that influence staking rewards include the number of coins being staked, how long the validator has been actively staking, and the number of coins staked on the network in total and their inflation rates.

Some blockchain networks decide staking rewards as a specified percentage. These rewards are distributed to validators as a form of compensation for inflation. This mechanism ensures that users spend their coins instead of holding them, which can increase the utilization of the cryptocurrency. As validators can calculate the exact staking reward they can expect, a predictable reward schedule can be more appealing to some individuals than a probabilistic chance of receiving a block reward. Since this is public information, it might incentivize more people to participate in staking, leading to an increase in network security and participation.

For example, in the case of the Cosmos Network, the staking rewards are calculated based on a complex formula that takes into account the total amount of ATOMs staked, the network inflation rate, the percentage of ATOMs staked, and the unbonding period. On the other hand, the Tezos network offers an estimated 6% staking reward to its users. The actual reward depends on the number of tokens staked and the network's inflation rate. Other networks such as Ethereum and Cardano are also moving towards the Proof of Stake consensus mechanism, and their staking rewards are determined based on various factors.

The Benefits & Challenges of Staking Pools for Crypto Investors

A staking pool is a group of coin holders who pool their resources to increase their chances of validating blocks and receiving rewards. The members of the pool combine their staking power and share the rewards proportionally to their contributions to the pool. However, setting up and managing a staking pool can be challenging and time-consuming, requiring expertise and technical know-how. Therefore, staking pools are generally more effective on networks with high technical or financial barriers to entry. To compensate for the expenses associated with operating the pool, most pool providers charge a fee for the staking rewards distributed to participants.

Apart from cost-sharing benefits, staking pools also offer additional flexibility for individual stalkers. Typically, staking requires a fixed period of stake lockup, with a specified withdrawal or unbinding time set through the protocol. Additionally, participants of a staking pool must maintain a substantial minimum balance to disincentivize malicious behaviour. However, most staking pools require a low minimum balance and impose no extra withdrawal times. Therefore, becoming a member of a staking pool can be more beneficial for newer users rather than staking solo.

For instance, if an individual wants to stake on the Cardano network, they can join a staking pool to increase their chances of earning rewards without the need for technical knowledge or high upfront costs. Staking pools distribute rewards among members according to their contributions to the pool, and participants can enjoy the benefits of staking without worrying about the technical aspects of setting up and maintaining a node. Additionally, participants can withdraw their stake at any time without incurring any penalties. However, staking pools charge a fee for their services, which can range from 0% to 20% of the staking rewards distributed to participants. *(Ref. No.: 142- 146)*

Benefits of Cold Staking in Cryptocurrency

Cold staking is a term used in the context of cryptocurrency staking, which refers to the process of staking coins in a wallet that is not connected to the internet. The concept of cold staking is gaining popularity among cryptocurrency enthusiasts who are looking for ways to secure their assets while also participating in staking activities. One of the primary advantages of cold staking is that it provides an additional layer of security against potential cyber-attacks. By keeping their staking funds offline, users can significantly reduce the risk of losing their assets to hackers or other malicious actors. Instead of keeping their coins on an exchange or a hot wallet, which are connected to the internet, users can store their assets in a hardware wallet or an air-gapped software wallet.

Cold staking can be performed directly from a hardware wallet or an air-gapped software wallet. Some blockchain networks support cold staking, allowing users to stake their coins while maintaining the security of their funds offline. However, it's important to note that if the stakeholder moves their coins out of cold storage, they will stop receiving rewards. Cold staking is especially beneficial for large stakeholders who want to ensure maximum protection of their funds while still supporting the network. By using cold staking, they can participate in staking activities without risking their assets to potential cyber threats. In addition, cold staking can be an attractive option for long-term investors who want to hold onto their coins for an extended period of time, while also earning staking rewards. *(Ref. No.: 145- 158)*

Staking Pools on Centralized Exchanges

Staking pools offer users a way to participate in staking without having to run a node or maintain technical requirements. Some staking pools are offered by centralized exchanges, which can be a convenient option for users who want to take advantage of the exchange's features while still earning staking rewards. By keeping their PoS coins in a staking pool on an exchange, users can enjoy the security and convenience of holding their coins on a centralized exchange, while also participating in staking.

There may be fees associated with staking on an exchange, but the exchange will take care of the technical requirements and distribute staking rewards to users. These rewards are generally distributed at the beginning of each month, and users can check their staking rewards under the Historical Yield tab on the staking page of the respective project. Additionally, staking on an exchange can provide users with extra benefits, such as the ability to easily trade their staked coins or use them as collateral for loans. However, it is important to note that staking on an exchange also carries some risks. In the event of a security breach or hack, users' staked coins may be at risk. Additionally, users do not have complete control over their coins while they are staked on an exchange, as the exchange may have control over the withdrawal process. Therefore, users should carefully consider the risks and benefits before staking on an exchange and should only use reputable and trusted exchanges for their staking needs.

Benefits of Cryptocurrency Staking:

Cryptocurrency staking has become increasingly popular in recent years due to its many benefits. The following are some of the key advantages of cryptocurrency staking:

1. **Easy to Earn Interest:** One of the primary benefits of cryptocurrency staking is that it provides a straightforward way to earn interest on your cryptocurrency holdings. By staking your coins, you can earn rewards for participating in the network, without having to do any complicated mining.

2. **No Need for Equipment:** Unlike cryptocurrency mining, which requires specialized equipment and technical expertise, staking can be done with any standard computer or even a mobile phone. This makes staking accessible to a wider range of individuals, including those who may not have the resources or technical knowledge to mine cryptocurrencies.

3. **Supporting Blockchain Network Security:** By staking your coins, you are helping to secure the blockchain network. Staking involves holding a certain number of coins as collateral, which helps to prevent malicious actors from attacking the network. In addition, staking incentivizes users to act in the best interest of the network, as they stand to lose their collateral if they engage in any malicious behaviour.

4. **Environmentally Friendly:** Staking is much more environmentally friendly than cryptocurrency mining. This is because mining requires a lot of energy and computing power, which can have a significant impact on the environment. In contrast, staking requires much less energy and computing power, making it a more sustainable way to participate in cryptocurrency networks.

Risks of Staking Crypto

When staking cryptocurrency, there are several risks that one should be aware of.

1. Crypto prices are volatile and can rapidly decrease, potentially offsetting any interest earned from staking.
2. Staking often requires locking up assets for a set time, leaving them illiquid and unavailable for selling or trading.
3. An unstaking period may be required, which can last seven days or longer, adding another layer of illiquidity.
4. The greatest risk associated with staking is a decline in cryptocurrency prices.

Closing Thoughts

Proof of Stake and staking options provide an excellent opportunity for anyone interested in participating in the consensus and governance of a blockchain network. As staking becomes increasingly accessible, the barriers to entry in the blockchain ecosystem are getting lower. However, it is essential to remember that staking is not without risk. Locking up finances in a smart contract is vulnerable to bugs, so it is crucial to do thorough research and use high-quality software and hardware wallets. Despite the risks, staking is an easy way to earn passive income by merely holding coins, and it is a promising trend for the future of blockchain. It is important to weigh the risks and benefits before deciding to stake, and to stay informed about the latest developments in the industry.

Chapter 7: Cryptography & Encryption

"The main advantage of blockchain technology is supposed to be that it's more secure, but new technologies are generally hard for people to trust, and this paradox can't really be avoided."

— Vitalik Buterin

The Fascinating History of Cryptography

Cryptography is the process of creating codes and ciphers to ensure secure communication. It is a crucial component in the development of cutting-edge cryptocurrencies and blockchains. The strategies used in cryptography today are the result of a long and fascinating history that dates back to ancient times. Symbol replacement, the simplest form of cryptography, was used in ancient Egyptian and Mesopotamian scripts. The earliest known instance of cryptography was found in the tomb of an Egyptian noble, Khnumhotep II, about 3,900 years ago. Symbol replacement was not used to conceal information, but rather to enhance its linguistic appeal. Similarly, in Mesopotamia about 3,500 years ago, a scribe used cryptography to hide a formula for pottery glaze that was written on clay tablets.

As time passed, cryptography became more important for safeguarding sensitive information, especially military secrets. For example, in the Greek city of Sparta, messages were encrypted using cylinders of specific sizes. The message could not be deciphered until it was wrapped around a similar cylinder by the recipient. Coded messages were also used by spies in ancient India, dating back to the 2nd century BC. The Romans developed some of the most advanced cryptography in the ancient world. Their Caesar cipher involved shifting the letters of an encrypted message by a certain number of places down the Latin alphabet. A recipient could decode the message if they knew the system and the number of places to shift the letters.

During the Middle Ages and the Renaissance, cryptography continued to develop. Cryptanalysis, the science of breaking codes and ciphers, began to catch up to the still relatively primitive technology of cryptography. Al-Kindi, a renowned Arab mathematician, developed a technique called frequency analysis around 800 AD that rendered substitution ciphers vulnerable to decryption. In 1465, Leone Alberti developed the polyalphabetic cipher, which was designed to counter Al-Kindi's frequency analysis technique. A message is encoded using different alphabets in a polyalphabetic cipher, making it much more secure.

In the Renaissance period, new techniques of encoding data were developed, such as binary encoding, which was invented by Sir Francis Viscount St. Albans in 1623. The history of cryptography shows how this technology has evolved and adapted over time to meet the needs of secure communication. Today, cryptography is an essential tool for protecting information in the digital age.

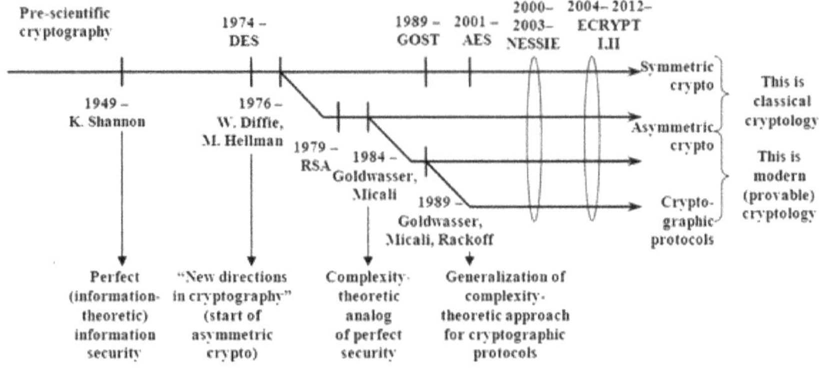

Figure 7.1: History of Cryptography

The Impact of Modern Computing on Cryptography

In more recent centuries, the technology of cryptography continued to progress and improve. One significant advancement was proposed by Thomas Jefferson in the 1790s with his invention of the cipher wheel. The cipher wheel consisted of 36 rings of letters on moving wheels that could be used to achieve complex encoding. This idea was so advanced that it served as the foundation for American military cryptography until as late as the Second World War. During World War II, the Enigma machine was an example of analog cryptography used by the Axis powers. Similar to the wheel cipher, this device used rotating wheels to encode a message, making it virtually impossible to read without another Enigma. However, early computer technology was eventually used to break the Enigma cipher, and the successful decryption of Enigma messages turned out to be a critical factor in the eventual Allied victory. With the advent of computers, cryptography entered a new era of rapid advancement. One of the most important cryptographic advances was the development of public-key cryptography in the 1970s. Public-key cryptography, also known as asymmetric cryptography, allows for secure communication over a public channel without the need for a shared secret key. This was a significant breakthrough that allowed for secure electronic communication over the internet. Today, cryptography is essential to the security of modern digital communication, including cryptocurrencies and blockchains. Cryptographic algorithms and protocols are continually being developed and improved to stay ahead of the latest threats from cybercriminals and state-sponsored attackers. The ongoing development of cryptography is critical to ensuring the privacy and security of individuals and organizations in the digital age.

Computerized Cryptography & the Importance of Public Key Cryptography

Cryptography has undergone significant advancements with the advent of computers, surpassing the analog era in terms of complexity and security. Today, 128-bit mathematical encryption is the standard for securing sensitive information on devices and computer systems. In the early 1990s, quantum cryptography was being developed to enhance encryption security. Cryptography has also played a vital role in enabling cryptocurrencies, which use advanced cryptographic strategies like hash functions, digital signatures, and public-key cryptography to ensure secure data storage and authentication of financial transactions. The Elliptical Curve Digital Signature Algorithm (ECDSA) is a specialized form of cryptography that underpins the security of cryptocurrency systems like Bitcoin.

Public key cryptography, also known as asymmetric cryptography, is a critical framework used in modern cyber security. Unlike symmetric cryptography that uses a single key to encrypt and decrypt data, public key cryptography utilizes a pair of keys, one private and one public, to encrypt and decrypt data. This provides a unique set of capabilities that can be used to address challenges inherent in other cryptographic techniques. Public key cryptography has become a critical component of the growing cryptocurrency ecosystem and is an essential element of modern cyber security. Its use in securing data transmission, authentication, and digital signatures has made it a powerful tool in the fight against cyber threats. As long as sensitive data requires protection, cryptography will continue to evolve, and public key cryptography will play an increasingly significant role in safeguarding it.

How does Public-Key Cryptography Work?

Public-key cryptography, also known as asymmetric cryptography, is a method of encryption that uses two mathematically related keys - a public key and a private key - to secure the transmission of information over a network. In this system, the public key is used to encrypt the data, and only the holder of the corresponding private key can decrypt it. The public key can be safely shared without compromising the security of the private key, which is kept secret by the owner. Each key pair is unique, ensuring that a message encrypted using a public key can only be decrypted by the individual possessing the corresponding private key.

Asymmetric encryption algorithms generate key pairs that are mathematically coupled, making it extremely difficult to compute a private key from its public counterpart. The longer length of the keys, typically between 1,024 and 2,048 bits, ensures a high level of security. One of the most commonly used algorithms for asymmetric encryption is the RSA (Rivest-Shamir-Adleman) scheme. In the RSA system, keys are generated using a modulus that is arrived at by multiplying two large prime numbers. The modulus generates two keys - a public key that can be shared and a private key that must be kept secret.

RSA was first described in 1977 by Rivest, Shamir, and Adleman and remains a major component of public-key cryptography systems. It is widely used for secure data transmission over the internet, including digital signatures, SSL (Secure Sockets Layer) certificates, and secure email. Public-key cryptography is an essential element of modern cybersecurity, as it provides a secure method of communication between parties without the need for a shared secret key. It is widely used in various applications, including secure online transactions, online banking, e-commerce, and secure communication between government agencies.

In summary, public-key cryptography provides a secure method for transmitting data over a network by using two mathematically related keys, a public key, and a private key. The keys are unique and cannot be computed from one another, ensuring the security of the information being transmitted. The RSA scheme is a widely used asymmetric encryption algorithm that generates secure key pairs for secure data transmission over the internet.

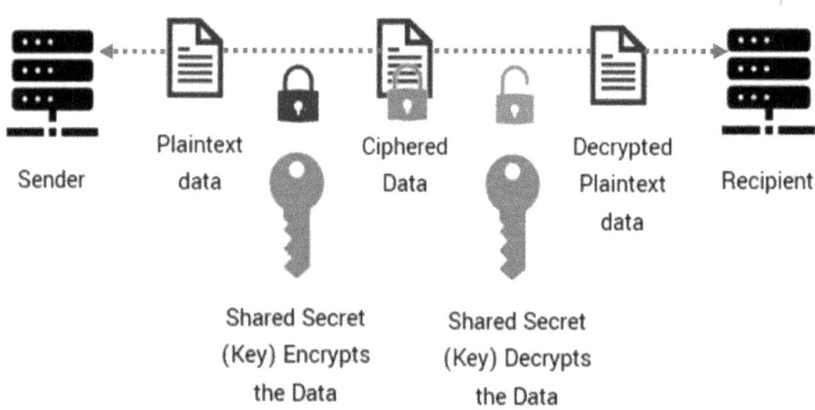

Figure 7.2: Private Key Encryption (Symmetric)

How Good is the Public-Key Cryptography as an Encryption Tool?

Public-Key Cryptography (PKC) is considered to be a highly effective encryption tool due to its unique approach to encryption and decryption. In traditional symmetric cryptography, the same key is used for both encryption and decryption. However, this poses a challenge in securely transmitting the key over an insecure connection. In contrast, PKC solves this challenge by using two separate and mathematically related keys - a public key and a private key.

The public key is widely distributed and used by anyone who wants to send an encrypted message to the owner of the private key. The owner of the private key, and only the owner, can decrypt the message using their private key. This way, the private key can be kept secret and secure without the need for secure transmission.

Asymmetric encryption algorithms generate key pairs that are mathematically coupled, and their key lengths are typically much longer than those used in symmetric cryptography. This makes it extremely difficult to compute a private key from its public counterpart. For example, the RSA algorithm, which is a widely used asymmetric encryption algorithm, uses a modulus to generate public and private keys that are mathematically related.

One of the key advantages of PKC is its ability to establish secure communication channels over insecure networks. For example, PKC is widely used to secure internet communication through the use of SSL/TLS protocols. It is also used in email communication through the use of S/MIME encryption.

Overall, PKC is considered to be a highly effective encryption tool due to its ability to securely transmit information without the need for secure key exchange, and its ability to establish secure communication channels over insecure networks.

In PKC, the security of the encryption relies on the fact that it is extremely difficult to compute the private key from the public key. This is because the public and private keys are mathematically related through a complex mathematical function, such as the RSA algorithm. The strength of the encryption depends on the length of the keys, with longer keys making it increasingly difficult to derive the private key from the public key.

For example, let us consider the RSA algorithm, which is one of the most widely used PKC algorithms. RSA keys are typically between 1,024 and 4,096 bits in length. To derive the private key from the public key in an RSA system, an attacker would need to factor the modulus, which is the product of two large prime numbers used to generate the keys. Factoring the modulus is a computationally intensive task, and the larger the key size, the more difficult it becomes.

To illustrate this point, let us take an RSA key with a length of 2,048 bits. The number of possible keys for a 2,048-bit RSA key is approximately 2^{2048}, which is an astronomically large number (approximately 3×10^{616}). Even with the most powerful computers available today, it would take an impossibly long time to compute the private key from the public key through brute force.

Therefore, the mathematical complexity of PKC algorithms, such as RSA, demonstrates that they are highly secure encryption tools, making them a critical component of modern cybersecurity.

Generating Digital Signatures

Asymmetric Cryptography algorithms, also known as Public Key Cryptography, have numerous applications in securing data communications. One of these applications is generating digital signatures to authenticate data. A digital signature is a mathematical scheme that provides authenticity, integrity, and non-repudiation to digital documents or messages.

To generate a digital signature, a hash is created using the data contained within a message, which is a fixed-sized output that represents the input data. The hash is then encrypted using the sender's private key, which ensures the integrity of the hash by making it tamper-evident. The encrypted hash, or digital signature, is then transmitted alongside the message.

The recipient of the message can then use the sender's public key to decrypt the digital signature and obtain the original hash. If the original hash matches the hash of the received message, the message is authenticated as originating from the sender and has not been altered in any way during transmission.

In some cases, digital signatures and encryption are used in combination, known as digital envelope. In this scenario, the hash itself may be encrypted as part of the message. The message is first encrypted using a symmetric key, and then the encrypted symmetric key is encrypted using the recipient's public key. This way, only the recipient can decrypt the symmetric key using their private key, and subsequently decrypt the message. Finally, the recipient can verify the digital signature using the sender's public key to ensure the authenticity of the message.

It is important to note that not all digital signature techniques involve encryption methods. Some digital signature schemes use only hashing functions to generate the signature, and these signatures are called "Hash-based signatures."

Digital signatures are essential in ensuring the authenticity, integrity, and non-repudiation of digital data. They are particularly important when transmitting sensitive or confidential information, as they provide a high level of confidence in the authenticity of the data.

Limitations of Public-Key Cryptography (PKC)

Public-Key Cryptography (PKC) is a widely used cryptographic technique for enhancing computer security and verifying message integrity. However, it also has several limitations. One of the limitations of PKC is that the complex mathematical functions involved in encryption and decryption can make asymmetric algorithms quite slow, particularly when dealing with large amounts of data. This can pose challenges for systems that require fast and efficient encryption and decryption processes.

Furthermore, PKC is dependent on the idea of keeping the private key a secret. If the private key is shared or exposed, the security of all messages encrypted with its corresponding public key will be compromised. Therefore, it is crucial to ensure that the private key remains confidential and is only accessible by authorized individuals. Another limitation is the risk of accidentally losing the private key. If the private key is lost, it becomes impossible for users to access the encrypted data. In such cases, recovery of the data becomes difficult, if not impossible, without the private key.

Moreover, PKC is vulnerable to attacks such as brute force attacks, in which an attacker systematically tries all possible combinations of keys until the correct one is found. This vulnerability means that encryption keys need to be long and complex enough to resist such attacks, making the encryption process even slower.

In addition, PKC is subject to man-in-the-middle (MITM) attacks, where an attacker intercepts and alters communications between two parties. To prevent MITM attacks, a trusted third party, such as a certificate authority, is required to verify the identity of the communicating parties and issue digital certificates to them.

In conclusion, while PKC provides many benefits for secure communication, it is important to be aware of its limitations and vulnerabilities. These limitations should be taken into consideration when implementing PKC-based systems, and appropriate measures should be taken to mitigate their impact.

Applications of Public-Key Cryptography (PKC)

Public-Key Cryptography (PKC) is widely used in cutting-edge cybersecurity systems to provide secure communication for sensitive information. For example, PKC can be used to encrypt emails, keeping their contents private. The Secure Sockets Layer (SSL) protocol, which enables secure connections to websites, also uses asymmetric cryptography.

PKC has also been explored as a way to provide a secure digital voting environment, enabling voters to participate in elections from their personal computers. Furthermore, PKC plays a crucial role in blockchain and cryptocurrency technology. When a new cryptocurrency wallet is set up, a pair of keys is generated - a Public Key and a Private Key. The Public Key is used to generate the wallet address, which can be shared with others securely, while the Private Key is used to create digital signatures, verify transactions, and must be kept secret.

In cryptocurrency applications, PKC is used for a specific algorithm to verify transactions known as the Elliptic Curve Digital Signature Algorithm (ECDSA). Unlike many PKC systems used for cybersecurity, the ECDSA does not use encryption to create digital signatures. This means that blockchain technology does not require encryption, as the digital signature itself provides the necessary security.

Once a transaction is confirmed by verifying the hash contained in the digital signature, it can be introduced to the blockchain ledger, ensuring that only the person with the Private Key associated with the corresponding cryptocurrency wallet can circulate the funds. It is important to note that the use of PKC in cryptocurrency applications is unique from those used for cybersecurity purposes. The use of PKC in blockchain and cryptocurrency technology has enabled secure, decentralized transactions, transforming the financial landscape.

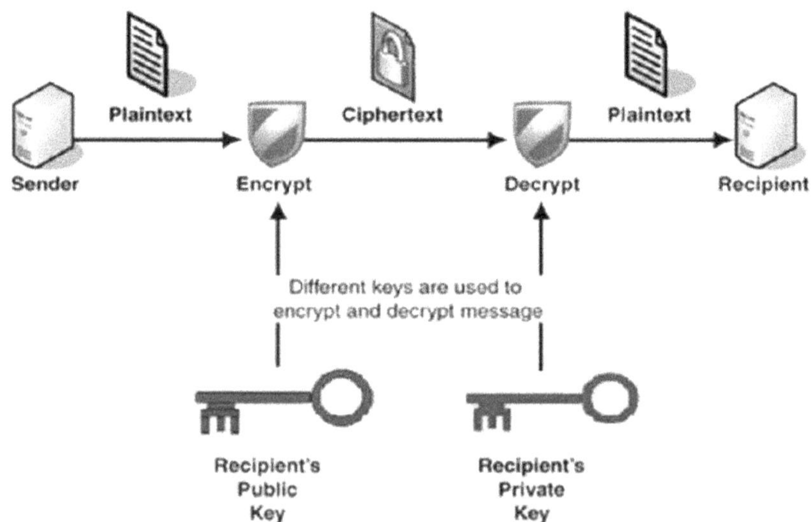

Figure 7.3: Diagram of Public-Key Cryptography (PKC)

Symmetric vs. Asymmetric Encryption

In the world of cryptography, there are two main fields of study: symmetric and asymmetric cryptography. Although symmetric encryption is commonly associated with symmetric cryptography, asymmetric cryptography is divided into two primary use cases: asymmetric encryption and digital signatures. Asymmetric encryption involves the use of a public key and a private key. The public key is widely distributed and can be used to encrypt messages, while the private key is kept secret and is used for decryption. This approach enables secure communication between parties without the need for a shared secret key.

Digital signatures, on the other hand, are created using a hash function that generates a unique fingerprint of a message. The hash is then encrypted using the sender's private key, producing a digital signature. The recipient can verify the signature by decrypting it using the sender's public key, ensuring that the message was not tampered with during transmission and originated from the expected sender.

Thus, we can distinguish between these two groups of cryptographic systems: symmetric cryptography, which primarily uses symmetric encryption, and asymmetric cryptography, which includes both asymmetric encryption and digital signatures. By understanding these different approaches to cryptography, we can better appreciate the methods used to protect sensitive information in a variety of contexts, from secure communication to digital transactions.

Consequently, these categories can be depicted as:

- Symmetric Key Cryptography
- Symmetric Encryption
- Asymmetric Cryptography (or Public-Key Cryptography)
- Asymmetric Encryption (or Public-Key Encryption)
- Digital Signatures (may or may not encompass Encryption)

Encryption algorithms are generally categorized as either symmetric or asymmetric encryption. The main difference between these two approaches is that symmetric encryption algorithms use a single key for both encryption and decryption, while asymmetric encryption uses two separate but related keys. This seemingly simple distinction has significant implications for the functionality and use of each encryption strategy.

Symmetric encryption is often used in scenarios where the parties involved in communication share a common secret key that is used to encrypt and decrypt messages. This type of encryption is relatively fast and efficient, making it suitable for use in scenarios where large amounts of data need to be encrypted or decrypted quickly. However, the use of a shared secret key introduces security risks, as the key must be securely distributed among all parties involved in communication. *(Ref. No.: 159- 167)*

Asymmetric encryption, on the other hand, uses a public key and a private key to encrypt and decrypt messages. The public key is widely distributed and can be freely shared with anyone, while the private key is kept secret and known only to the owner. This approach provides stronger security than symmetric encryption, as the private key can be kept confidential and not shared with others. Asymmetric encryption is commonly used in scenarios where secure communication between parties is required, such as online transactions or email communication. However, the use of two keys for encryption and decryption can result in slower performance compared to symmetric encryption.

Symmetric Encryption

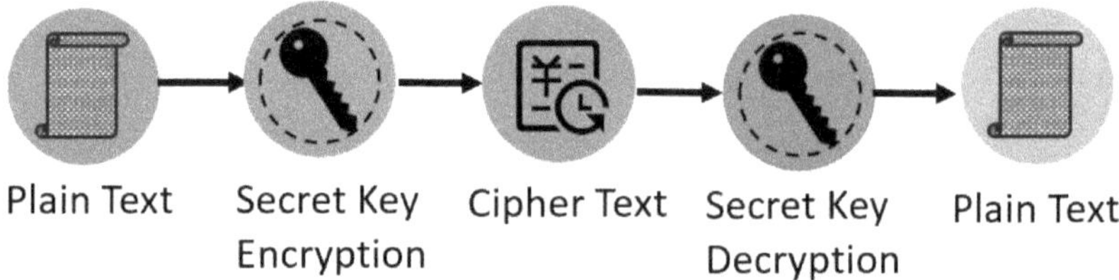

Asymmetric Encryption

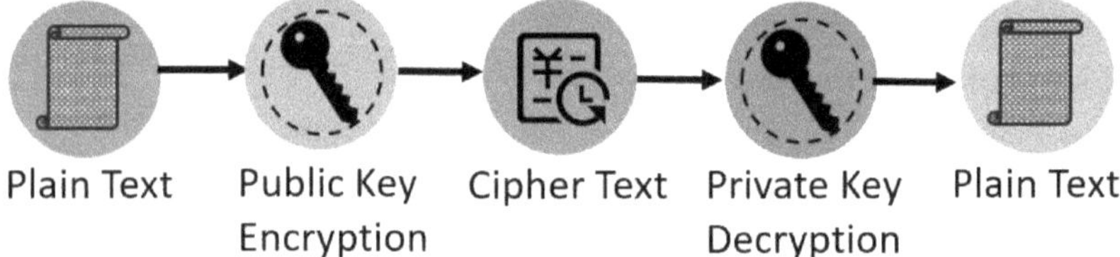

Figure 7.4: Symmetric vs. Asymmetric Encryption

Understanding Encryption Keys

Encryption keys are an essential component of encryption algorithms, which are divided into two main categories: symmetric and asymmetric encryption. These two forms of encryption differ significantly in how they use keys to encrypt and decrypt data.

Symmetric encryption algorithms use the same key for both encryption and decryption. This means that if Alice wants to send a secure message to Bob, she would use the same key to encrypt the message and share that key with Bob. However, if a malicious actor were to intercept the key, they could decrypt the message and access the encrypted information.

On the other hand, asymmetric encryption algorithms use two keys that are mathematically related but distinct: a public key and a private key. The public key is used for encryption, and it can be freely shared with anyone. The private key, which is kept secret, is used for decryption.

For instance, if Alice wants to send a secure message to Bob using an asymmetric scheme, she would encrypt the message with Bob's public key. Only Bob can decrypt the message using his private key, which only he has access to. Even if a malicious actor intercepts the message and obtains Bob's public key, they cannot decrypt the message without the corresponding private key.

Asymmetric encryption provides a higher degree of security than symmetric encryption because the private key must be kept secret, whereas the shared key in symmetric encryption is more vulnerable to interception. Asymmetric encryption is commonly used for secure communication over the internet, such as in email encryption and secure website connections.

Encryption algorithms use mathematical functions to generate keys and to encrypt and decrypt data. Symmetric encryption algorithms use a single key for both encryption and decryption. For example, the popular Advanced Encryption Standard (AES) algorithm uses a symmetric key to encrypt and decrypt data.

Let us say that the key for the AES algorithm is represented by the variable "k". If Alice wants to send a message to Bob using the AES algorithm, she would first encrypt the message using the key "k", resulting in ciphertext. Bob would then use the same key "k" to decrypt the ciphertext and recover the original message.

In mathematical terms, we can represent this process as:

ciphertext = AES_encrypt (plaintext, k) plaintext = AES_decrypt (ciphertext, k)

Asymmetric encryption algorithms, on the other hand, use two different but mathematically related keys for encryption and decryption. One key is used for encryption and is called the public key, while the other key is used for decryption and is called the private key.

Let us consider the popular RSA algorithm as an example. RSA generates two keys, a public key, and a private key, both of which are related mathematically. The public key consists of a pair of numbers (e, n), while the private key consists of a pair of numbers (d, n), where n is a product of two large prime numbers and e and d are mathematical functions of n.

To encrypt a message using RSA, Alice would first obtain Bob's public key (e, n). She would then use this public key to encrypt the message, resulting in ciphertext. Bob would then use his private key (d, n) to decrypt the ciphertext and recover the original message.

In mathematical terms, we can represent this process as:

$$ciphertext = plaintext^e \bmod n \quad plaintext = ciphertext^d \bmod n$$

Here, the "^" symbol represents exponentiation, and "mod n" represents the modulus operation, which gives the remainder after division by n.

Functional Differences Between Symmetric & Asymmetric Encryption

Symmetric and asymmetric encryption have a functional difference when it comes to key length and security. The length of the key in bits is directly related to the level of security provided by the cryptographic algorithm. In symmetric encryption, the keys are randomly chosen and their lengths are usually set at either 128 or 256 bits, depending on the required degree of security.

On the other hand, asymmetric encryption requires a mathematical relationship between the public and private keys, which means that there may be a mathematical pattern between the two. This pattern can potentially be exploited by attackers to break the encryption, so asymmetric keys need to be much longer to provide an equivalent level of security.

The difference in key length is so significant that a 128-bit symmetric key and a 2048-bit asymmetric key provide roughly similar levels of security. This means that asymmetric encryption is generally considered to be more secure than symmetric encryption, but at the cost of slower performance due to the complex mathematical functions involved in the encryption and decryption processes.

Advantages & Disadvantages of Symmetric & Asymmetric Encryption

Advantages of Symmetric Encryption:

1. Faster than asymmetric encryption due to the simpler algorithm.
2. Requires less computational power, making it more efficient for large amounts of data.
3. Can be implemented in hardware, making it even faster and more efficient.
4. Better suited for use cases where the same parties need to encrypt and decrypt data frequently and securely, such as in internal communication systems.

Disadvantages of Symmetric Encryption:

1. The same key is used for both encryption and decryption, so key distribution is a major concern.
2. The key must be securely shared among all parties who need to access the encrypted data.
3. If the key is compromised, all data encrypted with that key is vulnerable.
4. Limited in terms of scalability, since each additional user requires the distribution of a new key.

Advantages of Asymmetric Encryption:

1. Uses separate keys for encryption and decryption, eliminating the need for key distribution.
2. The public key can be widely distributed, reducing the risk of interception.
3. Offers better security, since even if the public key is intercepted, it cannot be used to decrypt the message.
4. Suitable for applications where secure communication between many parties is required, such as in online transactions.

Disadvantages of Asymmetric Encryption:

1. Slower than symmetric encryption due to the more complex algorithm and longer key lengths.

2. Requires more computational power, which can be a disadvantage for large amounts of data or resource-limited devices.
3. More difficult to implement in hardware, which may affect performance.
4. Longer key lengths required to provide equivalent security, which can be challenging to manage and store.

Examples of Symmetric Encryption

Popular examples of symmetric encryption include the:

1. **Data Encryption Standard (DES):** DES is one of the oldest and widely used symmetric key encryption algorithms. It uses a 56-bit key to encrypt and decrypt data. Despite being a widely used algorithm, it has some security vulnerabilities, and it is recommended to use it with caution.
2. **Triple Data Encryption Standard (Triple DES):** Triple DES is an advanced version of DES that uses three keys instead of one. It offers a higher level of security than DES and is more difficult to crack. However, it is slower than DES due to the additional rounds of encryption.
3. **Advanced Encryption Standard (AES):** AES is currently the most popular symmetric key encryption algorithm used worldwide. It is used in a variety of applications, including wireless security, internet security, and file encryption. AES uses 128, 192, or 256-bit key lengths, making it more secure than DES.
4. **International Data Encryption Algorithm (IDEA):** IDEA is a symmetric key block cipher that was designed to replace DES. It is faster than DES and more secure than Triple DES, and uses 128-bit keys. IDEA is widely used in financial transactions and is considered one of the strongest encryption algorithms.
5. **TLS/SSL protocol:** TLS/SSL is a protocol that is widely used to secure internet communications. It is based on asymmetric encryption, where a public key is used to encrypt data, and a private key is used to decrypt it. TLS/SSL is used in web browsing, email, and many other applications that require secure communication over the internet.

Applications of Symmetric & Asymmetric Encryption:

Symmetric Encryption:

1. Symmetric encryption is extensively used in modern computer systems due to its greater speed and efficiency.

2. The Advanced Encryption Standard (AES) is widely used to encrypt sensitive and classified information by the United States government.
3. Other examples of symmetric encryption algorithms include the Data Encryption Standard (DES), Triple DES, and the International Data Encryption Algorithm (IDEA).
4. Symmetric encryption is commonly used in securing data stored on computer hard drives, USB drives, and cloud servers.

Asymmetric Encryption:

1. Asymmetric encryption is commonly used in scenarios where multiple users need to encrypt and decrypt data, such as in encrypted email communications.
2. In asymmetric encryption, the public key can be freely shared while the private key must be kept secure.
3. The Transport Layer Security/Secure Sockets Layer (TLS/SSL) protocol uses asymmetric encryption to secure web communication.
4. Asymmetric encryption is also used in digital signatures to ensure the authenticity and integrity of electronic documents.

Hybrid Encryption Systems

Hybrid cryptographic systems are commonly used in many applications, which combine both symmetric and asymmetric encryption techniques. For instance, the SSL and TLS cryptographic protocols, designed to secure communication on the internet, use hybrid encryption. The SSL protocol is no longer considered secure and its use should be discontinued, but the TLS protocol is regarded as safe and widely used by major web browsers.

1. Hybrid systems refer to cryptographic techniques that combine both symmetric and asymmetric encryption methods to obtain the benefits of each approach.
2. The SSL and TLS protocols are used to secure online communication between web servers and clients, such as web browsers.
3. SSL was widely used in the past but has been replaced by TLS, which is considered safer and more reliable.
4. The TLS protocol uses a hybrid approach to encryption, where a symmetric key is generated for a session and used to encrypt the data, while the asymmetric encryption technique is used for key exchange and authentication.

5. The use of hybrid encryption allows for a more secure and efficient method of encryption, as it leverages the strengths of both symmetric and asymmetric encryption techniques.

Cryptocurrency Encryption & End-to-End Encryption (E2EE)

Cryptocurrency Encryption:

Cryptocurrency wallets use encryption techniques to provide improved levels of security to end-users. While encryption algorithms are implemented when users set up a password for their crypto wallets, it is a common misconception that blockchain systems make use of asymmetric encryption algorithms. In reality, not all digital signature techniques make use of encryption processes, even if they present public & private keys. The RSA is one instance of an algorithm that can be used for signing encrypted messages, but the digital signature algorithm utilized by Bitcoin (named ECDSA) does not use encryption at all.

End-to-End Encryption (E2EE):

In our digital communication era, our messages are recorded and stored in a central server, and users hardly communicate directly with their peers. In this scenario, end-to-end encryption (E2EE) may be the solution for users who do not want their message to be read by the server responsible for passing the message between the sender and the receiver. End-to-end encryption is a technique for encrypting communications between sender & receiver such that they are the only parties who can decrypt the data. E2EE's origins could be traced back to the 1990s when Phil Zimmerman released Pretty Good Privacy (referred to as PGP). PGP stands for Pretty Good Privacy, an encryption software application designed to provide privacy, security, and authentication for online communication systems. End-to-end encryption ensures that the message is encrypted before being sent to the server, and only the intended recipient can decrypt the message.

Enhancing Message Security with End-to-End Encryption (E2EE)

Once the message has been passed along to the recipient's device, it is typically stored on that device and displayed in the recipient's chat log within the messaging application. In some cases, the message may be stored temporarily on the central server for delivery purposes, but it is eventually removed once it has been successfully delivered.

The problem with this process is that messages are transmitted and stored in an unencrypted format, which means that anyone with access to the central server or the recipient's device could potentially read the message. This includes not only the intended recipient but also any third parties who may be able to intercept or access the message in transit or on the device itself. Furthermore, even if the messaging platform itself uses encryption to protect data in transit between the user's device and the central server, the messages may still be stored in an unencrypted format on the server or the recipient's device. This means that if the server or device is compromised, an attacker may be able to gain access to the messages.

End-to-end encryption (E2EE) addresses these issues by encrypting messages in a way that only the sender and recipient can decrypt them. With E2EE, the message is encrypted on the sender's device before it is transmitted to the central server. The central server does not have access to the encryption key, which means it cannot read the contents of the message. When the message is received by the recipient's device, it is decrypted using a key that only the recipient has access to. This means that even if the central server is compromised, the attacker will not be able to read the message.

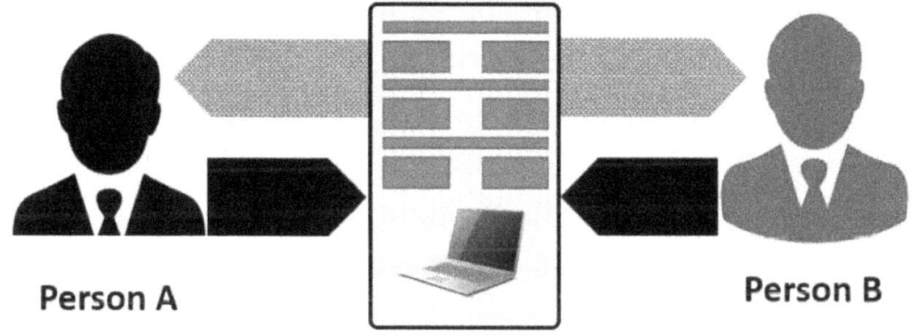

Figure 7.5: Message Encryption & Decryption

(Users A and B communicating. They must pass data through the server (S) to reach each other.)

The model where a server handle most of the data transmission is commonly known as the client-server model. In this setup, a user's phone, acting as the client, is not heavily involved in the data transmission process. Instead, the server takes care of most of the work. However, this also means that the carrier company acts as a middleman between the sender and the recipient. To secure the data transmission, the information exchanged between A <-> S & S <-> B in the diagram is often encrypted. One of the encryption methods used is Transport

Layer Security (TLS), which provides a secure connection between clients and servers, preventing outsiders from intercepting the message during transmission. Although TLS and similar security measures can prevent unauthorized access to the data, the server can still read the information. This is where encryption plays a vital role. If the information from node A is encrypted with a cryptographic key belonging to node B, the server cannot read or access it. Without End-to-End Encryption (E2EE) methods, the server may store the data in a database alongside millions of others, which could result in disastrous implications for end-users, as demonstrated by large-scale data breaches.

End-to-End Encryption (E2EE) & its Implementation in Software Applications.

End-to-End Encryption (E2EE) is a method of encrypting communications between two parties, such that only the sender and the recipient can decrypt and read the data. This technique guarantees that nobody, not even the server, can access the user's communications. E2EE is implemented in various software applications such as WhatsApp, Signal, and Google Duo, to secure the transmission of sensitive information such as text messages, emails, documents, and video calls. The process of E2EE involves encrypting the data on the sender's device before it leaves their device and transmitting the encrypted data to the recipient's device. The recipient's device then decrypts the data, so it can be read. This entire process is done automatically, without the need for user intervention.

One crucial aspect of E2EE is that the encryption and decryption keys are only known to the sender and the recipient. No one else, including the server, has access to these keys. This is what distinguishes E2EE from other encryption methods, such as those used in HTTPS and TLS. While HTTPS and TLS encrypt data between the client and the server, the server can still read the data. On the other hand, E2EE encryption ensures that the data remains confidential, even if it is intercepted by a third party. E2EE is achieved using cryptographic algorithms and keys. Here is a simplified explanation of how it works using equations and code snippets:

Encryption:

When a user sends a message, the message is encrypted using a cryptographic key generated on the user's device. This key is unique to the user and the recipient, and it is never shared with the server or any other third party.

Encryption can be represented mathematically as follows:

$$C = E\ (K,\ P)$$

Where C is the ciphertext, E is the encryption algorithm, K is the encryption key, and P is the plaintext message.

In code, encryption might look something like this:

```
def encrypt_message(key, message):
    encrypted = encryption_algorithm(key, message)
    return encrypted
```

Decryption:

When the recipient receives the message, they use their own unique decryption key to decrypt the message. Only the intended recipient has access to this key, which is never shared with anyone else.

Decryption can be represented mathematically as follows:

$$P = D\ (K,\ C)$$

Where P is the plaintext, D is the decryption algorithm, K is the decryption key, and C is the ciphertext.

In code, decryption might look something like this:

```
def decrypt_message(key, encrypted):
    decrypted = decryption_algorithm(key, encrypted)
    return decrypted
```

With E2EE, the message is only decrypted on the recipient's device, and it is never stored on the server or any other third-party device in plain text. This makes it extremely difficult for anyone to intercept or read the message, even if they have access to the server or the network.

The Diffie-Hellman Key Exchange: Safely Sharing Secrets

The Diffie-Hellman key exchange is a cryptographic technique developed by Whitfield Diffie, Martin Hellman, and Ralph Merkle. It allows two parties to generate a shared secret in a potentially hostile network without physically swapping keys. The exchange can take place on insecure forums without compromising the messages that follow.

To understand the concept better, let us use an analogy of paint colours. Suppose Alice and Bob are in separate hotel rooms at opposite ends of a hallway and want to share a particular color of paint without anyone else finding out. Unfortunately, the hotel is swarming with spies, so they cannot enter each other's rooms.

So, they agree to interact in the hallway and choose a common paint color, let us say yellow. They get a can of yellow paint, divide it between themselves, and go back to their respective rooms. In their rooms, they blend in a secret paint that nobody knows about. Alice uses a shade of blue, and Bob uses a shade of red.

The spies cannot see the secret colours they are using, but they become aware of the ensuing combinations because Alice and Bob go out of their respective rooms with their blue-yellow/red-yellow concoctions. They swap these combos out in the open, and it does not matter if the spies see them because they may not be able to determine the precise shade of the colours introduced in. Alice takes Bob's mix, and Bob takes Alice's, and they go back to their rooms again. Now, they blend their secret colours back in. Alice combines her secret shade of blue with Bob's red-yellow mix, giving a red-yellow-blue blend, while Bob combines his secret shades of red with Alice's blue-yellow blend, giving a blue-yellow-red blend. Both combinations have the same colours, so they should look identical. It is important to note that this analogy only explains the concept of the Diffie-Hellman key exchange. The actual mathematics behind this system is much more complex and makes it even harder to guess the shared secret.

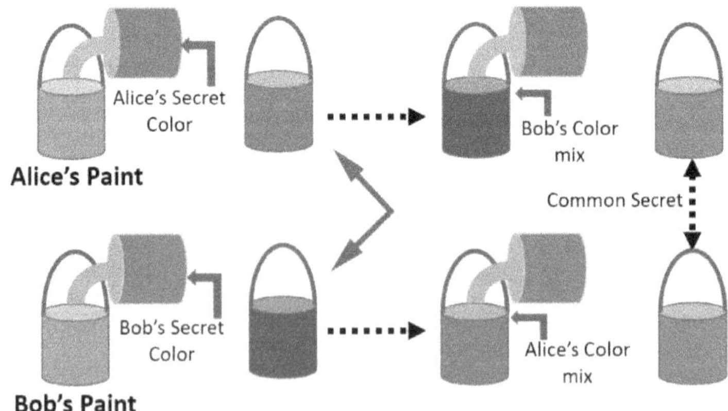

Figure 7.6: Diffie-Hellman Key Exchange

End-to-End Encryption in Secure Messaging

After the Diffie-Hellman key exchange, the parties involved can now use the shared secret to establish a symmetric encryption technique. Although more advanced encryption techniques may be added for increased security, these details are typically hidden from users. When two users connect on an E2EE application, their devices can effectively handle the encryption and decryption of messages, as long as there are no immediate software vulnerabilities. This means that regardless of who the user is, whether a hacker, a service provider, or even law enforcement, intercepting an E2EE message will result in a garbled value, rendering it completely unreadable.

In simpler terms, E2EE provides secure communication between two parties by generating a shared secret that is used to encrypt messages. This process ensures that no one can intercept and read the messages, even if they manage to access them during transmission. Thus, E2EE is an important security feature in messaging applications, as it ensures that users can communicate without fear of interception or surveillance.

Pros & Cons of End-to-End Encryption (E2EE)

End-to-End Encryption (E2EE) stands as a commendable instrument for the augmentation of confidentiality and security; nevertheless, it does not elude certain drawbacks. It is imperative to embark on a detailed examination of both the merits and demerits inherent in E2EE.

Pros of End-to-End Encryption (E2EE)

1. End-to-End Encryption (E2EE) stands as an invaluable asset in fortifying confidentiality and bolstering security. It garners fervent support from privacy advocates on a global scale. Moreover, E2EE possesses the attribute of seamless integration with software applications, rendering this technology accessible to a broad spectrum of users, including those proficient in mobile device usage.

2. End-to-End Encryption (E2EE) plays a pivotal role in safeguarding against the extraction of substantial message content data by hackers, even in the event of a security breach within a company whose user base relies on E2EE. In such scenarios, the most that malicious actors might gain access to is a portion of metadata. This marks a significant advancement in securing encrypted messages, affording protection to sensitive communications and vital identity documents, the

compromise of which could potentially result in severe repercussions for individuals.

Cons of End-to-End Encryption (E2EE)

1. Some people view the sheer value proposition of E2EE as problematic since nobody can access a user's messages without the associated key. Competitors argue that this enables criminals to communicate securely, knowing that governments and tech companies cannot decrypt their communications.
2. Law-abiding individuals might not need to keep their messages and phone calls secret, which is a sentiment echoed by many politicians who support regulation that would allow backdoor access to encrypted systems. However, this would defeat the purpose of E2EE.
3. E2EE applications are not 100% secure. Messages are obfuscated when relayed from one device to another, but they are visible on the endpoints, which are the laptops or smartphones at each end.
4. There are other threats that still exist, such as a stolen or compromised device, or a man-in-the-middle attack where an attacker intercepts messages and also the key to decrypt them. To mitigate this, many applications integrate a security code feature, such as a string of numbers or a QR code, to ensure that a third party is not snooping on communications. *(Ref. No.: 168- 182)*

Closing Thoughts

In conclusion, Public-Key Cryptography (PKC) occupies a pivotal role in the contemporary digital landscape, serving as an indispensable tool for safeguarding a myriad of digital systems. Its versatile application extends across diverse domains, encompassing computer security and the verification of cryptocurrency transactions. PKC operates on the bedrock of asymmetric cryptography, artfully harnessing pairs of public and private keys to orchestrate the intricate ballet of data encryption and decryption. This paradigm elegantly resolves the security conundrums posed by symmetric ciphers, offering a potent shield against potential threats.

The enduring utility of PKC is underscored by its extensive tenure in the digital realm, continually evolving to meet the demands of emerging applications and programs. Its dynamic flexibility and adaptability render it a priceless asset within the ever-evolving landscape of computer security. While it is acknowledged that asymmetric encryption might entail a trade-off in speed compared to its symmetric counterpart, its steadfast commitment to delivering heightened security remains unwavering.

Within the realm of digital security, both symmetric and asymmetric encryption shoulder indispensable roles, each uniquely suited to distinct applications. The swiftness and efficiency of symmetric encryption are the trusted workhorses for enciphering substantial volumes of data, while the methodical and security-focused nature of asymmetric encryption is the stalwart guardian of confidential communication.

As the discipline of cryptography charts a course through the shifting currents of technological evolution, it must contend with the emergence of ever-more sophisticated threats. Thus, it is highly probable that both symmetric and asymmetric cryptographic systems will continue to be indispensable sentinels of digital security.

In this landscape, the advent of freely accessible End-to-End encryption (E2EE) tools heralds a promising development. Esteemed messaging platforms such as Apple's iMessage and Google's Duo, alongside the proliferation of privacy-conscious software, offer robust fortifications against the nefarious designs of cyber assailants. While it is imperative to acknowledge that E2EE is not infallible, it diligently mitigates the risks associated with the exposure of sensitive data.

In harmony with E2EE, the ensemble of digital privacy safeguards includes the likes of Tor, Virtual Private Networks (VPNs), and cryptocurrencies. These tools form an integral part of the contemporary arsenal for preserving digital privacy. Encouragingly, these resources are becoming increasingly accessible, thus augmenting the overall security of digital communication and transactions. In sum, the harmonious amalgamation of PKC, symmetric and asymmetric encryption, and E2EE tools stands as a formidable bastion of digital security in the face of an ever-evolving technological landscape.

Chapter 8: Blockchain Applications: Charity, Supply Chain, Healthcare & Finance

"When it comes to early adoption, invisible technologies are often the hardest for humans to relate with. It takes a lot of time and stories to gain a narrative that becomes viable. This is blockchain technology's problem too."

— Olawale Daniel

Exploring the Potential of Blockchain Technology

Blockchain technology, first conceived in 1991, gained significant attention with the development of Bitcoin in 2009. Bitcoin was created by an unknown person or group under the pseudonym Satoshi Nakamoto, and the technology has had a profound impact on the world of finance. Blockchain functions as a distributed ledger that records and protects digital data through the use of cryptography. While most people associate blockchain with cryptocurrencies, the technology's decentralized and highly secure nature makes it useful for many other industries as well.

As the cryptocurrency space grows and blockchain-based solutions improve, it is becoming increasingly important to understand how this innovative technology can be applied in unique scenarios. One of the major advantages of blockchain technology is that it eliminates the need for trust and expensive security measures, which can improve efficiency and reduce costs. Additionally, the decentralized nature of blockchain networks makes them ideal for creating transparent databases that are visible to all participants. This means that blockchain technology has the potential to create a distributed yet unified record, which could enhance performance and security in industries such as charity, supply chain management, and healthcare. One of the most significant benefits of blockchain technology is its ability to increase trust in digital transactions. Because blockchain transactions are secured by complex cryptography and recorded on a distributed ledger, they are almost impossible to tamper with. This makes blockchain an attractive option for companies and organizations that need to share sensitive data or perform transactions with parties they do not know or trust. As the technology continues to evolve, it is likely that blockchain will become even more useful and versatile, providing new solutions to old problems in a wide range of industries. Therefore, understanding the principles behind blockchain technology and its potential applications is becoming increasingly important for businesses and individuals who want to stay ahead of the curve in an ever-changing digital world.

Use of Blockchain in Charity

Charitable organizations often face challenges when it comes to managing resources, maintaining transparency in operations, and ensuring effective governance. Fortunately, blockchain technology presents an opportunity to improve the process of receiving and managing funds for these organizations. There are already several noteworthy examples of blockchain being used in the charity sector, such as the Blockchain Charity Foundation (BCF). This non-profit organization is dedicated to achieving sustainable development goals that combat poverty and inequality by leveraging blockchain-powered philanthropy on a global scale. By harnessing the power of blockchain, charitable organizations can enhance their operations and make a greater impact in the communities they serve.

Use of Blockchain in the Supply Chain

The current supply chain management system faces several challenges in terms of transparency and efficiency. The traditional system relies heavily on trust, which often results in a lack of proper integration between different parties involved in the process. However, with the advent of blockchain technology, there is a solution to these challenges. By leveraging the distributed ledger technology, the entire process of creating and distributing goods within a supply chain network can be tracked and monitored in a secure and transparent manner. Blockchain can also provide a secure and tamper-proof way of recording relevant data, ensuring the authenticity of the products, and improving transparency of payments and transportation. With the integration of blockchain technology into the supply chain, businesses can improve their efficiency, minimize errors, and build trust between all parties involved. This could potentially revolutionize the delivery chain industry, making it more reliable and secure.

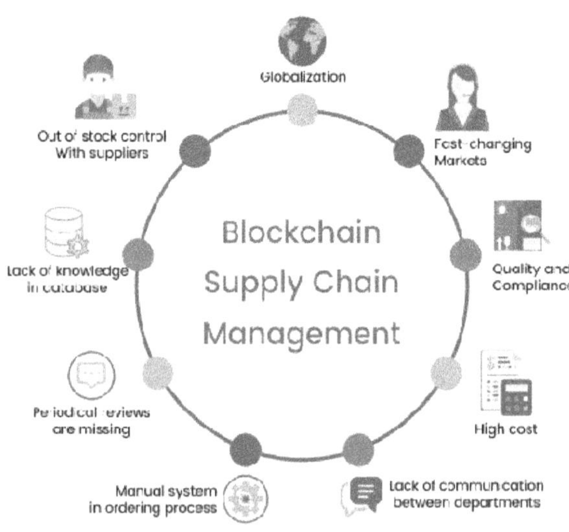

Figure 8.1: Blockchain in Supply Chain Management

Use of Blockchain in Healthcare Sector

The healthcare industry is confronted with an array of formidable challenges that have implications for its efficiency, accuracy, and administrative burden. Among these challenges are operational inefficiencies, data inaccuracies, and the burdensome paper-based processes that still persist. Blockchain technology presents a promising panacea for these ills, offering a multitude of potential use cases to transform the healthcare landscape. One such application lies in the meticulous tracking of pharmaceuticals as they traverse the intricate labyrinth of the supply chain. By employing blockchain's immutable ledger, the provenance and journey of drugs can be transparently recorded, minimizing the risks of counterfeit medications entering the market. This not only safeguards patient safety but also bolsters the pharmaceutical industry's integrity. Furthermore, the management of patient information, a cornerstone of healthcare, can be elevated to new heights through blockchain technology. The decentralized nature of blockchain networks ensures that patient records are securely stored, accessible only to authorized personnel. This not only streamlines the process of sharing crucial medical data among healthcare providers but also empowers patients with greater control over their health information. In an era where data breaches loom as an ever-present threat, the healthcare sector stands as a prime target due to the sensitivity of the information it safeguards. *(Ref. No.: 183- 192)*

Hospitals, in particular, grapple with the constant spectre of cyberattacks. Blockchain's robust security features can offer these institutions a fortified defence against such malicious incursions. By decentralizing data storage and employing cryptographic techniques, blockchain fortifies the defences of healthcare facilities, shielding sensitive patient data from malevolent actors. Companies and innovators are actively exploring the integration of blockchain technology for the secure storage of digital health data. The potential benefits are manifold, as these solutions have the capacity to reduce overall operational costs, enhance the accuracy of information, and fortify privacy measures. In essence, blockchain's implementation in the healthcare sector holds the promise of streamlining operations, curbing errors, and providing an impregnable bulwark around sensitive medical data.

The cumulative effect of embracing blockchain in healthcare translates into improved patient outcomes, more efficient healthcare services, and a sector better equipped to meet the demands of the digital age. As the industry navigates these transformative possibilities, it stands on the precipice of a new era where the promise of blockchain technology can usher in a healthcare paradigm that is secure, precise, and patient-centric.

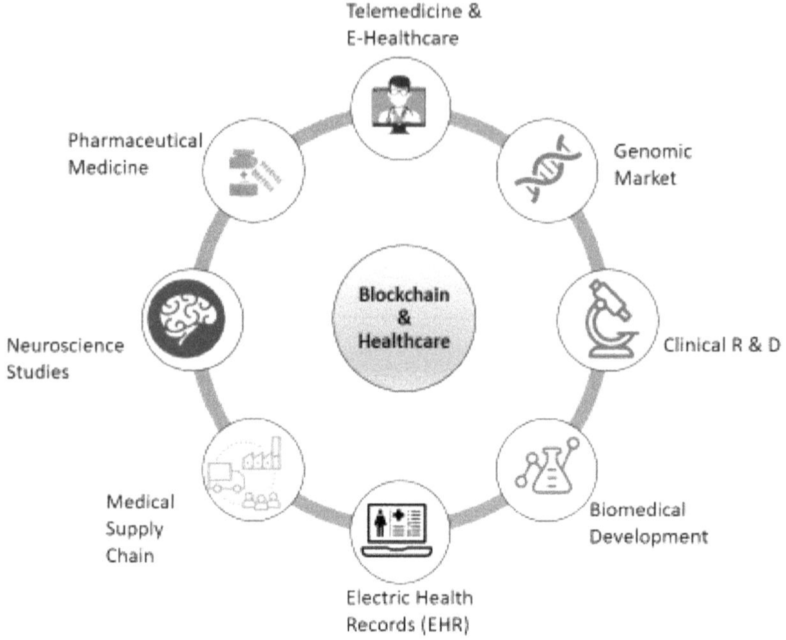

Figure 8.2: Blockchain in Healthcare

Use of Blockchain in Royalty Payments

The digital age has brought about many opportunities for artists, musicians, and content creators to reach a wider audience. However, it has also brought its fair share of challenges, such as digital piracy, unfair agreements with third-party agencies, and non-payment of royalties. These issues can make it difficult for creators to receive the full compensation they deserve for their work. Blockchain technology offers a potential solution to these problems. By creating an immutable and transparent record of who is renting, buying, and using their content, artists can ensure they receive proper compensation. Smart contracts, which are essentially self-executing digital contracts, can also be used to facilitate payments and ensure that creators receive the royalties they are due. Through the use of blockchain technology, artists can have more control over their work and receive fair compensation for their efforts.

Use of blockchain in Governance

The use of blockchain technology has the potential to significantly improve governance across various sectors, by providing a more democratic, transparent, and secure approach to managing networks and operations. By leveraging blockchain-based systems, it is possible to eliminate voting fraud and the need for trust in elections and other constitutional processes, thus ensuring the accuracy and legitimacy of results. In addition, blockchain can serve as a powerful tool against corruption, enhancing data integrity and traceability in areas ranging from tax collection to the distribution of financial aid. The decentralized and highly secure nature of blockchain technology offers a unique advantage in addressing governance challenges, leading to greater accountability and trust in institutions.

Payment Solutions & Decentralized Applications

Blockchain technology has revolutionized the way we send money across borders. Traditional centralized banks and payment systems are often costly and slow, but blockchain-based solutions have proven to be much more efficient. Cryptocurrencies can be sent to anyone around the world quickly and cheaply, without the need for intermediaries. However, blockchain technology is not limited to just sending and receiving money. Decentralized applications built on blockchain (dApps) allow users to interact with each other directly without the need for middlemen, resulting in reduced fees, greater incentives, and increased transaction efficiency. This also gives users more control over their data, unlike centralized websites and apps. Blockchain-based solutions also have the potential to automate and streamline numerous other industries, removing the need for centralized systems and intermediaries. As Vitalik Buterin, the co-founder of Ethereum, said, "blockchain technology automates away the centre." This allows for greater democratization, transparency, and control in various sectors, ultimately benefiting consumers and businesses alike.

Application of Blockchain in Internet of Things (IoT)

The integration of blockchain technology and the Internet of Things (IoT) is a natural match. As IoT networks are often used to gather data from sources that are dispersed across various locations, blockchain, being a decentralized technology, allows organizations to maintain an immutable and transparent ledger of IoT devices, the data they collect, and their interactions with each other. With its security features and cryptocurrency applications, blockchain provides an ideal platform for Machine-to-Machine (M2M) transactions. By integrating blockchain with IoT, it ensures accountability, data accuracy, and security. The marriage of blockchain and IoT has numerous benefits. It can help in creating an efficient and secure network of IoT devices that can communicate and share data with each other without relying on a central authority. By using blockchain technology, companies can create tamper-proof records of device interactions and data exchanges, which can help in detecting any potential tampering attempts. Moreover, by using smart contracts, blockchain technology can automate many of the processes involved in IoT networks, such as data sharing and processing. Several companies have already invested significant resources in building blockchain-powered IoT networks. For instance, IOTA is a blockchain-based platform that is specifically designed for IoT devices. It allows IoT devices to communicate and transact with each other using a cryptocurrency called MIOTA. Similarly, Bosch, one of the world's largest manufacturing companies, has also developed a blockchain-powered IoT platform that helps in managing and securing IoT devices. In summary, blockchain technology provides an ideal platform for managing IoT networks securely, transparently, and efficiently.

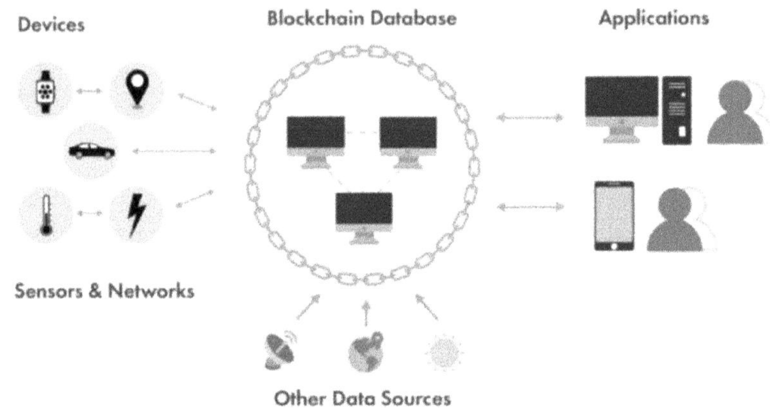

Figure 8.3: Application of Blockchain in IoT Platform

Exploring the Benefits of Blockchain Technology in Healthcare

Blockchain technology is often associated with cryptocurrencies such as Bitcoin, but its potential applications are not limited to financial transactions. Blockchain's distributed architecture and cryptographic security make it a promising tool for storing and safeguarding sensitive data in various industries, including healthcare. In the healthcare sector, blockchain technology can offer several benefits. One of the critical features that make blockchains suitable for storing medical data is their immutability. As most blockchain systems are designed as distributed databases that use cryptography to protect and record information, it is difficult for anyone to tamper with or alter the data without the approval of all network participants. This property allows for the creation of incorruptible databases for medical records.

Furthermore, the peer-to-peer architecture of blockchains ensures that all copies of a patient's record are synchronized with one another, even if they are stored on different computers. Each network node holds a replica of the entire blockchain, and they communicate regularly to ensure that data is updated and authentic. Decentralization and data distribution are also essential aspects of blockchain technology. It is worth noting that blockchains can be distributed but not necessarily decentralized, depending on how nodes are distributed and the overall architecture. In healthcare, blockchain networks are typically built as private networks, unlike public ones that are commonly used as cryptocurrency ledgers. Private blockchain networks in healthcare require permission and are controlled through a smaller number of nodes, unlike public blockchains where anyone can join and contribute to the improvement of the system. The use of blockchain in healthcare has several potential applications, including secure sharing of patient data between healthcare providers, tracking pharmaceutical supply chains, and monitoring clinical trials' data management. In summation, blockchain technology is an invaluable resource for the healthcare sector, promising secure and efficient solutions for data storage and sharing. Its adoption in healthcare represents a profound shift towards enhanced patient care, fortified data privacy, and unassailable data security. Nonetheless, as with any transformative technology, the integration of blockchain in healthcare mandates careful consideration of its potential advantages and challenges. With a judicious approach, blockchain holds the promise of revolutionizing healthcare operations, ushering in an era where data is not merely a digital asset but an incorruptible testament to patient well-being.

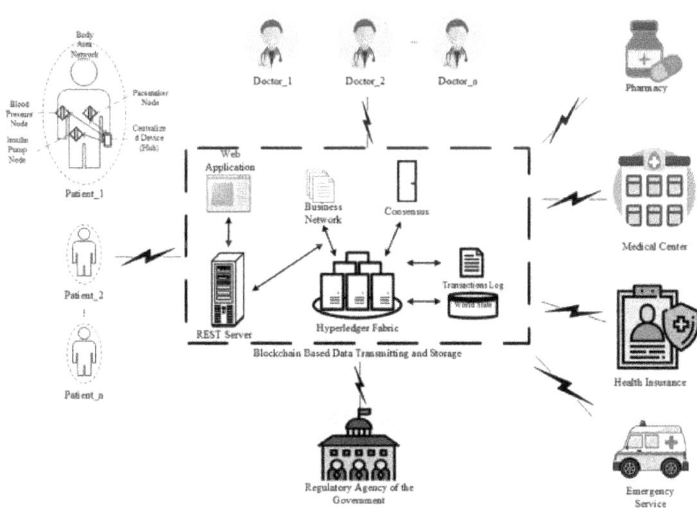

Figure 8.4: Advantages of Blockchain in Healthcare

The Potential of Blockchain in Healthcare: Creating Secure & Unified Databases

Increased Security

One of the most promising applications of blockchain technology in the healthcare sector is the creation of secure and unified databases that use a distributed peer-to-peer system. The immutability of blockchain ensures that data corruption is no longer a concern, making it possible to effectively register and track the medical records of numerous patients. Unlike traditional databases that rely on a centralized server, distributed systems offer higher levels of security for data exchange while reducing administrative expenses. Furthermore, the decentralized nature of blockchain technology makes it less vulnerable to technical failures and external cyber-attacks, which frequently compromise valuable data in the healthcare sector. The security provided by blockchain networks is particularly beneficial for hospitals and nursing homes that are often targets of hacker invasions and ransomware attacks.

The potential advantages of blockchain in healthcare extend beyond data security. The technology can also improve patient care by enabling secure sharing of medical records between healthcare providers, ensuring accurate and up-to-date information is available to all involved parties. Blockchain can also be used to track pharmaceutical supply chains, enhancing transparency, and reducing the risk of counterfeit drugs entering the market.

Improving Interoperability in Healthcare with Blockchain-Based Medical Records

Interoperability is a key challenge in healthcare, with different clinics, hospitals, and service providers using a range of data storage systems that often do not communicate effectively with one another. This lack of interoperability can lead to difficulties in sharing patient files and other medical records, which can have serious implications for patient care.

One of the key benefits of blockchain-based medical records is their ability to improve interoperability in the healthcare sector. By creating a distributed peer-to-peer system that permits authorized parties to access a unified database of patient files or even medicine distribution records, blockchain technology can resolve the problem of technological variations in data storage systems. Rather than attempting to interface with each other's internal storage systems, healthcare service providers can work collectively on a single blockchain-based database, enabling them to share patient information securely and efficiently. This enhanced interoperability can lead to improved patient care by ensuring that healthcare providers have access to accurate and up-to-date information about their patients.

Furthermore, blockchain-based medical records can enable secure sharing of patient data between healthcare providers, improving care coordination and reducing the risk of errors and duplications. This can lead to more efficient and effective treatment for patients, particularly those with complex medical conditions who require care from multiple providers.

Accessibility & Transparency

In addition to facilitating the sharing of healthcare data between providers, blockchain systems have the potential to provide patients with greater accessibility and transparency over their personal health information. By allowing patients to access and verify the accuracy of their medical records, blockchain technology can improve patient engagement and involvement in their own care. Blockchain-based medical records can enable patients to track any changes made to their records and request verification of those changes, ensuring the accuracy and integrity of their health data. This added layer of security can protect against human errors and intentional falsifications, providing patients with greater confidence in the information being used to inform their care. Furthermore, blockchain-based systems can increase transparency in healthcare by enabling patients to control who has access to their medical records and for what purposes. This can help to build trust between patients and healthcare providers by ensuring that patient data is being used in a responsible and ethical manner.

Reliable Supply Chain Management

Blockchain technology offers a reliable solution for pharmaceutical companies to track their products throughout the manufacturing and distribution process, helping to combat the serious problem of drug counterfeiting. By recording each stage of the supply chain on an immutable ledger, blockchain systems can provide an audit trail that verifies the authenticity and integrity of pharmaceutical products.

In addition to blockchain, Internet of Things (IoT) devices can be used to monitor and measure factors such as temperature and humidity during storage and transportation, providing an added layer of verification and quality control.

This combination of blockchain and IoT technology can ensure that drugs are being stored and transported in optimal conditions, and that the products delivered to patients are safe, effective, and genuine.

By improving supply chain management with blockchain, pharmaceutical companies can reduce the risk of counterfeit drugs entering the market, protecting both patient safety and brand reputation. Moreover, blockchain can enhance transparency and accountability throughout the supply chain, making it easier to identify and address issues such as product recalls and shortages.

Insurance Fraud Protection

The use of blockchain technology can also contribute to reducing medical insurance fraud, a problem that imposes a significant cost on the American healthcare system, estimated to be around $68 billion annually. By leveraging the immutable nature of blockchain records, insurance providers can verify the authenticity and accuracy of medical claims and billing information. This can help prevent common types of fraud, such as charging for services that were never provided or billing for unnecessary procedures. Through the use of blockchain, fraudulent activities can be detected more quickly and accurately, reducing the overall cost of fraud to the healthcare industry.

In essence, blockchain's decentralized and secure nature can provide insurance companies with a transparent and reliable way of tracking medical claims and payments, minimizing the risk of fraudulent activity. This, in turn, could result in more affordable and accessible healthcare for patients, as well as a more sustainable and efficient healthcare system for providers.

Clinical Trials Recruiting

Blockchain technology has various applications in the healthcare industry, including the potential to improve the quality and effectiveness of clinical trials. One of the main challenges in clinical trials is identifying suitable patients who could benefit from the drugs being tested. By utilizing medical records stored on blockchains, clinical trial recruiters can quickly and easily identify potential participants. This recruitment system can significantly enhance the enrolment of clinical trials, as many patients are unaware of ongoing drug trials, and as a result, miss out on the opportunity to participate. Once patients are identified and recruited, blockchain technology can be used to ensure the integrity of the data collected during the trial. The immutability and transparency of blockchain technology make it an ideal solution to prevent any tampering or manipulation of data during the clinical trial process. By using blockchain, researchers can ensure that the data collected is trustworthy and can be used to make informed decisions about the efficacy and safety of new drugs.

Benefits and Limitations of Blockchain Technology in Healthcare

Blockchain technology can speed up medical credentialing, which is the process of verifying a healthcare professional's qualifications, experience, and credentials. This is accomplished by creating a secure network to store and verify credentials in real-time, without relying on direct human references.

Here are some key benefits of blockchain for patients:

1. Empowers patients to undertake ownership of their medical data
2. Supports consent mechanisms prohibiting healthcare providers from accessing information without patient permission
3. Enables patients to participate in research and otherwise monetize their data without intermediaries
4. Collects and stores data from wearable devices in a secure manner

However, despite these advantages, blockchain technology has some potential limitations that must be addressed before it can be widely adopted in the medical sector.

These limitations include:

1. Compliance with data regulations, such as the Health Insurance Portability and Accountability Act (HIPAA) in the US, is necessary for companies to ensure patient data privacy and security. US-based companies would need to set up custom-designed blockchain record systems to comply with HIPAA, with improved privacy features and limited accessibility.
2. Initial costs and speed of implementation are significant barriers for healthcare providers to adopt blockchain solutions. The initial investment required for companies to implement blockchain can be high, and distributed systems like blockchain can be slower than centralized ones in terms of transactions per second.
3. Large databases like those containing CT scans or MRIs could take longer to transmit and synchronize data in a big blockchain network with numerous nodes compared to centralized systems.

Despite these challenges, blockchain technology has the potential to revolutionize the medical industry by providing a more secure, efficient, and patient-centric approach to data management.

Supply Chain Management: Enhancing Efficiency & Transparency through Blockchain Technology

A supply chain refers to the network of individuals and businesses involved in the creation and distribution of a particular product or service, from the initial suppliers to the end-users and customers. The current supply chain management system often includes the suppliers of raw materials, manufacturers, logistics businesses, and retailers. However, this system is plagued by inefficiencies, lack of transparency, and difficulties in integrating all parties involved.

The conventional supply chain model struggles to maintain a consistent and efficient system, which negatively impacts the profitability and final retail price of companies. Blockchain technology can provide novel approaches to record, transmit, and share data, addressing some of the most pressing problems faced by supply chains.

Using blockchain technology, products, and materials, as well as money and data, can move seamlessly throughout the various stages of the supply chain. Some of the benefits of implementing blockchain technology in supply chain management include:

1. Improved transparency and traceability of products and materials
2. Increased efficiency in managing inventory and logistics
3. Enhanced security and fraud prevention through immutable records
4. Greater collaboration and communication between all parties involved in the supply chain

However, there are also potential limitations to implementing blockchain technology in supply chain management, such as high initial costs and the need for compliance with existing data regulations. Despite these challenges, blockchain technology has the potential to revolutionize supply chain management and create a more efficient and transparent system.

The Advantages of using Blockchain for the Supply Chain

The inherent properties of blockchain, such as decentralization and immutability, make it an ideal fit for supply chain networks. In a blockchain system, data blocks are cryptographically linked in a sequence, creating a tamper-proof and unalterable ledger. This ensures that the information stored on the blockchain is secure and cannot be modified without consensus from the entire network.

Blockchain technology is not limited to recording cryptocurrency transactions, and its application to the supply chain can provide several advantages. By integrating blockchain into the supply chain network, companies can ensure that data related to product movement, quality, and compliance is secure and accurate. This can lead to increased transparency, efficiency, and traceability throughout the supply chain network.

Furthermore, since the blockchain system operates without intermediaries, it can reduce transaction costs and speed up settlement times. The decentralized nature of blockchain also means that the network is resilient to cyber-attacks and can continue to function even if some of the nodes fail.

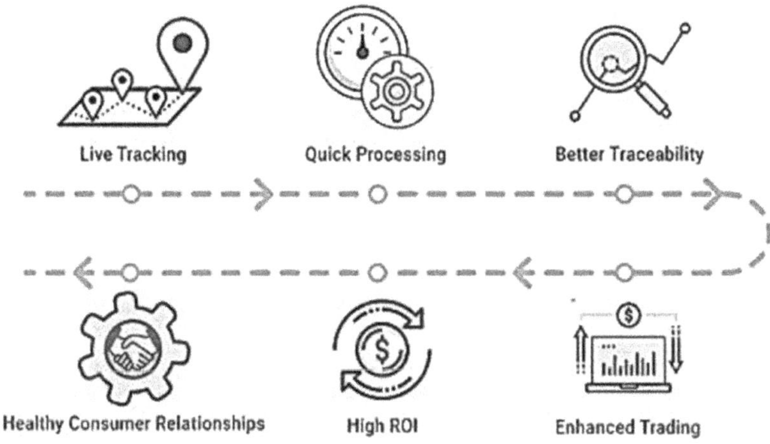

Figure 8.5: Use case of Blockchain in Supply Chain Management

Transparent & Immutable Records

The usage of blockchain technology in supply chain management can result in transparent and immutable records. Multiple companies and establishments can use the blockchain system to record and share data about the location and ownership of their materials and products. The decentralized nature of the blockchain system ensures that all parties involved can see the recorded data, and the data cannot be altered or tampered with, making it easier to determine the accountable party in case of any issues or errors in the supply chain.

The Benefits of Implementing Blockchain in Supply Chain Networks:

Cutting Costs:

To reduce waste and inefficiencies in supply chain networks, companies can use blockchain technology to track their materials and products more accurately. This is especially important in industries with perishable goods. By improving tracking and data transparency, companies can identify areas of waste and implement cost-saving measures. Additionally, blockchain technology can eliminate fees associated with financial institutions and payment processors, which can eat into profit margins.

Blockchain for Interoperable Supply Chain Data:

One of the biggest challenges in supply chain management is integrating data across multiple members of the process. Blockchain, as a distributed system, provides a unique and transparent data repository where each node of the network contributes to adding new data and verifying its integrity. All parties in the supply chain have access to the same information, ensuring that one business enterprise can easily confirm the data being broadcasted by the other.

Replacing Electronic Data Interchange with Blockchain

Many businesses rely on Electronic Data Interchange (EDI) systems to share information with each other. However, these systems often send data in batches rather than in real-time. This can lead to delays in information sharing, especially in cases where rapid pricing changes or missing freight need to be addressed. With blockchain, data is updated regularly and can be quickly distributed to all nodes of the network in real-time, improving efficiency and reducing delays.

Digital Agreements & Document Sharing on Blockchain

Having a single version of the truth is critical for any type of supply chain document sharing. Blockchain allows essential documentation and contracts to be associated with transactions and virtual signatures, ensuring that all participants have access to the original version of the agreements and documents. The blockchain guarantees immutability of files and agreements can only be modified if all concerned parties reach a consensus. This way, corporations can reduce the time spent on paperwork and legal negotiations and focus on developing new products and promoting business growth.

Overcoming Challenges to Blockchain Adoption in Supply Chain Management

Blockchain technology has significant potential for the supply chain industry, but there are also challenges to its adoption that need to be addressed.

Deploying New Systems

Implementing blockchain-based systems can be a daunting task for companies that already have existing systems in place. They may not be able to adapt their current infrastructure and business strategies to a blockchain-based environment, which can cause disruptions in operations and resources diverted from other projects. As a result, upper management may be hesitant to invest in this technology before seeing its significant adoption by other essential players in their industry.

Getting Partners On-Board

It is also crucial for partners involved in the supply chain to be willing to embrace blockchain technology. While companies can still benefit from using blockchain for only part of the process, they cannot take full advantage of its benefits when there are holdouts. Additionally, transparency is not something that all companies may desire. It may take some time and effort to convince all parties involved to adopt blockchain technology fully.

Change Management

Adopting a blockchain-based system requires companies to promote its adoption to their employees. Employees must be trained on the technical aspects of blockchain technology and ways to improve its application in the corporate world. Companies should conduct training programs to address new features or innovations in blockchain-based technology, which can require time and resources. Change management should be a crucial part of the adoption process to ensure a smooth transition and address any resistance to change.

Blockchain & Digital Identity Management

Digital identity management and verification is a promising use case of blockchain technology, which offers more secure techniques for storing, transferring, and verifying sensitive data. The need for secure identity management has become more crucial than ever before, as personal data breaches have become more frequent and widespread. In 2018 alone, billions of people were affected by such data breaches across the globe. Centralized databases, which are commonly used for identity management, are vulnerable to cyberattacks and data breaches. In contrast, blockchain technology offers a decentralized and tamper-proof data storage mechanism that can significantly improve the security and privacy of personal information. By using blockchain-based digital identity systems, individuals can control their personal data and provide selective access to third parties. This way, individuals can protect their privacy and reduce the risk of identity theft. Moreover, blockchain-based digital identity systems can help address several challenges associated with traditional identity verification systems. For instance, blockchain technology can enable instant verification of identity, eliminating the need for lengthy and cumbersome verification processes. This can be especially useful in scenarios where fast and efficient identity verification is required, such as in border control, voting, or financial transactions.

How Blockchain Technology Can Enhance Digital Identification Systems

Blockchain technology can provide a powerful solution to enhance the security of digital identification systems. When data is stored on a blockchain network, it is protected by numerous nodes that maintain the network. This means that the authenticity of the data is ensured by a batch of claims from multiple users, making it very difficult for a single entity to manipulate or tamper with the data. In digital identification systems, this can provide a highly secure method of storing and verifying sensitive information, such as personal identification details or biometric data. The nodes of the network can be controlled by authorized companies or government establishments, who are responsible for verifying and validating the digital records. Each node can "cast a vote" regarding the authenticity of the data, ensuring that the files can be used as official documents with increased levels of security.

Function of Cryptography in Blockchain-based Identification Systems

Cryptography plays a crucial role in maintaining the security of blockchain-based identification systems. Instead of directly sharing sensitive information, digital data can be shared and verified through various cryptographic techniques, including hashing functions, digital signatures, and Zero-Knowledge Proofs. Using hashing algorithms, any document can be converted into a hash, which is a unique string of letters and numbers representing the data used to create it.

This hash acts as a digital fingerprint and can be authenticated by authorized entities providing digital signatures, confirming the validity of the document.

Moreover, Zero-Knowledge Proofs enable the sharing and authentication of credentials or identities without revealing any data about them. This means that even though the data is encrypted, its authenticity can still be verified. For example, an individual could use Zero-Knowledge Proofs to prove they are of legal age to enter a club or drive a car without revealing their exact date of birth. This technology ensures that sensitive information is protected while maintaining the integrity and authenticity of digital identities.

Self-Sovereign Identity

Self-sovereign identity refers to a concept where individuals have complete control over their personal data, which is stored in private wallets like crypto wallets. This approach allows users to determine how and when their data is shared, such as using a private key to sign transactions and prove ownership of credit card credentials. Blockchain technology, which is often associated with cryptocurrencies, can also be used to securely share, and validate personal documents and signatures. For example, a government agency can provide approval of an individual's investor status, which can then be verified by a broker through a Zero-Knowledge proof protocol without revealing any specific personal information about the investor's net worth or income. Self-sovereign identity provides individuals with greater autonomy and privacy over their personal information.

Why do we need Blockchain for Identity?

Blockchain technology offers a solution to the challenges that exist in traditional identity management systems. These challenges include limited accessibility, data insecurity, and fraudulent identities. Current systems are inaccessible to some populations, which can create a range of issues, including discrimination and exclusion from social and economic activities. Blockchain's decentralized and secure platform ensures universal access to identity management services. Additionally, data insecurity is a major concern that can lead to identity theft, fraudulent transactions, and other forms of criminal activity. Blockchain technology provides enhanced security, privacy, and control over personal data through encryption and decentralization. Lastly, fraudulent identities are a significant problem, but blockchain enables a trustless and tamper-proof system that reduces the risk of fraudulent identities.

What are the Benefits of Decentralized Identity?

Decentralized identity is becoming increasingly popular in response to concerns over data security and privacy. Regulations like the EU General Data Protection Regulation (EU GDPR) have strengthened identity standards, which requires modern identity solutions. Governments are looking towards distributed ledger technology to provide identities to the unidentified and protect citizens' personally identifiable information. Blockchain technology offers several benefits to decentralized identity systems. Firstly, it provides a decentralized public key infrastructure (DPKI), allowing users to control their public keys and authenticate their identity without relying on a centralized authority. Secondly, decentralized storage ensures that personal data is not stored in a single location, reducing the risk of data breaches. Finally, blockchain technology provides manageability and control over personal data, allowing users to determine how their data is accessed and used.

Advantages & Limitations of Blockchain-based Digital Identification Systems

Cryptography and blockchain technologies can provide significant benefits to digital identification systems. Firstly, users can have better control over how and when their personal data is used, reducing the risks associated with storing sensitive information in centralized databases. Additionally, blockchain networks can offer better levels of privacy through the use of cryptographic systems. Zero-knowledge proof protocols allow customers to prove the validity of their documents without sharing information about them. Secondly, blockchain-based digital ID systems may be more reliable than conventional ones. The use of digital signatures makes it relatively easy to verify the source of a claim made about a user, and blockchain systems can efficiently protect all forms of information against fraud.

However, as with many use cases of blockchain, there are some potential limitations to using the technology for digital identification systems. Synthetic identity theft, which involves combining legitimate information from specific individuals to create a completely new identity, is a significant challenge. Although every piece of data used to create a synthetic identity is accurate, some systems may recognize the fake ones as authentic. This problem can be mitigated using digital signatures to prevent made-up combinations of documents from being accepted as information on a blockchain. Another factor to consider is the possibility of 51 percent attacks, which are more likely in small blockchain networks. A 51 percent attack can reorganize a blockchain, changing its information, and is especially concerning in public blockchains where anyone can join the process of verifying and validating blocks. Private blockchains can reduce the probability of such attacks by including only trusted entities as validators, but this would constitute a more centralized and less democratic model.

How Blockchain Will Impact the Banking Industry

The traditional role of banks as intermediaries in the global financial system can be disrupted by blockchain technology. Currently, banks rely on their internal ledgers to manage and coordinate the monetary system, which is not transparent and forces trust in the banks and their outdated infrastructure. With blockchain technology, it is possible to create a trustless, borderless, and transparent system that is accessible to everyone, potentially cutting out the middlemen in the banking industry. This can promote faster and cheaper transactions, increase access to capital, provide higher data security, enable trustless agreements through smart contracts, make compliance smoother, and more. Moreover, the interaction between the various financial building blocks made available through blockchain can create entirely new types of financial services.

Blockchain technology has the potential to revolutionize the banking industry by providing a transparent, decentralized, and secure financial system. One of the significant advantages of blockchain is the ability to facilitate faster and cheaper transactions. With blockchain, banks can streamline cross-border payments, reducing the need for intermediaries and minimizing transaction times and fees. This also provides greater access to capital for individuals and businesses who were previously underserved due to the high costs associated with traditional banking methods.

Another benefit of blockchain technology is its ability to ensure data security. The decentralized nature of blockchain provides a secure platform for transactions and data storage, reducing the risk of fraud, data breaches, and cyber-attacks. Additionally, blockchain enables the creation of smart contracts, which can enforce trustless agreements between parties, reducing the need for intermediaries and increasing efficiency.

Furthermore, blockchain technology can simplify compliance processes by providing a transparent record of all transactions, making it easier for banks to adhere to regulatory requirements. This can also lead to greater accountability and transparency in the banking industry, providing consumers with greater confidence in the financial system.

What are the main Advantages of Blockchain for Banking & Finance?

Blockchain technology offers several benefits for the banking and finance industry, including:

1. **Security:** Blockchain eliminates single points of failure, reducing the need for intermediaries to hold data and increasing security.
2. **Transparency:** By homogenizing shared processes, blockchain creates a single source of truth for all network participants, enhancing transparency.
3. **Trust:** Transparent ledgers make it easier for parties to collaborate and come to agreements, building trust.
4. **Programmability:** Smart contracts enable the automation of business processes, increasing efficiency and reducing the risk of human error.
5. **Privacy:** Privacy technologies enabled by blockchain allow for selective sharing of information between businesses, ensuring data confidentiality.
6. **Performance:** Blockchain networks are designed to support high transaction volumes while maintaining interoperability, creating an interconnected web of blockchains.

Efficient & Secure Payment Settlement through Blockchain

Traditional banking systems have long been associated with delayed transactions, high costs, and extensive verification processes. As technology has advanced, these systems have struggled to keep up with the demands of an interconnected world. Blockchain technology, on the other hand, offers a faster and cheaper method of payment settlement, available 24/7, without geographical restrictions, and with robust security guarantees. By utilizing a decentralized network that eliminates intermediaries, transactions can be processed in real-time, reducing the time and costs associated with traditional banking transactions.

Fundraising Directly on the Blockchain Network

Crowdfunding has been a popular method for entrepreneurs to raise money from the public, but it still often involves going through intermediaries such as crowdfunding platforms. Blockchain technology has made it possible to bypass these intermediaries altogether and raise funds directly on the blockchain network. Initial Coin Offerings (ICOs) and Initial Exchange Offerings (IEOs) are two such fundraising methods that use blockchain technology to enable companies to sell tokens in exchange for funding. These tokens can represent various assets such as equity, utility, or security, and the funds raised can be used to develop new products or services. One of the primary advantages of fundraising through ICOs or IEOs is that it can be done without the need for banks or other financial institutions. This can potentially lower the costs of fundraising by avoiding fees charged by banks for facilitating business securitization and Initial Public Offerings (IPOs).

However, it is important to note that the ICO market is still largely unregulated and comes with significant financial risks for potential investors. The relative ease of setting up an ICO has led to instances of fraud and scams, where projects have raised significant amounts of funds without delivering on their promises. Therefore, investors must conduct thorough research and due diligence before investing in any ICO or IEO.

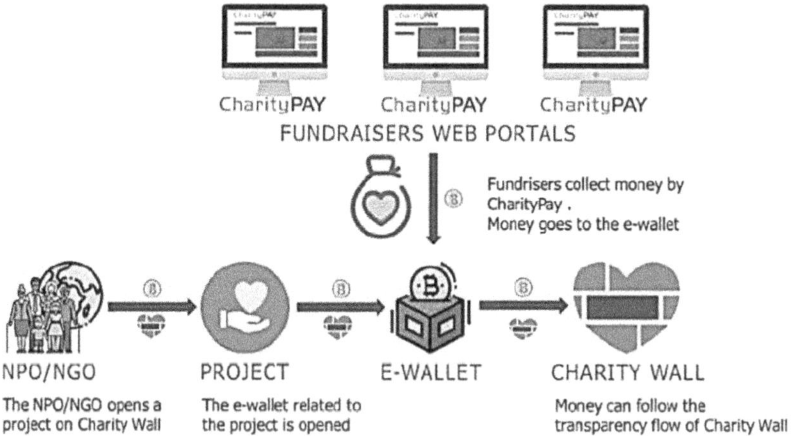

Figure 8.6: Fundraising on Blockchain Network

Asset Tokenization on the Blockchain

Asset tokenization on the blockchain is a process of converting physical assets into digital tokens and recording them on a blockchain network. This technology enables assets like real estate, art, and commodities to be bought and sold in a more efficient and cost-effective way by providing a transparent and secure platform for asset ownership transfer.

The traditional method of buying and selling securities and other assets involves multiple intermediaries, making the process complex and expensive. With blockchain technology, the need for intermediaries is eliminated, and assets can be tokenized, making them easier to trade and manage.

Tokenization offers several benefits, such as fractional ownership, which allows investors with limited capital to invest in expensive assets. Additionally, tokenization can make the process of asset ownership transfer more accessible, faster, and cheaper. It also provides transparency, reduces the risk of fraud, and offers more liquidity for assets that are traditionally illiquid, such as real estate. However, asset tokenization is not without challenges. One of the significant concerns is the lack of regulation. Tokenization may lead to regulatory issues that need to be addressed, as securities and assets are typically subject to legal and regulatory requirements. Another challenge is the need for standardized tokenization frameworks that can be widely adopted across various industries.

Lending Money using the Blockchain

The lending industry has long been dominated by banks and other financial institutions, giving them the power to charge high-interest rates and limit access to capital based on credit scores. This has resulted in a time-consuming and costly process for borrowers. However, with the advent of blockchain technology, a new lending ecosystem has emerged, known as Decentralized Finance (DeFi), which aims to create a more inclusive financial system by placing all financial applications on blockchains.

DeFi enables peer-to-peer lending, allowing anyone to borrow and lend money in a secure, affordable, and straightforward manner, without arbitrary restrictions. This system provides more competition for traditional banks, which may lead them to offer better terms to their customers to remain competitive. Moreover, DeFi can serve as an alternative to traditional financial institutions, as it allows individuals to bypass the need for intermediaries, thereby enabling a more decentralized, transparent, and accessible financial system.

Blockchain-based lending offers several advantages, including affordability, accessibility, security, and transparency, which could ultimately disrupt the traditional lending industry, benefiting borrowers and investors alike.

Blockchain's Impact on Global Trade Finance

The process of engaging in international trade can be a cumbersome and time-consuming task due to the numerous international policies and guidelines that are imposed on importers and exporters. The manual processes involved in keeping track of products as they move through each stage also require hand-written documentation and ledgers, which can be error-prone and inefficient. Blockchain technology provides a solution to these challenges by enabling trade finance participants to create a shared ledger that accurately tracks goods as they move across the globe. This shared ledger helps to simplify and streamline the complicated world of trade finance, ultimately saving importers, exporters, and other organizations significant amounts of time and money. With blockchain technology, the entire trade finance process can be automated and digitized, eliminating the need for intermediaries, such as banks, to verify transactions. This not only saves time but also reduces costs and minimizes the risk of fraud. Additionally, blockchain technology can provide greater transparency into the movement of goods, enabling all parties involved in the trade to have a clear view of the entire process. Furthermore, blockchain technology can also offer traceability, enabling importers and exporters to track the origin of goods and ensure that they meet regulatory compliance standards. This can be particularly important in industries such as food and pharmaceuticals, where safety and quality control are critical. Blockchain technology can revolutionize the world of international trade finance by simplifying and streamlining the entire process, saving time and money, reducing risk and fraud, and providing greater transparency and traceability.

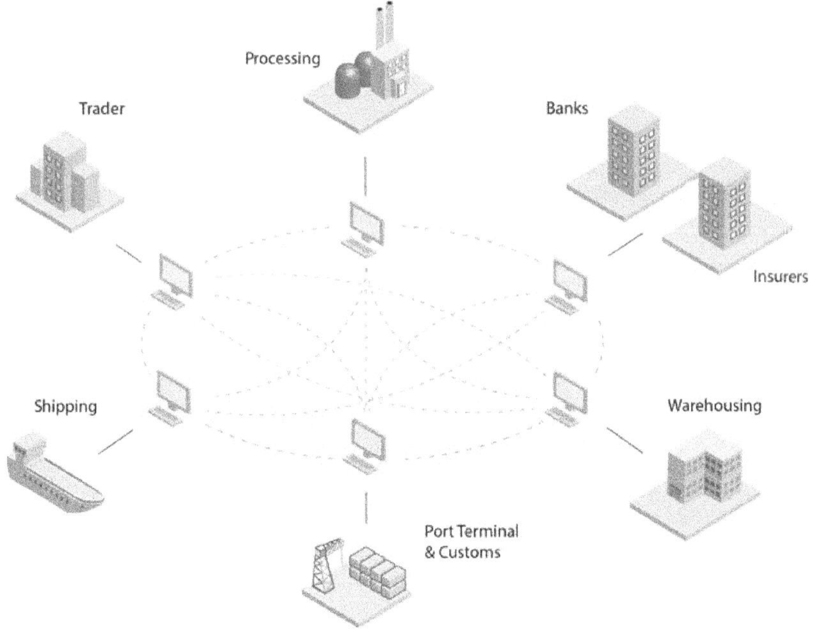

Figure 8.7: Blockchain's impact on global trade finance

Safer Agreements through Smart Contracts

Contracts are intended to safeguard individuals and corporations when they enter into agreements, but this protection comes with a hefty price tag. Drafting a contract necessitates a significant amount of manual work from legal experts due to the intricate nature of contracts. Smart contracts, on the other hand, allow for the automation of agreements through blockchain-based, tamper-proof, and deterministic code. Money can be held in escrow and only released when certain conditions of the agreement are met. Smart contracts considerably lessen the need for trust in reaching an agreement, reducing the likelihood of financial agreements going wrong and decreasing the possibility of ending up in court. In essence, smart contracts enable parties to conduct business with one another in a more secure and transparent manner, without requiring the involvement of a third-party intermediary. As a result, they provide a cost-effective and efficient alternative to traditional legal contracts.

Data Integrity & Security Facilitated by Blockchain

In the realm of financial institutions, data is the lifeblood of operations, and safeguarding it against compromise or fraud is paramount. Trusted intermediaries often play a pivotal role in data sharing, yet this introduces an inherent risk of data breaches. Furthermore, many financial institutions still rely on archaic paper-based storage methods that not only drain financial resources but also operate with marked inefficiency. Blockchain technology emerges as a compelling solution to these challenges, introducing streamlined processes that automate data verification, reporting, and management. It offers a digital transformation by digitizing Know Your Customer (KYC) and Anti-Money Laundering (AML) data, along with transaction histories. What's more, blockchain facilitates real-time authentication of financial documents, ensuring that the integrity and accuracy of these documents remain beyond question. The adoption of blockchain translates into multifaceted benefits for financial institutions. It significantly minimizes operational risks, curbs the potential for fraudulent activities, and slashes the exorbitant costs associated with data management. This is achieved through the decentralized and immutable nature of blockchain, which ensures that all data resides in a tamper-proof and transparent ecosystem. The result is heightened transparency and accountability, redefining how financial data is handled and authenticated.

In essence, blockchain-based solutions empower financial institutions to not only streamline their operations but also to fortify their data's integrity and security. This, in turn, fosters increased trust and confidence among clients, as the immutable ledger of blockchain ensures that financial data is not just stored but safeguarded with an unwavering commitment to data integrity and security. In a financial landscape defined by data, blockchain stands as a formidable ally, paving the way for enhanced efficiency, reduced risks, and a future where clients can trust in the sanctity of their financial interactions. *(Ref. No.: 193- 206)*

Closing Thoughts

Blockchain technology is a decentralized distributed ledger that can provide networks and companies with improved security, transparency, accountability, and efficiency. This technology enhances privacy, eliminates the need for trust, and constructs a network of value where users can perform borderless peer-to-peer transactions. Blockchain technology and cryptocurrencies are not only here to stay, but also have the strength to convert all forms of industries and aspects of life, from finance, agriculture, big data to government voting and law. The healthcare sector has several promising use cases for blockchain networks, from creating and sharing immutable medical records to increasing transparency in the pharmaceutical supply chain. Though there are some technical, logistical, and regulatory challenges, the implementation of these systems will probably play a significant role in the future of medical data storage and transfer. Several big players of the supply chain enterprise are already embracing blockchain-based distributed systems, and we are likely to see global supply chain platforms leveraging blockchain technology to streamline the way companies share data as merchandise and materials move around. Blockchain technology has the potential to transform businesses in lots of unique ways, from production and processing to logistics and responsibility, eliminating areas of inefficiency that are common in traditional management models. Despite the drawbacks and limitations, blockchain technology has outstanding potential to revolutionize the way digital data is verified, stored, and shared. While many organizations and startups are already exploring the possibilities, there is still a lot to be done. We will definitely see more services focused on digital ID management in the coming years, and blockchain is likely to be a crucial part of it. The banking and financial industry is one of the major sectors that will be impacted by blockchain, with potential use cases ranging from real-time transactions to tokenization of assets, lending, smoother global trade, more robust digital agreements, and many more. Solving all the technological and regulatory hurdles required to fully realize the potential of this new financial infrastructure seems only to be a matter of time. A banking and financial stack based on a trustless, transparent, and borderless base layer is liable to be effective in enabling a more open and interconnected financial system.

Chapter 9: Applications of Blockchain in the Mining Industry

"Blockchain is so wide that you cannot know everything – you are just a few moments away from meeting a geek who knows so much than you do."

— Olawale Daniel

Potential of Blockchain Technology in Addressing Challenges of the Mining Industry

The mining industry is a crucial sector in the global economy, with its products and services being used in various industries, such as manufacturing, construction, and technology. According to the United States Geological Survey (USGS), the global production value of non-fuel minerals was estimated to be approximately $91.4 billion in 2020, with China being the largest producer and consumer of non-fuel minerals. Moreover, the mining industry provides employment opportunities to millions of people worldwide, with approximately 7 million people being employed in the sector globally, according to the International Council on Mining and Metals (ICMM).

Despite its significant contribution to the global economy, the mining industry faces several challenges, including complex supply chains, high maintenance costs, and inefficient royalty payment systems. The industry's complex supply chains result in a lack of transparency and accountability, making it challenging to track the movement of minerals from their source to their final destination. Additionally, the high maintenance costs associated with mining equipment and machinery contribute to increased operational expenses, and reducing profitability. Finally, the inefficient royalty payment systems can result in disputes and mistrust between mining companies and governments, affecting the industry's sustainability.

Blockchain technology has the potential to address these challenges by improving transparency, efficiency, and security in the mining industry. By providing a transparent and immutable ledger, blockchain technology can enable better tracking of minerals throughout the supply chain, reducing the risk of fraud and unethical practices. Additionally, blockchain-based equipment monitoring systems can help reduce maintenance costs by detecting issues before they become more significant problems. Finally, blockchain-based royalty payment systems can improve trust and reduce disputes by providing an immutable record of royalty payments.

Mineral tracking & supply chain management are critical components of the mining industry, but they often lack transparency and accountability. However, blockchain technology has the potential to address these challenges, as it provides a transparent and immutable ledger that tracks minerals from their source to their destination. Here are some statistics to illustrate how blockchain technology can improve mineral tracking and supply chain management in the mining industry:

1. According to the World Economic Forum, blockchain technology can enable better traceability of minerals and metals, reducing the risk of unethical practices such as child labor and conflict minerals.
2. The adoption of blockchain technology in mineral tracking could increase transparency and traceability of minerals and metals, which is essential for building consumer trust. A survey by Deloitte found that 65% of consumers are willing to pay a premium for products that demonstrate social and environmental responsibility.
3. A blockchain-based mineral tracking system can provide real-time data on the environmental and social impacts of mining activities. This can help companies identify and mitigate negative impacts and improve sustainability.
4. A blockchain-based mineral tracking system can also improve supply chain efficiency by reducing paperwork and manual processes, increasing the speed of transactions, and reducing the risk of errors.
5. According to a report by BCG, blockchain technology can reduce supply chain costs by up to 5% and increase supply chain transparency by up to 30%.
6. The use of blockchain-based mineral tracking systems can also benefit governments by enabling better monitoring of mining activities and reducing the risk of illegal mining.

The mining industry relies heavily on equipment and machinery to extract minerals from the earth. However, this equipment is subject to wear and tear, which can result in costly maintenance and downtime. Furthermore, mining equipment failures can pose serious safety risks to workers. Blockchain technology has the potential to address these challenges by enabling more efficient and effective equipment maintenance and monitoring.

Equipment Maintenance & Monitoring Challenges in the Mining Industry

Mining equipment and machinery require regular maintenance to ensure they operate efficiently and effectively. However, the maintenance of this equipment can be time-consuming and costly. Equipment downtime can result in significant financial losses, and manual maintenance processes can be inefficient and prone to errors.

Moreover, mining equipment failures can pose serious safety risks to workers. Faulty equipment can cause accidents, resulting in injuries or even fatalities. Traditional maintenance and monitoring systems may not be effective in detecting equipment malfunctions or deviations from normal operating conditions in real-time, which can result in increased safety risks.

Blockchain Technology in Equipment Maintenance & Monitoring

Blockchain technology has the potential to address these challenges by enabling more efficient and effective equipment maintenance and monitoring in the mining industry. Here are some potential applications of blockchain technology in equipment maintenance and monitoring:

1. **Real-time Equipment Monitoring:**
   Blockchain technology can be used to create a real-time monitoring system for mining equipment. Sensors can be installed on equipment to collect data on equipment performance, such as temperature, pressure, and vibration. This data can be stored on a blockchain, providing a transparent and immutable record of equipment performance.

2. **Predictive Maintenance:**
   By analyzing data collected through blockchain-based equipment monitoring systems, machine learning algorithms can be trained to predict when equipment failures are likely to occur. Predictive maintenance can help mining companies schedule maintenance activities proactively, reducing downtime and maintenance costs.

3. **Smart Contracts:**
   Smart contracts can be used to automate equipment maintenance schedules and trigger maintenance activities when specific conditions are met. For example, a smart contract could automatically trigger a maintenance activity when a sensor detects that a machine's vibration exceeds a certain threshold. This can reduce manual labor and improve the efficiency of maintenance activities.

4. **Improved Safety:**
   Blockchain-based equipment monitoring systems can improve safety in the mining industry by providing real-time alerts and notifications when equipment malfunctions or deviates from normal operating conditions. This can enable mining companies to take proactive measures to prevent accidents and ensure worker safety.

Benefits of Blockchain Technology in Equipment Maintenance & Monitoring

The adoption of blockchain-based equipment maintenance and monitoring systems can benefit mining companies in various ways, including:

1. Reduced maintenance costs and downtime
2. Improved equipment performance and longevity
3. Enhanced safety and risk management
4. Increased transparency and accountability

According to a report by Accenture, blockchain technology can help mining companies reduce maintenance costs by up to 20% and increase equipment uptime by up to 10%.

Moreover, the adoption of blockchain technology in equipment maintenance and monitoring can benefit the mining industry as a whole by improving sustainability. By reducing downtime and maintenance costs, mining companies can improve operational efficiency and reduce their environmental footprint.

Royalty Payments:

The current royalty payment system in the mining industry is plagued with challenges such as payment delays, administrative errors, and disputes between mining companies and governments. This leads to a lack of trust and accountability in the system, which can discourage investments and hinder economic growth. Blockchain technology can address these challenges by providing a transparent, secure, and efficient royalty payment system.

1. **Streamlining Royalty Payment Systems**
   Blockchain technology can streamline royalty payment systems by providing a decentralized and secure platform for payments. Smart contracts, which are self-executing contracts with the terms of the agreement between buyer and seller being directly written into lines of code, can be used to automate royalty payments based on predefined rules and criteria. Smart contracts can also reduce administrative costs by automating the payment process and eliminating the need for intermediaries such as banks and payment processors.
   One of the benefits of blockchain-based royalty payment systems is that they are transparent and secure. Every transaction is recorded on the blockchain, providing an immutable record of all royalty payments. This eliminates the possibility of fraud and ensures that payments are made to the rightful owners. Moreover, the transparency of blockchain technology can promote trust and accountability in the royalty payment system, which can increase investments and economic growth.

2. **Automating Royalty Payments**
   Blockchain technology can automate royalty payments based on predefined rules and criteria. Smart contracts can be programmed to

automatically calculate royalties based on the quantity and value of the minerals extracted, as well as any applicable taxes and fees. This eliminates the need for manual calculations and reduces the likelihood of errors and disputes.

Moreover, blockchain-based royalty payment systems can provide real-time updates on the status of royalty payments. This ensures that mining companies and governments have access to accurate and up-to-date information about the payments, reducing the risk of misunderstandings and disputes. This can also promote transparency and accountability in the royalty payment system, increasing trust and confidence in the system.

3. **Improving Trust and Reducing Disputes**

   Blockchain-based royalty payment systems can improve trust and reduce disputes between mining companies and governments. The transparency and security of blockchain technology can provide a clear and immutable record of all royalty payments, reducing the likelihood of disputes and misunderstandings. Additionally, blockchain technology can enable mining companies and governments to track royalty payments from the point of extraction to the point of sale, providing an accurate and complete picture of the royalties owed and paid.

Managing Data Challenges in the Mining Industry

Data Management in Mining:

The mining industry generates vast amounts of data every day from a wide range of sources, including geospatial data from exploration, geological data from mining operations, environmental data from monitoring, production data from processing plants, and equipment data from sensors and machines.

To effectively manage this data, mining companies must employ various data management techniques, such as data modelling, data integration, data warehousing, and data analytics. Data modelling involves creating a logical representation of the data to understand the relationships between different types of data. Data integration involves combining data from different sources to create a unified view of the data. Data warehousing involves storing large amounts of data in a centralized repository, while data analytics involves using statistical and machine learning techniques to gain insights from the data.

However, managing data in the mining industry is not without its challenges. One of the most significant challenges is the volume of data generated by mining operations. Mining companies must collect, process, and store large volumes of data from multiple sources, often in real-time, which can be costly and complex.

Another challenge is data privacy and security. Mining companies must comply with various data privacy regulations to protect sensitive information, such as employee and customer data, from unauthorized access, theft, or loss. They must also implement robust cybersecurity measures to protect against data breaches, hacking, and other cyber threats.

Finally, data integrity is a critical challenge in mining data management. Mining data is often complex and heterogeneous, which can lead to errors and inconsistencies. To ensure data integrity, mining companies must implement data quality controls, such as data cleansing, data profiling, and data validation, to identify and correct errors and ensure that data is accurate, complete, and consistent. *(Ref. No.: 320- 336)*

Challenges in Mining Data Management:

Managing data in the mining industry presents a multitude of unique challenges that can affect the efficiency, safety, and profitability of mining operations. In the following section, we will delve deeper into these challenges to gain a better understanding of their impact and how they can be addressed.

1. **Data Privacy & Security:** Mining companies deal with sensitive information that is critical for their business operations. Such information includes geological data and financial information that can provide insights into the quantity, quality, and location of minerals and help in making financial decisions. Therefore, it is essential for mining companies to ensure the confidentiality and security of these data. Unauthorized access, theft, or misuse of sensitive information can lead to significant financial losses, legal consequences, and reputational damage. To protect sensitive information, mining companies implement various security measures, such as access control, encryption, firewalls, and intrusion detection systems. Access control involves limiting access to sensitive data to authorized personnel only, such as data scientists and geologists. Encryption is used to protect data in transit or at rest, making it unreadable to unauthorized persons. Firewalls and intrusion detection systems are used to monitor network traffic and detect and prevent unauthorized access attempts. Additionally, mining companies must have policies and procedures governing the handling, storage, and disposal of sensitive information.

These policies should outline how data are collected, used, and shared, as well as how they are secured and protected against unauthorized access or disclosure.

The following are some specific data privacy and security challenges that mining companies face:

- Protecting confidential data from unauthorized access, theft, or data breaches.
- Managing access to data, especially in remote or inaccessible mining sites.
- Ensuring that third-party vendors comply with data privacy and security policies.

2. **Data Integrity:** Ensuring the accuracy and consistency of data is crucial for mining operations as it can have a significant impact on the decision-making process. Accurate data is required to determine the quality and quantity of minerals that can be extracted from a mining site, which can influence the mining process's planning and execution. Moreover, consistency in data collection and analysis is essential to ensure that the mining operation's progress is accurately tracked and compared to the initial projections. Inaccurate or inconsistent data can lead to costly mistakes and losses, which can have a significant impact on the overall profitability of the mining operation. To ensure data accuracy and consistency, mining companies use various data management techniques such as data validation, data cleansing, and data normalization. Data validation involves the process of verifying the data to ensure that it is accurate and consistent with the expected format. Data cleansing is used to correct or remove errors, inconsistencies, or duplicates from the data set. Data normalization is used to ensure that data is consistent across different data sets and is in the expected format.

The following are some specific data integrity challenges that mining companies face:

- Ensuring that data is captured and recorded accurately, consistently, and in a timely manner.
- Managing data quality, including data verification, validation, and cleaning.
- Avoiding data duplication or redundancy.

3. **Large volumes of Data:** Mining operations generate vast amounts of data throughout the mining process, from exploration to mineral

processing and beyond. This data includes geological data, financial data, equipment data, and operational data, among others. The volume of data generated can be enormous, making it challenging to store, manage, and analyze. The data can be in various formats, including images, videos, and text, and can be distributed across different locations, including remote sites. Furthermore, the variety of data generated by mining operations can make it challenging to integrate and analyze. Mining companies must manage this data effectively to make informed decisions, identify trends, and optimize mining operations. Failure to manage data effectively can lead to lost opportunities, inefficient operations, and increased costs.

To manage the vast amounts of data generated by mining operations, mining companies use various data management techniques and technologies. These techniques and technologies include data warehousing, data mining, and big data analytics. Data warehousing involves collecting and storing data in a central repository for easy access and analysis. Data mining involves extracting useful information from large data sets using machine learning algorithms. Big data analytics involves analyzing large and complex data sets to identify patterns and trends.

The following are some specific challenges associated with managing large volumes of data:

- Storing and managing large volumes of data, including unstructured data such as images, videos, and audio files.
- Ensuring that data is accessible and available in a timely manner, especially in remote or inaccessible mining sites.
- Managing data processing and analysis, including data visualization, machine learning, and other advanced analytics techniques.

4. **Legal & Regulatory Compliance:** Mining companies must comply with complex legal and regulatory frameworks when it comes to data management. These frameworks are designed to protect the privacy and security of sensitive data and ensure that companies operate in an ethical and responsible manner. Non-compliance with these regulations can result in significant legal penalties, reputational damage, and loss of trust among stakeholders.

One of the most important legal frameworks that mining companies must comply with is data protection laws. These laws regulate the collection, use, storage, and sharing of personal and sensitive data. They require companies to implement appropriate security measures to protect data against unauthorized access, theft, or misuse. Failure to comply with these laws can lead to significant financial penalties and legal consequences.

Additionally, mining companies must comply with regulations specific to the mining industry. For example, they must comply with environmental regulations that require them to monitor and report on their environmental impact. They must also comply with health and safety regulations that require them to ensure the safety of their workers and implement appropriate measures to prevent accidents and injuries.

The following are some specific challenges associated with legal and regulatory compliance:

- Ensuring compliance with data privacy regulations, such as the General Data Protection Regulation (GDPR).
- Complying with data retention and data destruction requirements.
- Ensuring compliance with industry-specific regulations, such as the U.S. Securities and Exchange Commission (SEC) regulations for mining companies.

Exploring Data Privacy Challenges in the Mining Industry: Importance, Protection & Regulations

This section of Data Privacy aims to conduct a comprehensive analysis of the importance of preserving confidentiality in the mining industry and the specific challenges related to protecting sensitive data. One key factor to consider in this regard is the use of cryptography for secure communication and storage of confidential information. Cryptography involves mathematical algorithms that transform plaintext data into ciphertext, which only authorized parties with the appropriate key can decipher. In the context of blockchain technology, the ERC20 standard comprises a set of regulations that dictate the secure and transparent transfer and management of tokens on the Ethereum network. It utilizes cryptographic techniques to ensure the security of transactions.

1. **Importance of data privacy:** The mining industry deals with sensitive data, including personal information, financial data, and trade secrets. The following are some specific reasons why data privacy is important in the mining industry:
   - Protecting confidential information from competitors and unauthorized access.

- Building trust with stakeholders, including investors, employees, and customers.
- Complying with legal and regulatory frameworks.

2. **Challenges of data privacy in the mining industry:** Mining companies face specific challenges when it comes to data privacy, including the following:
   - Managing access to data, especially in remote or inaccessible mining sites.
   - Ensuring that third-party vendors comply with data privacy policies.
   - Protecting against cyber threats, including data breaches and unauthorized access.

3. **Legal and regulatory frameworks:** The mining industry is subject to various legal and regulatory frameworks related to data privacy. The following are some specific frameworks that mining companies must comply with:
   - **The General Data Protection Regulation (GDPR):** This regulation applies to mining companies operating within the European Union and regulates the processing and transfer of personal data.
   - **The U.S. Securities and Exchange Commission (SEC) regulations:** These regulations require mining companies to disclose information related to their operations, including data privacy policies.
   - **Industry-specific regulations:** Mining companies may be subject to additional regulations related to data privacy, depending on the countries and regions in which they operate.

Securing Mining Industry Data: Challenges, Measures & Protection

Data security is critical for businesses operating in the mining industry, as it involves the management and storage of sensitive information, including customer data, financial information, and intellectual property. As the industry continues to rely heavily on digital technology and data, it is crucial to ensure that data is protected from cyber threats, including hacking, data breaches, and other malicious activities. In this section, we will discuss the importance of data security in the mining industry and the challenges associated with protecting data from cyber threats. Additionally, we will explore the various security measures that can be put in place to protect data, such as encryption and access controls.

Importance of Data Security in the Mining Industry

The mining industry is highly competitive, and mining companies must keep sensitive data confidential to protect their interests. Companies in this industry need to protect their data from cyber threats to ensure that they remain competitive and maintain their market position. The importance of data security in the mining industry includes:

1. **Protecting intellectual property:** Mining companies invest a lot of resources in research and development to develop new technologies, techniques, and processes. They need to protect their intellectual property from competitors who may steal their ideas.
2. **Maintaining customer trust:** Mining companies have access to sensitive customer data, including personal information, financial data, and confidential business information. They need to ensure that this data is protected to maintain customer trust.
3. **Compliance with regulations:** The mining industry is heavily regulated, and companies must comply with various laws and regulations related to data privacy and security.

Challenges Associated with Protecting Data in the Mining Industry

The mining industry faces various challenges in protecting data from cyber threats, including:

1. **Advanced persistent threats:** Cybercriminals use sophisticated techniques to gain access to mining company systems, steal data, and cause damage.
2. **Lack of awareness:** Many mining companies do not have adequate knowledge of cybersecurity threats and how to protect themselves against them.
3. **Insider threats:** Mining companies face threats from employees who may intentionally or unintentionally expose sensitive data to unauthorized individuals.

Security Measures that can be put in place to Protect Data

To protect data in the mining industry, various security measures can be put in place, including:

1. **Encryption:** Encryption is the process of converting data into an unreadable format to prevent unauthorized access. Encryption can be used to protect data in transit and data at rest.
2. **Access controls:** Access controls are mechanisms that restrict access to data to authorized users only. Access controls can be implemented using techniques such as authentication, authorization, and accounting.
3. **Firewalls:** Firewalls are security devices that control access to a network by analyzing incoming and outgoing traffic. They can be used to prevent unauthorized access to a mining company's network.
4. **Regular backups:** Regular backups can help mining companies recover lost or stolen data quickly. Backups should be stored in a secure location and tested regularly to ensure that they are functional.

Ensuring Mining Industry Data Integrity: Challenges & Measures

Data integrity is crucial for businesses operating in the mining industry, as it involves ensuring the accuracy and reliability of data. Inaccurate data can have severe consequences for mining companies, including financial losses, damage to reputation, and regulatory non-compliance. In this section, we will discuss the importance of data integrity in the mining industry and the challenges associated with ensuring the accuracy and reliability of data. Additionally, we will explore the various measures that can be put in place to ensure data integrity, such as data validation and data quality checks.

Importance of Data Integrity in the Mining Industry

The mining industry is heavily dependent on accurate and reliable data for decision-making, resource planning, and operational efficiency. The importance of data integrity in the mining industry includes:

1. **Resource management:** Accurate data is essential for determining the location, quantity, and quality of mineral resources, which is critical for effective resource management.
2. **Safety:** Mining companies rely on data to identify and mitigate safety risks. Inaccurate data can result in hazardous working conditions, accidents, and injuries.
3. **Compliance:** The mining industry is subject to various regulations and standards related to data integrity. Companies must ensure that their data meets these requirements to avoid regulatory sanctions and penalties.

Challenges Associated with Ensuring Data Integrity in the Mining Industry

Mining companies face various challenges in ensuring the accuracy and reliability of their data, including:

1. **Data complexity:** Mining data is complex and includes a vast array of information sources, including geological, geophysical, and engineering data. Managing and validating this data can be challenging.
2. **Human error:** Data can be compromised by human error, such as data entry mistakes, incorrect calculations, and incorrect interpretations.
3. **Data manipulation:** Data can be intentionally manipulated to produce misleading or false results, which can have severe consequences for mining companies.

Measures that can be put in place to Ensure Data Integrity

To ensure data integrity in the mining industry, various measures can be put in place, including:

1. **Data validation:** Data validation involves verifying the accuracy and completeness of data. This can be achieved by comparing data from different sources, performing data consistency checks, and identifying and resolving data anomalies.
2. **Data quality checks:** Data quality checks involve assessing the quality of data against specific criteria, such as completeness, accuracy, consistency, and timeliness. These checks can be performed using automated tools and techniques.
3. **Data governance:** Data governance involves establishing policies, procedures, and standards for managing and ensuring the quality and integrity of data. This includes defining roles and responsibilities, establishing data ownership, and implementing data management processes.

Effective Data Governance in Mining Industry: Challenges & Standards

Data governance is a critical component of effective data management in the mining industry. It involves the process of managing and ensuring the quality and integrity of data across an organization. In this section, we will discuss the importance of data governance in the mining industry, the challenges associated with managing data effectively, and the various frameworks and standards that can be used to guide data governance in the mining industry.

Importance of Data Governance in the Mining Industry

Effective data governance is essential for mining companies for the following reasons:

1. **Compliance:** The mining industry is subject to various regulations and standards related to data management, such as data privacy regulations, data security requirements, and data quality standards. Effective data governance ensures that mining companies comply with these regulations and standards.
2. **Operational efficiency:** Data governance enables mining companies to manage their data more efficiently, improving the accuracy and reliability of their data. This, in turn, helps to enhance operational efficiency, reduce costs, and improve decision-making.
3. **Reputation:** Mining companies are under increasing pressure to demonstrate their commitment to ethical and responsible business practices. Effective data governance helps to ensure that data is managed in an ethical and transparent manner, which can enhance a company's reputation and credibility.

Challenges Associated with Managing Data Effectively

Mining companies face various challenges in managing their data effectively, including:

1. **Data complexity:** Mining data is complex and includes a vast array of information sources, including geological, geophysical, and engineering data. Managing this data can be challenging, requiring specialized skills and expertise.
2. **Data silos:** Data can be stored in multiple locations and in different formats, creating data silos. This can make it difficult to access and manage data effectively.
3. **Data security:** Mining data is often sensitive and confidential, requiring robust security measures to protect it from cyber threats, such as hacking and data breaches.

Frameworks for Data Governance in the Mining Industry

Various frameworks and standards can be used to guide data governance in the mining industry, including:

1. **ISO 8000:** This standard provides a framework for managing data quality and is relevant to all industries, including mining. It includes guidelines for data modelling, data exchange, and data quality assessment.
2. **Global Mining Standards & Guidelines:** This is a collection of guidelines and best practices developed by the mining industry to promote best practices in various areas, including data governance. The guidelines cover topics such as data management, data security, and data privacy.
3. **Mining Association of Canada's (MAC) Towards Sustainable Mining (TSM):** TSM is a framework developed by the MAC to promote sustainable mining practices. It includes a data management component that requires mining companies to establish data management policies, processes, and practices to ensure the quality and integrity of their data. *(Ref. No.: 330- 342)*

Closing Thoughts

To manage their data efficiently and comply with regulatory requirements, mining companies must prioritize effective data governance. This involves implementing frameworks and standards such as ISO 8000, Global Mining Standards and Guidelines, and TSM. By adhering to these standards, mining companies can ensure that their data is of high quality and integrity, which is crucial for decision-making and overall operational efficiency. Furthermore, complying with regulatory requirements helps mining companies avoid potential legal consequences and reputational damage.

ISO 8000 is an international standard that provides guidelines for data quality management, including data exchange, data interoperability, and data integration. Global Mining Standards and Guidelines (GMSG) is a consortium of mining industry stakeholders that work together to develop best practices and guidelines for mining companies. The Towards Sustainable Mining (TSM) initiative is a program developed by the Mining Association of Canada that provides a framework for mining companies to manage their social, environmental, and economic impacts.

Overall, implementing effective data governance through these frameworks and standards is critical for mining companies to maintain data quality and integrity, improve operational efficiency, and comply with regulatory requirements.

Chapter 10: Smart Contracts in the Mining Industry

"Data privacy is an illusion because you are not in control. Any information you don't want out there shouldn't be shared anywhere on and off the internet."

— Olawale Daniel

How Smart Contracts are Revolutionizing the Mining Industry

The mining industry is a highly complex and capital-intensive sector that is faced with a multitude of challenges, including high operational and transactional costs, regulatory compliance issues, logistical complexities, and the challenge of maintaining stakeholder trust. These challenges are further compounded by the inherent risks associated with mining operations, such as accidents, natural disasters, and geopolitical instability. In recent years, there has been a growing interest in the potential of blockchain technology to address some of the key challenges facing the mining industry. Blockchain is a distributed ledger technology that enables secure and transparent transactions between parties without the need for intermediaries. Smart contracts, a self-executing computer program that runs on a blockchain, are an innovative solution that can automate many of the key processes involved in mining.

Smart contracts offer several benefits for the mining industry. Firstly, they can help to automate payment processes, which are often complex and time-consuming. By using pre-programmed rules, smart contracts can automatically execute payments when certain conditions are met, such as the delivery of goods or completion of a project. This can help to reduce the transactional costs associated with manual payment processes and improve cash flow for mining companies and their suppliers.

Secondly, smart contracts can help to enforce regulations, which is a critical issue for the mining industry. Compliance with environmental, health, safety, and labor regulations are essential to maintain the social license to operate and mitigate reputational risks. By using pre-programmed rules, smart contracts can ensure that all relevant regulations are being adhered to and automatically report on compliance. This can help to reduce the costs associated with regulatory compliance and enhance the transparency and accountability of mining operations.

Thirdly, smart contracts can help to streamline logistics, which is another key challenge facing the mining industry. The mining supply chain is complex and involves multiple stakeholders, including suppliers, contractors, and logistics providers. By using pre-programmed rules, smart contracts can automate the tracking and management of supply chain data, including the delivery of goods and the payment of invoices. This can help to reduce the logistical complexities and transactional costs associated with manual supply chain management.

Smart Contracts in Mining:

1. Automating Payment Processes:

The mining industry involves a vast network of suppliers, contractors, and other stakeholders that work together to extract minerals and other resources from the earth. These stakeholders need to be paid for their services, and the payment processes involved in mining can be quite complex. The use of smart contracts can help to automate these processes and make payments more efficient and transparent. A smart contract is a computer program that runs on a blockchain network. It is a self-executing contract that is programmed to automatically execute when certain conditions are met. In the case of the mining industry, smart contracts can be used to automate the payment processes for suppliers, contractors, and other stakeholders.

One of the key benefits of using smart contracts in the mining industry is that they can help to ensure prompt and accurate payments. When a supplier delivers goods or services, a smart contract can be programmed to automatically release payment once certain conditions are met. For example, the contract could be set up to release payment as soon as the goods are received and verified as meeting the required specifications. Smart contracts can also help to eliminate disputes and reduce the need for intermediaries. Because the terms of the contract are pre-programmed and stored on the blockchain, there is no need for intermediaries to mediate disputes or oversee payments. This can help to reduce costs and streamline the payment process. Another benefit of using smart contracts in the mining industry is that they can help to increase transparency. Because the terms of the contract are stored on the blockchain, all parties involved can see the terms of the agreement and the status of the payment process. This can help to reduce the risk of fraud and increase trust between parties.

Overall, the use of smart contracts in the mining industry can help to automate payment processes, increase transparency, reduce disputes, and streamline the payment process. As blockchain technology continues to evolve and become more widely adopted, it is likely that we will see more and more industries adopting smart contracts to help automate their payment processes.

2. Enforcing Regulations:

The mining industry is subject to a wide range of regulations that are designed to protect the environment, ensure worker safety, and promote responsible mining practices. Compliance with these regulations can be challenging, as they are often complex and involve multiple stakeholders, including government agencies, local communities, and environmental groups.

Smart contracts offer a potential solution to the challenge of regulatory compliance in the mining industry. A smart contract is a self-executing contract that is programmed to automatically execute when certain conditions are met. In the context of regulatory compliance, smart contracts can be used to ensure that mining companies are adhering to all relevant regulations by automating the process of monitoring and reporting on compliance.

For example, a smart contract could be used to track a mining company's carbon emissions and ensure that it is complying with environmental regulations. The contract could be programmed to automatically collect data on the company's emissions, compare it to regulatory standards, and report any violations to the relevant authorities. This would help to ensure that the mining company is meeting its obligations under environmental regulations, and reduce the risk of fines or other penalties.

Smart contracts can also help to enforce other types of regulations in the mining industry, such as those related to worker safety or community engagement. For example, a smart contract could be used to ensure that a mining company is providing adequate training to its workers, or that it is engaging with local communities in a responsible and transparent manner. By automating the process of monitoring and reporting on compliance, smart contracts can help to reduce the risk of regulatory violations and improve the overall sustainability of mining operations. In addition to improving compliance with regulations, smart contracts can also help to increase transparency and accountability in the mining industry.

3. Streamlining Logistics:

The mining industry involves a complex network of suppliers, contractors, and other stakeholders, which can make it challenging to manage the logistics of the industry effectively. There are many different stages in the mining process, from exploration and extraction to transportation and processing. Each stage involves multiple players, and it can be challenging to ensure that all parties are working together efficiently.

Smart contracts can help to streamline logistics in the mining industry by automating the tracking and management of supply chain data. For example, when ore is extracted from a mine, it needs to be transported to a processing plant where it can be refined and processed. This process involves multiple parties, including trucking companies, shipping companies, and processing plants. Each of these parties needs to be notified when a delivery is made and needs to be paid based on the terms of their contract.

A smart contract can be used to automate this process, ensuring that all relevant stakeholders are notified when deliveries are made, and payments are made automatically based on the terms of their contracts. This can help to eliminate the need for manual tracking and management of supply chain data, reducing the risk of errors and delays.

By streamlining logistics in this way, smart contracts can help to increase efficiency and reduce costs in the mining industry. This, in turn, can help to improve profitability and competitiveness, making it easier for mining companies to operate in an increasingly competitive global market.

How Blockchain Can Help Modernize the Mining Industry

The mining industry faces significant challenges in managing the supply chain of minerals such as cobalt, which is experiencing a surge in global demand due to its use in lithium-ion batteries. The industry relies on outdated, paper-based processes, resulting in high costs, a lack of transparency, and the opportunity for fraud and unsafe working conditions. To address these challenges, IBM has introduced two blockchain-based projects designed specifically for the mining industry. The first project is a consortium that includes Ford Motor Company, LG Chem, and other partners on a blockchain network aimed at enhancing mineral traceability and provenance transparency. Tracing the source of cobalt is difficult because, despite the majority of it originating from the Democratic Republic of the Congo, the metal passes through several companies and countries before reaching manufacturers. The use of blockchain technology in this consortium allows for the creation of an immutable, secure, and transparent digital ledger that can track the journey of minerals through the supply chain from the mine to the end buyer. This ensures that all parties involved in the process, including manufacturers and end consumers, have access to accurate information about the origin of the minerals used in their products, thereby enhancing the transparency and traceability of the mining supply chain. The second project is a partnership with MineHub Technologies, which is creating an online platform that utilizes blockchain to reduce paper documentation and manual processes in the mining and metals industry. This will lead to greater transparency and collaboration in the mining supply chain, resulting in cost savings, improved security, and a faster downstream mineral travel speed. The first use case, currently under development on the MineHub platform, manages ore from GoldCorp's Penasquito Mine in Mexico throughout its journey to the market. Overall, the use of blockchain technology in the mining industry has the potential to modernize the supply chain, making it more efficient, secure, and transparent. By enabling greater transparency and collaboration, blockchain technology can reduce costs, increase efficiency, and ultimately improve the safety and sustainability of the mining industry.

Inefficiencies in Conventional Mining Supply Chain

The path of minerals and metals along the supply chain is burdened by manual processes and complexity, which creates inefficiencies and presents opportunities for fraud. The complexity arises from the involvement of multiple parties, such as banks and smelters, each of whom must verify a significant number of paper documents related to mineral and ore traceability, rights management, regulatory compliance, or trade finance.

The inefficiencies of these manual processes cost the mining industry a significant amount of money, while the lack of visibility and data transparency presents numerous challenges. The situation is made more challenging by the fact that minerals like cobalt are often found in challenging conditions, requiring additional measures to ensure responsible and sustainable mining practices. Tracking the origins of a particular mineral becomes even more challenging when relying on paper-based processes.

Overall, the mining industry needs to find ways to modernize and digitize their processes to increase efficiency and reduce fraud. There is a need for greater transparency, data visibility, and traceability across the entire mining supply chain. Such measures would help in promoting responsible and sustainable mining practices while increasing operational efficiency and reducing costs.

Modernizing Outdated Trading in the Mining Industry

In today's world, trading in the mining industry is conducted manually, where buyers have to contact sellers directly to place orders. The sellers then need to verify any existing contracts manually, without any electronic recording or automation, which heavily relies on intermediaries. This manual process is time-consuming, and the lack of transparency and automation often creates opportunities for fraud.

To address these issues, the MineHub Platform is introducing smart contracts that automate most of these manual processes. By eliminating intermediaries, the platform enables quicker mineral sales, creating substantial value for all participants in the supply chain. This automation also reduces the time it takes for minerals and metals to move between points A and B, making the entire process more efficient.

In addition to improving the efficiency of the supply chain, the MineHub Platform helps minimize the possibility of commodity trading fraud, making the trading process more transparent and secure. This will provide confidence to all stakeholders in the mining industry, from miners and smelters to end customers, and increase mining industry financing.

Transforming Mining Operations with IoT, Wireless Communication, and Blockchain Technology

The mining industry has been grappling with several challenges related to safety, efficiency, and sustainability in its demanding environments. One of the major concerns is the need for robust tracking and communication systems to ensure the well-being of workers, monitor mining operations, and maintain a smooth supply chain. In recent years, there have been significant advancements in technology that have shown promising potential for addressing these challenges. The integration of Internet of Things (IoT) devices and wireless communication has improved data collection, real-time monitoring, and communication within mining sites. IoT sensors can be deployed to monitor various aspects of mining operations, such as equipment performance, environmental conditions, and worker safety.

Despite the progress made by IoT and wireless communication, there are still limitations to consider. One critical issue is the security and integrity of the collected data. As the mining industry relies heavily on accurate and reliable information, any tampering or unauthorized access to data can lead to serious consequences. Additionally, mining operations often take place in remote and challenging terrains, which can make maintaining a seamless and reliable wireless connection a significant challenge.

This is where blockchain technology comes into play as a potentially revolutionary solution. Blockchain is a decentralized and immutable ledger that ensures data transparency, security, and traceability. By combining blockchain with IoT and wireless communication, mining companies can achieve enhanced transparency, data integrity, and security throughout the entire mining process. One of the most significant advantages of blockchain technology is its ability to improve the traceability of minerals from the point of extraction to the end market. With blockchain, each step of the mining supply chain can be recorded, creating an auditable and unalterable trail. This level of traceability fosters responsible and sustainable mining practices, as stakeholders can verify the origin, quality, and ethical sourcing of minerals with confidence. Moreover, the integration of blockchain with IoT devices can enhance the reliability and credibility of the data collected. As IoT sensors gather information, it can be securely stored and verified on the blockchain, reducing the risk of data manipulation or errors. This reliability is crucial for informed decision-making processes, safety protocols, and compliance requirements within the mining industry.

Another exciting application of blockchain in mining is the use of smart contracts. Smart contracts are self-executing contracts with predefined rules and conditions. These contracts can automate various processes within the mining supply chain, including payment distributions, royalty agreements, and contractual obligations. By removing the need for intermediaries, smart contracts streamline operations, reduce administrative burdens, and increase overall efficiency.

Furthermore, cybersecurity is a paramount concern in the mining industry. With valuable data and critical infrastructure at stake, mining companies are prime targets for cyberattacks. Blockchain's decentralized nature and cryptographic security mechanisms offer enhanced protection against cyber threats, making it more difficult for malicious actors to compromise sensitive data or disrupt mining operations.

The convergence of IoT, wireless communication, and blockchain technology presents a transformative opportunity for the mining industry. By leveraging the strengths of these technologies, mining operations can achieve enhanced safety, transparency, efficiency, and sustainability. Blockchain's potential to secure data, ensure traceability, and automated processes can revolutionize how mining companies operate, paving the way for a more responsible, digitally-enabled future in the mining sector.

IoT & Wireless Communication in the Mining Industry

Blockchain technology has emerged as a ground-breaking solution for various industries, and its integration with IoT is revolutionizing the way we perceive and utilize technology. IoT, which stands for the Internet of Things, has become the new standard for combining different technologies and aspects from diverse methods. It brings together pervasive computing, sensing technologies, embedded devices, ubiquitous computing, internet protocols, and communication technologies to create systems where the digital and physical worlds converge and continuously interact. In the realm of IoT, smart objects serve as the fundamental building blocks. By equipping everyday objects with IoT capabilities, they are transformed into intelligent entities capable of collecting information from their surroundings, controlling, and interacting with the physical world. These smart objects are interconnected through the internet, enabling seamless information and data exchange between them. The vast number of interconnected devices and the abundance of accessible data in the IoT ecosystem open up numerous opportunities for creating valuable services that can benefit the environment, individuals, the economy, and society as a whole. However, certain industries face unique challenges that can be mitigated through the application of IoT. For instance, working in underground coal mines is a complex and high-risk endeavour due to factors such as long and narrow tunnels, intricate structures, and the presence of potentially explosive gases.

To address these challenges and enhance safety measures, the mining industry has started embracing IoT solutions. These solutions focus on improving communication and enabling timely responses to unusual activities within mining operations. However, the mining industry has been relatively slow in adopting technological advancements due to infrastructural limitations in areas such as communication, information exchange, data management, and storage. While some research efforts have concentrated on specific issues within the mining industry, such as ventilation monitoring, personal and fleet management, accident analysis, and tailings dam monitoring, a comprehensive IoT architecture specifically tailored to the mining industry's needs is still lacking.

In order to overcome the existing challenges and maximize the potential of IoT in mining, it is crucial to address issues related to data processing, interoperability, and remote control throughout all stages of mining operations, from exploration to reclamation. By adopting an Industrial IoT (IIoT) architecture, mining companies can create safer working environments for their employees, achieve operational efficiencies, enable seamless integration between modern and traditional devices and systems, automate processes to reduce human intervention, and establish real-time underground surveillance by converging Information Technology (IT) with Operational Technology (OT). Implementing IIoT in the mining industry requires overcoming various open research challenges. These challenges include mobility management, digital twins, scalability, and edge virtualization within the IIoT framework. Additionally, the monitoring of underground engineering construction is essential for ensuring safety. The current status of underground monitoring systems revolves around data acquisition, transmission, and processing. Optical-fibre sensors are gaining prominence in data acquisition, while wireless transmission based on Zigbee technology is becoming a popular choice for data transmission. In terms of data analysis and processing, various models, including the Neural Network (NN) model, are continuously evolving to improve prediction accuracy.

The integration of IoT with the mining industry involves utilizing a wide range of technologies to collect data. Wireless Sensor Networks (WSNs) play a crucial role in monitoring and sensing services within mining operations. WSNs consist of interconnected micro sensors that communicate wirelessly and provide extensive monitoring capabilities. WSNs have been applied in underground coal mines, leveraging their distributed system for monitoring, and addressing the unique communication challenges presented in such environments. These nodes within the WSN do not require direct communication with a central station or tower but interact with their local peers. By leveraging sensors, communication technologies, microcomputers, and automatic detection systems, this monitoring system enables real-time data analysis, remote control, image gathering, and security information monitoring.

Other innovative approaches to monitoring and securing coal mines include the utilization of web-based data transmission systems, wearable smart-jackets for miners, and blockchain technology. Web-based data transmission systems facilitate the monitoring of various parameters in coal mines, such as light detection, humidity and temperature conditions, fire detection, and gas leakage. These systems utilize sensors and process the sensor data to generate actionable insights for mine exploration and safety. Wearable smart-jackets are designed to ensure the safety of coal miners by monitoring vital health parameters, detecting hazardous gases, tracking location, and transmitting data through Wi-Fi shields.

Blockchain technology, with its inherent characteristics of distributed networks and security, presents itself as a promising solution for the mining industry. It can help mitigate cyber-attacks, ensure secure storage and data transmission, and address vulnerabilities and deficiencies in the current IoT infrastructure. However, challenges still remain in the implementation of blockchain technology within the mining industry, including issues related to cyber security, lack of standards, and structural limitations. *(Ref. No.: 316- 341)*

To realize the full potential of IoT in mining, it is essential to identify and address seismic hazards, enhance safety measures, and develop efficient communication systems. The grey-system theory has been utilized to identify seismic hazards in underground mines, showcasing better performance compared to conventional methods. Moreover, wearable devices and RFID technology have been explored to improve communication and safety in underground mines. Overall, the integration of IoT with the mining industry brings numerous benefits, including cost savings, improved productivity, enhanced fleet management, online sensor calibration and monitoring, real-time data alerts, and improved health and safety measures. The global IoT market is projected to experience exponential growth, generating vast amounts of data, and opening up new possibilities across various sectors, including mining. By embracing IoT technologies and addressing industry-specific challenges, the mining industry can unlock significant opportunities for efficiency, safety, and sustainability.

Blockchain-powered Tracking & Communication System

Blockchain technology offers various ways to enhance and solidify each process within the tracking and communication system:

Wall Receiver: by integrating blockchain into the wall receiver system, the tracking and monitoring of miners can be further strengthened. Blockchain can provide an immutable and transparent ledger to record the tracking information uploaded on the Wireless Sensor Network (WSN). This ensures the integrity and authenticity of the data, preventing any tampering or unauthorized modifications. Additionally, smart contracts on the blockchain can automate the verification and validation processes, ensuring that the tracking information is accurate and reliable. The decentralized nature of blockchain technology also enhances the security and resilience of the system, making it less susceptible to single points of failure or malicious attacks.

Cap-Lamp Transmitter: with blockchain integration, the cap-lamp transmitter can benefit from enhanced data security and privacy. The tracking and location information of miners can be securely stored and shared on the blockchain network, preventing unauthorized access, or tampering. Miners' identities and personal information can be encrypted and stored on the blockchain, ensuring their privacy is protected. Moreover, blockchain-based smart contracts can automate the tracking process, enabling real-time updates and alerts when miners enter restricted areas. The transparency and immutability of blockchain records also provide a reliable audit trail for compliance and regulatory purposes.

Access Point for the Wireless Network: Blockchain technology can play a vital role in the access point system by enabling secure and decentralized connectivity. Blockchain-based identity management systems can ensure that only authorized devices and sensors can connect to the wireless network. By leveraging blockchain's consensus mechanisms, the access point system can verify the authenticity and integrity of data collected from various sensors within the underground mines. This ensures the accuracy and reliability of the real-time data gathered. Additionally, blockchain-based data management solutions can provide a secure and transparent infrastructure for storing and sharing data, overcoming the challenges of communication, data management, and storage within the mining environment.

Wireless-Voice Communication Device: Integrating blockchain into the wireless-voice communication device enhances the security, reliability, and traceability of communication within underground coal mines. Blockchain's decentralized architecture ensures that voice communication data is distributed across multiple nodes, making it resistant to single points of failure or unauthorized access. The immutability of blockchain records provides a tamper-proof history of voice communications, ensuring the integrity of the messages exchanged. Smart contracts on the blockchain can automate emergency alerts and notifications, enabling timely responses to ensure miner safety. Additionally, blockchain-based authentication mechanisms can verify the identity of individuals participating in voice communication, preventing unauthorized access, and maintaining the confidentiality of sensitive information.

By leveraging the capabilities of blockchain technology, the tracking and communication system can be significantly enhanced in terms of data integrity, security, privacy, and reliability. It provides a robust infrastructure to ensure the safety of miners and efficient operations within underground mines.

Blockchain Technology in Underground Mining: Enhancing Wireless Communication, Monitoring, and Safety

Wireless Communication

The Wireless Access Point (WAP) refers to a device that enables different types of wireless networks to connect to the local area network (LAN) and access resources. By integrating wireless networks into the existing wired infrastructure, facilitates seamless connectivity. For instance, in underground mines, a router equipped with Wi-Fi capabilities serves as an access point for wireless data communication. These access points, often referred to as Wi-Fi-compatible devices, enable internet connectivity and support various mining activities. Access points play a crucial role in gathering real-time data from sensors and devices deployed in underground mines. The data collected can vary in complexity, with sensors used for blast control, toxic gas detection, and exploration, among other mining-related tasks. However, the integration of a fully automated system faces challenges such as communication infrastructure limitations, data management, and storage. Researchers have analyzed these challenges and proposed models tailored to specific mining sections, incorporating technologies like wireless sensor networks (WSNs) and global data management.

Wireless-Voice Communication systems serve critical functions such as monitoring miners' locations, displaying path histories, sending emergency alerts to control rooms, issuing warnings about restricted areas, localizing trapped miners, and broadcasting voice messages to underground miners.

1. **Decentralized Location Monitoring:** Blockchain technology enables the creation of a decentralized and tamper-proof ledger, which records real-time data regarding miners' locations within the mine. Through Wireless-Voice Communication systems, miners' movements can be continuously tracked and updated on the blockchain, providing mine operators and control rooms with accurate and transparent information about their whereabouts.

2. **Immutable Path Histories:** As miners traverse through the underground maze, the communication system can log their path histories on the blockchain. This permanent record ensures that the history of movements and activities remains immutable, allowing for reliable analysis and assessment of mining operations, safety protocols, and efficiency measures.

3. **Emergency Alert System:** In the event of an emergency, such as a cave-in or hazardous gas leakage, Wireless-Voice Communication systems can instantly trigger emergency alerts. These alerts are transmitted to the blockchain network, which propagates the information across all authorized nodes. This decentralized approach guarantees that crucial notifications reach the control rooms and relevant personnel promptly, enabling rapid response and rescue operations.

4. **Restricted Area Warnings:** Certain sections of underground mines may be designated as restricted areas due to safety concerns or ongoing operations. Through blockchain-enabled communication systems, warnings about restricted zones can be transmitted to miners' devices. These warnings are securely stored on the blockchain, ensuring that miners receive consistent and tamper-proof notifications about areas they should avoid.

5. **Localization of Trapped Miners:** In unfortunate scenarios where miners become trapped or lost, the Wireless-Voice Communication

system can aid in their localization. Miners can use their communication devices to send distress signals, which are recorded on the blockchain with their precise location details. This information aids rescue teams in pinpointing the trapped miners' whereabouts and expediting rescue efforts.

6. **Broadcasting Voice Messages:** The ability to broadcast voice messages through the Wireless-Voice Communication system is a powerful tool for disseminating important instructions, updates, or safety protocols to all underground miners simultaneously. With blockchain technology, these broadcast messages are time-stamped and securely stored, creating a reliable communication trail for future reference and audit.

Mine Environment Monitoring

Mine Environment Monitoring, involves the use of sensors to monitor parameters such as humidity, air velocity, and temperature in underground mines. Due to the harsh and risky nature of mining tunnels, reducing human presence is desirable. Researchers have presented an inspection robot prototype controlled by an operator via radio, assisting maintenance staff during inspections. The robot can identify hotspots and store its position with images. Additionally, temperature and electric current monitoring systems have been implemented in conveyor drive units, enhancing operational supervision. Radio-controlled systems ensure the safety of operators. This type of robot can be categorized as a mobility monitoring system for assessing the spatial distribution of underground infrastructure.

Here's how blockchain can be incorporated into the project to enhance data management, security, and collaboration:

1. **Blockchain Data Storage:** The project can utilize blockchain as a decentralized and immutable data storage solution for collecting sensor data from different monitoring systems. Each sensor reading, such as humidity, air velocity, and temperature, can be timestamped and recorded as a transaction on the blockchain. Smart contracts can be employed to define data structures and ensure the validity of incoming data.

2. **Smart Contract Integration:** Smart contracts play a vital role in automating certain processes and enforcing rules within the blockchain network. For instance, a smart contract can be designed to trigger an emergency alert if certain environmental parameters exceed predefined thresholds. Additionally, smart contracts can govern access control,

ensuring that only authorized personnel have permission to interact with specific data on the blockchain.

3. **Internet of Things (IoT) Integration:** To collect real-time data from various sensors, IoT devices can be deployed in the mining environment. These devices can be equipped with sensors for measuring humidity, air velocity, temperature, and other relevant parameters. The data collected by these IoT devices can then be securely transmitted to the blockchain for storage and analysis.

4. **Blockchain Consensus Mechanism:** Choosing an appropriate consensus mechanism is crucial for the project's success. Depending on the specific requirements, a consensus algorithm like Proof of Work (PoW) or Proof of Stake (PoS) can be adopted. PoW ensures data integrity by requiring miners to solve complex mathematical puzzles, while PoS allows validators to create new blocks based on their stake in the network.

5. **Permissioned Blockchain:** Given the sensitive nature of mine environment monitoring data, a permissioned blockchain can be preferred over a public blockchain. This approach allows only authorized participants, such as maintenance staff, operators, and supervisors, to be part of the network. This ensures that data access is limited to trusted stakeholders, enhancing security and privacy.

6. **Data Visualization & Analytics:** Blockchain data can be combined with data visualization and analytics tools to gain meaningful insights from the sensor readings. Interactive dashboards can display real-time environmental data, historical trends, and spatial distribution of parameters. Advanced analytics can help identify patterns and anomalies, aiding in predictive maintenance and optimizing mine operations.

7. **Integration with Inspection Robots:** The inspection robot prototype can be enhanced with blockchain capabilities to securely store images and data collected during inspections. These robot-generated reports can be recorded as transactions on the blockchain, creating a permanent and tamper-proof record of inspection activities and identified hotspots.

8. **Cross-Platform Communication:** The project may involve different devices and systems, such as IoT devices, inspection robots, and radio-controlled monitoring systems. Interoperability among these devices can be facilitated through standardized communication protocols and data formats. Blockchain can serve as a common platform for seamless data sharing and collaboration among these disparate systems.

Revolutionizing Gas-Monitoring in Underground Mines

Blockchain technology presents an unparalleled opportunity to revolutionize the Gas-Monitoring System deployed in underground mines. By integrating blockchain, we can reinforce the security, immutability, and transparency of the gas concentration data, promoting safety and accountability within the mining industry. Let us delve into the technical aspects, including code snippets and other technical details, to better understand the advantages:

1. **Real-time Gas Monitoring with Sensors:** The gas monitoring system employs sensors to measure gas concentrations in the mine environment. These sensors upload real-time data to a wireless network for continuous surveillance. Below is a simplified Python code snippet illustrating how data from the sensors can be transmitted to the network:

   In python:

```python
import requests
def upload_gas_data_to_network(gas_data):
    api_endpoint = "https://gas-monitoring-network-api/upload"
    headers = {'Content-Type': 'application/json'}
    response = requests.post(api_endpoint, json=gas_data, headers=headers)
    return response.status_code
# Sample gas concentration data
gas_data = {
    'carbon_monoxide': 35.6,
    'methane': 18.2
}
status_code = upload_gas_data_to_network(gas_data)
print(f"Data upload status: {status_code}")
```

2. **Zigbee-based Wireless Sensor Networks:** To establish seamless communication between the sensors and the coal mine safety monitoring systems, Zigbee-based wireless sensor networks are employed. Zigbee ensures efficient and reliable data transmission among the devices. Here is a brief code snippet to demonstrate a simple Zigbee communication setup:

In python:

```python
from zigbee_library import Zigbee

def configure_zigbee_network():
    # Code to configure the Zigbee network
    pass

def receive_gas_data_from_sensors():
    zigbee = Zigbee()
    gas_data = zigbee.receive_data()
    return gas_data

configure_zigbee_network()
received_gas_data = receive_gas_data_from_sensors()
print("Received gas data:", received_gas_data)
```

3. **iBeacons for Precise Miner Identification:** Innovative solutions like iBeacons are proposed to enable precise tracking of miners within the mine environment. iBeacons emit Bluetooth signals that can be detected by mobile devices or dedicated receivers. Here is a conceptual explanation of iBeacon implementation:

In python:

```
Miner's Helmet (with iBeacon) -> iBeacon emits Bluetooth signal -> Receiver device detects the signal and identifies the miner -> Miner's location and identity are recorded and monitored.
```

4. **Service-Oriented Architecture (SOA) for Optimal Functionality:** The system adopts a service-oriented architecture for optimal functionality and scalability. SOA enables modular design and loosely coupled services, ensuring the system's flexibility and extensibility.
5. **Blockchain Integration for Enhanced Security & Transparency:** By incorporating blockchain technology, we can bolster the integrity of the gas monitoring system. Below are some key technical aspects of how blockchain enhances the system:

Blockchain Transaction for Gas Data: Each gas concentration reading is recorded as a transaction on the blockchain. Smart contracts can be employed to automate data recording and verification processes. Here is a high-level representation of a blockchain transaction:

In Yaml:

> Gas Data Transaction:
> Transaction Hash: 0xabcdef...
> From: Sensor Address 1
> To: Mining Company's Blockchain Wallet
> Gas Concentration Data: {'carbon_monoxide': 35.6, 'methane': 18.2}
> Timestamp: 2023-07-28 12:00:00

Immutability & Tamper Resistance: Once recorded on the blockchain, gas concentration data becomes immutable. No entity can alter or delete past readings, ensuring the data's authenticity and preventing tampering attempts.

Transparent Data Verification: Blockchain's transparent nature allows all stakeholders to validate and verify gas concentration data. Miners, safety inspectors, and management can independently access the blockchain and cross-check the accuracy of the data. The convergence of cutting-edge gas monitoring technologies and blockchain provides the mining industry with a safer, more efficient, and transparent future. By safeguarding the lives of workers, ensuring regulatory compliance, and promoting ethical practices, this integrated system significantly enhances safety standards within the mining sector.

Strata Monitoring in Underground Mines

With the adoption of blockchain technology, the Strata Monitoring system stands to gain substantial improvements in data security, immutability, and transparency. The integration of blockchain fortifies the integrity of convergence measurements, ensuring reliable and tamper-resistant records for the mining industry. Now, let us delve into the technical intricacies, which include code snippets and other relevant technical details:

1. **Convergence Measurement with Linear Potentiometer and Piston:** The Strata Monitoring system utilizes a linear potentiometer and a piston for accurate convergence measurements. This combination enables precise detection of changes in the mine's strata. The data collected by the convergence measurement component is the foundation for subsequent blockchain transactions.

2. **Microcontroller for Data Coordination:** The system's microcontroller acts as the central processing unit and facilitates data acquisition, processing, and transmission. By employing smart contracts on the blockchain, the microcontroller can automate data recording and coordination processes. Here is a simplified representation of a smart contract to handle data coordination:

In solidity:

```solidity
pragma solidity ^0.8.0;

contract ConvergenceData {
    mapping(uint256 => bytes32) public
convergenceMeasurements;

    function
recordConvergenceMeasurement(uint256
timestamp, bytes32 dataHash) public {

convergenceMeasurements[timestamp] =
dataHash;
    }

    function
getConvergenceMeasurement(uint256
timestamp) public view returns (bytes32)
{
        return
convergenceMeasurements[timestamp];
    }
}
```

3. **ESP for Secure Communication and Data Transmission:** The ESP module plays a vital role in securely connecting the Strata Monitoring system to the blockchain network. It uses cryptographic techniques to ensure data privacy and authenticity during transmission. Below is a conceptual explanation of how the ESP module securely transmits convergence data to the blockchain:

As an alternative to ESP, the Strata Monitoring system can use LoRaWAN (Long Range Wide Area Network) for secure communication and data transmission. LoRaWAN is well-suited for underground mining environments due to its long-range capabilities and low power consumption. It operates on unlicensed frequency bands, allowing for reliable communication in challenging underground conditions. LoRaWAN also supports encryption and authentication mechanisms, ensuring data privacy and integrity during transmission.

In Rust:

Convergence Measurement Data -> Hashed and Encrypted -> Sent via Secure Connection to the Blockchain Node -> Blockchain Transaction Created -> Data Stored on the Blockchain.

4. **Load-Cell Sensor for Accurate Force Measurements:** The load-cell sensor is a critical component responsible for capturing and quantifying forces exerted on the piston during convergence measurements. These precise force readings are instrumental in determining convergence changes accurately.

By leveraging blockchain technology, the Strata Monitoring system ensures the security, transparency, and immutability of convergence measurements. Blockchain transactions record the convergence data as it is measured, establishing an immutable and transparent history of strata changes within the underground mines. In the context of underground mine communication challenges, blockchain technology can address some issues. For instance, the secure and tamper-resistant nature of blockchain ensures that communication data remains intact and unaltered, even in harsh environments. Additionally, blockchain's decentralized nature can help maintain communication networks in underground mines, even if certain nodes or devices experience connectivity issues. For India's vast coal reserves, an IoT-based wireless and real-time monitoring system can effectively manage and track hazard information, production data, and geological information. By integrating blockchain into this IoT system, data security and transparency can be further enhanced, ensuring the reliability of the information shared among relevant stakeholders. This convergence of cutting-edge technologies can significantly improve safety and productivity in the mining industry.

From a blockchain perspective, the research objectives can be summarized as follows:

1. Developing a blockchain-powered miners' tracking system and mine-wide communication infrastructure for underground coal mines. By integrating blockchain technology, the system aims to enhance data integrity, security, and reliability, enabling efficient tracking and communication among miners within the mine environment.

2. Designing and developing an IoT-enabled system that incorporates blockchain technology to monitor strata conditions and sub-surface mine environments. This integration facilitates real-time data collection, secure storage, and traceability, ensuring accurate monitoring and analysis of the mine's geological conditions.
3. Designing and developing a web-based decision-making system that leverages blockchain technology to predict sub-surface environments. By utilizing blockchain's capabilities, the system enhances data reliability, transparency, and security, enabling informed decision-making based on accurate predictions.

By adopting blockchain technology in various aspects of tracking, communication, mine environment monitoring, and strata monitoring, the underground mining industry can benefit from improved data integrity, security, reliability, and real-time monitoring capabilities. The integration of blockchain ensures the trustworthiness of the collected data, enhances transparency, and facilitates secure communication among stakeholders. Ultimately, these advancements contribute to safer and more efficient mining operations, providing valuable insights for decision-making processes within the industry.

Integration of Blockchain in IoT for Enhanced Connectivity and Data Management in Underground Mining

IoT encompasses two major elements: Sensors & RFID tags.

Sensors play a crucial role in providing solutions for IoT. These embedded devices detect various environmental parameters such as temperature and humidity, converting them into interpretable data for machines or humans. The collected electronic data can be further transmitted.

RFID (Radio Frequency Identification) tags enable wireless transfer of digital data through electromagnetic or radio waves. Consisting of an Integrated Circuit (IC), substrate, and an antenna, RFID technology facilitates automatic detection of objects, gathering information, and transmitting it to a computer. Passive RFID tags draw power from RFID readers, while active RFID tags have their own power source. This technology is used for data capture and automatic identification.

WSN (Wireless Sensor Network) is a self-configured wireless communication system employed for collecting mine information and monitoring underground mine conditions. Specialized transducers and sensor nodes positioned at various detection stations form WSNs. These nodes are equipped with transceivers, power supplies, transducers, and microcomputers. WSN applications include monitoring environmental factors such as temperature, air pressure, threat detection, humidity, and surveillance.

Middleware & Data Structure: IoT

In the realm of IoT, middleware plays a critical role as a service layer external to operating systems. It simplifies the development of complex distributed applications and provides a connection between applications, users, and data. Middleware offers various facilities and accessible environments through standard interfaces, ensuring the smooth operation of services.

The middleware layer includes applications, service composition, service management, and object abstraction. Applications refer to downloadable functionalities accessible to end-users, while service composition focuses on arranging services for building networking applications. Service management handles data services such as reading, writing, transformation, computation, and tracking within managed IoT scenarios. Object abstraction facilitates coordination and access to devices through standard language procedures. Security Operation Centers (SOC) enable interconnection and information exchange using web-based protocols.

The data structure in IoT is crucial for networking, sensing, data gathering, and secure information exchange. The structure encompasses multiple layers, including the application layer, processing layer, transmission layer, analysis layer, and perception layer. The application layer facilitates communication between IoT devices and networks, ensuring successful data transmission. The processing layer performs computation and data processing based on gathered sensor data. The transmission layer enables data access and communication within the IoT network. The analysis layer analyses collected data, selects relevant information, and eliminates irrelevant data for transmission. The perception layer gives meaning to data collected from sensors such as RFID tags, converting it into digital signals for further processing.

Digital Mine Features through IoT & Blockchain Technology

In the mining industry, various applications such as wireless communication, environmental monitoring, prediction systems, emergency response, and employee record management play crucial roles in ensuring safety, efficiency, and overall operational effectiveness. With the advent of blockchain technology, these applications can be further enhanced, addressing challenges related to data security, transparency, and trust within the mining ecosystem.

Blockchain technology holds tremendous potential to revolutionize mining applications through a range of transformative methods, as outlined below:

1. **Wireless communication:** Blockchain can be employed to secure and authenticate wireless communication within the mining environment. By incorporating blockchain into the wireless communication system, data transmission can be encrypted, and the integrity of the communication network can be ensured. Additionally, blockchain-based smart contracts can automate and enforce communication protocols, ensuring reliable and accountable wireless communication among miners.

2. **Environmental monitoring:** Blockchain can enhance the reliability and transparency of environmental monitoring data. Sensor data collected from the monitoring systems can be securely stored on the blockchain, creating an immutable record of the environmental conditions. This ensures the integrity and authenticity of the data, allowing stakeholders to trust and verify the accuracy of the collected information.

3. **Prediction system:** Blockchain technology can improve the prediction system by providing a tamper-proof and auditable record of the collected data. The data used for predictive analysis can be securely stored on the blockchain, ensuring its integrity, and preventing unauthorized modifications. Smart contracts can be utilized to automate data analysis and prediction processes, enhancing efficiency and accuracy.

4. **Emergency response:** Blockchain can facilitate seamless and transparent communication during emergency situations. By integrating blockchain into the emergency response system, relevant stakeholders can have real-time access to critical information, such as emergency alerts and response plans. This promotes coordination and collaboration among different entities involved in emergency response, leading to faster and more effective actions.

5. **Employee record management:** Blockchain can be utilized to securely manage and authenticate employee records. By storing employee information on the blockchain, it becomes tamper-proof and immutable, ensuring the accuracy and integrity of the records. Additionally, blockchain-based identity management solutions can enable secure and efficient verification of employee identities, enhancing the overall management of the workforce.

Wireless Sensor Networks (WSNs) are widely used in underground coal mines for various monitoring and tracking applications. However, the resource-limited environment of underground mines poses challenges related to energy efficiency. Alongside energy consumption, the deployment, localization, and routing algorithms present inherent difficulties for WSNs in such environments.

To overcome these challenges, Visible Light Communication (VLC) has emerged as a promising technique for efficient communication and illumination in underground coal mines. Researchers have been focused on understanding VLC channel shadowing, path loss, and the impact of matter on illumination intensity and VLC system performance. By incorporating VLC technology, communication and tracking systems can be significantly improved in underground mining operations.

Furthermore, the integration of blockchain technology with WSNs, VLC, and other systems holds tremendous potential. Blockchain technology can enhance security and prevent cyber-attacks through its distributed network and utilization of heterogeneous devices. By leveraging blockchain, the integrity and confidentiality of data transmitted within the mining environment can be ensured.

Intelligent mining technology, including unmanned equipment and advanced algorithms, has also been explored to enhance mining equipment efficiency, monitoring capabilities, and overall operational efficiency. This approach aligns with the objective of upgrading conventional industries through advanced technologies like artificial intelligence (AI), digital intelligence, and automation. The integration of blockchain technology in intelligent mining systems can provide secure and transparent data management, allowing stakeholders to access and verify mining-related information in a decentralized manner.

To further improve monitoring in underground coal mines, a distributed monitoring system based on WSN has been proposed. WSN nodes communicate directly with local peers, creating a wireless network with unique characteristics and challenges. The safety theory of WSN communication in underground coal mines and the main methods for distributed monitoring systems have been discussed, along with the functions and layout of WSN in an underground coal mine. Blockchain technology can be employed to ensure the immutability and integrity of the data collected by WSNs, providing a tamper-resistant and auditable record of mining operations.

Researchers have also analyzed Zigbee network node arrangements for monitoring underground communication and tracking. Performance evaluation considers metrics such as Packet Delivery Ratio (PDR), security, energy consumption, and end-to-end delay. The analysis reveals that the mesh topology provides better performance in terms of PDR, network security, and high throughput, while the cluster tree topology is preferred for minimum energy consumption and end-to-end delay. Blockchain technology can be integrated into these network arrangements to establish a decentralized and transparent network infrastructure.

Blockchain-Enabled Gas Monitoring System for Underground Mines:

The gas monitoring system used in underground coal mines, which employs sensors to monitor gas concentrations (including SO2, H2S, O2, and CO), can greatly benefit from the integration of blockchain technology. Blockchain offers several advantages that can enhance the efficiency and security of the system. One key aspect is the secure and transparent transmission of data.

By connecting the gas monitoring system to a blockchain network, the wireless network information and gas concentration data captured by the sensors can be securely and transparently uploaded to the blockchain. This ensures the integrity and immutability of the data, preventing unauthorized tampering or alteration. Furthermore, blockchain technology enables real-time monitoring of gas concentrations in underground mines. Authorized personnel can access the gas concentration data stored on the blockchain network, allowing for continuous monitoring and prompt responses to any abnormal or hazardous conditions. This real-time monitoring capability is crucial for maintaining a safe working environment for miners.

In addition to real-time monitoring, blockchain provides traceability and compliance benefits. Each data point recorded on the blockchain can be timestamped and associated with a unique identifier. This creates an auditable trail of gas concentration data, facilitating compliance with safety standards and regulatory reporting. Mining companies can demonstrate their adherence to safety regulations by providing auditors with access to the blockchain records, ensuring transparency and accountability.

Blockchain also facilitates data sharing and collaboration among stakeholders involved in gas monitoring. Authorized parties such as mining companies, safety inspectors, and regulators can securely access the gas concentration data stored on the blockchain. This promotes collaboration, transparency, and cooperative decision-making, leading to improved safety measures and effective risk management. Moreover, blockchain-based smart contracts can automate the generation of alerts based on predefined gas concentration thresholds. When gas levels exceed or fall below the specified limits, the smart contract can trigger alerts or notifications to relevant personnel via messages or emails. This automated alert system ensures prompt actions to mitigate potential risks and enhances the overall safety of the mining operation.

Another significant advantage of integrating blockchain technology is the simplification of compliance audits. Auditors can securely access the blockchain network to verify the integrity of the gas concentration data and ensure compliance with regulatory requirements. The immutability of blockchain records streamlines the auditing process, reduces administrative burdens, and increases trust among stakeholders.

By incorporating blockchain technology into the gas monitoring system in underground coal mines, the industry can benefit from enhanced data security, transparency, auditability, collaboration, and automation. These features contribute to improved operational efficiency, risk management, and overall safety in mining operations, ensuring the well-being of miners and the integrity of underground environments.

There are blockchain-based projects that focus on gas monitoring in various industries, including mining. One example is the Methane Data Platform developed by ConsenSys, a blockchain software technology company. The Methane Data Platform utilizes blockchain technology to enable the transparent and secure tracking of methane emissions in the oil and gas industry. While it may not specifically target underground coal mines, it demonstrates the application of blockchain in gas monitoring and emissions tracking. The platform leverages smart contracts and IoT devices to collect data from gas sensors and other monitoring equipment. This data is securely recorded on the blockchain, ensuring its integrity and immutability.

The decentralized nature of the blockchain network enables stakeholders, including regulators, operators, and auditors, to access and verify the gas monitoring data in a transparent manner. By implementing blockchain technology, the Methane Data Platform enhances data accuracy, provides real-time monitoring capabilities, and facilitates compliance with emission regulations. It enables stakeholders to collaborate, share data, and make informed decisions regarding gas monitoring and emissions reduction efforts.

1. Smart Contract for Gas Monitoring Data:

A smart contract is a self-executing contract with the terms of the agreement directly written into code. In the context of the Methane Data Platform, smart contracts are used to handle the recording and verification of gas monitoring data on the blockchain.

In solidity

```solidity
// Solidity smart contract example for methane data recording

pragma solidity ^0.8.0;

contract MethaneData {
    // Structure to store methane data readings
    struct MethaneReading {
        uint256 timestamp;
        uint256 methaneLevel;
        address sensorAddress;
        // Add more relevant data as needed
    }

    // Array to store methane readings
    MethaneReading[] public methaneReadings;

    // Function to add new methane data readings
    function addMethaneReading(uint256 _timestamp, uint256
_methaneLevel) public {
        MethaneReading memory newReading =
MethaneReading(_timestamp, _methaneLevel, msg.sender);
        methaneReadings.push(newReading);
    }

    // Function to get the total number of methane readings
    function getMethaneReadingCount() public view returns (uint256) {
        return methaneReadings.length;
    }
}
```

2.   IoT Device Communication with Smart Contract:

IoT devices, such as gas sensors, communicate with the smart contract on the blockchain to record the methane data readings.

In JavaScript

```javascript
// JavaScript code example for IoT device communication
with smart contract

const Web3 = require('web3');
const contractAbi = require('...'); // Replace with the actual
ABI of the MethaneData contract
const contractAddress = '0x...'; // Replace with the actual
address of the deployed contract

// Connect to the Ethereum node
const web3 = new
Web3('https://mainnet.infura.io/v3/YOUR_INFURA_API_
KEY');

// Create an instance of the smart contract
const methaneDataContract = new
web3.eth.Contract(contractAbi, contractAddress);

// Simulate data reading from the gas sensor
const timestamp = Math.floor(Date.now() / 1000);
const methaneLevel = 100; // Replace with the actual
methane level read from the sensor

// Send the data to the smart contract
methaneDataContract.methods.addMethaneReading(timest
amp, methaneLevel).send({ from:
'YOUR_IOT_DEVICE_ADDRESS' })
  .then((receipt) => {
     console.log('Methane reading added to the
blockchain:', receipt.transactionHash);
  })
  .catch((error) => {
     console.error('Error adding methane reading:', error);
  });
```

3. **Accessing & Verifying Methane Data on the Blockchain:**

Stakeholders, including regulators and auditors, can access and verify the methane data recorded on the blockchain.

In JavaScript

```javascript
// JavaScript code example for accessing and verifying methane data on
the blockchain

// Get the total number of methane readings
methaneDataContract.methods.getMethaneReadingCount().call()
  .then((count) => {
    console.log('Total methane readings:', count);
  })
  .catch((error) => {
    console.error('Error getting methane reading count:', error);
  });

// Retrieve a specific methane reading by index
const readingIndex = 0; // Replace with the desired index
methaneDataContract.methods.methaneReadings(readingIndex).call()
  .then((reading) => {
    console.log('Methane reading:', reading);
  })
  .catch((error) => {
    console.error('Error retrieving methane reading:', error);
  });
```

The provided illustrations showcase how smart contracts and IoT device communication are employed to record and retrieve methane data on the blockchain. For real-world application, it is essential to incorporate supplementary security measures, access controls, and data encryption to guarantee the integrity and confidentiality of the gas monitoring data.

Future Perspective: Blockchain-Enabled Strata Monitoring System

In the future, the integration of blockchain technology into the strata monitoring system promises to revolutionize the way underground mining operations ensure safety and stability. The system architecture would be designed to harness the full potential of blockchain's decentralized, secure, and transparent nature.

1. **Decentralized Data Collection & Storage:** The future strata monitoring system would employ a network of IoT devices, equipped with advanced sensors, for data collection. These devices would continuously gather data on critical parameters such as bed load, bed separation, convergence, and pillar stress. The data would then be securely recorded on the blockchain, ensuring its integrity and immutability. Miners, regulators, and other stakeholders would have access to this decentralized database, promoting transparency and fostering collaborative efforts.

2. **Smart Contracts for Autonomous Operations:** Smart contracts would play a pivotal role in automating certain functions within the strata monitoring system. These self-executing contracts would be programmed to execute predefined actions based on specific conditions. For example, when sensor data indicates a potential hazard, smart contracts could trigger immediate alerts and initiate safety protocols without human intervention. This autonomous capability would enhance response times and reduce risks in critical situations.

A smart contract written in Solidity to store and manage the strata monitoring data on the blockchain.

In solidity

```solidity
// Solidity smart contract for strata monitoring data storage

pragma solidity ^0.8.0;

contract StrataMonitoring {
    // Structure to store strata monitoring data
    struct StrataData {
        uint256 timestamp;
        uint256 bedLoad;
        uint256 bedSeparation;
        uint256 convergence;
        uint256 pillarStress;
        // Add more relevant data as needed
    }

    // Array to store strata monitoring data
    StrataData[] public strataData;

    // Function to add new strata monitoring data
```

```
            function addStrataData(uint256 _timestamp, uint256
        _bedLoad, uint256 _bedSeparation, uint256 _convergence,
        uint256 _pillarStress) public {
            StrataData memory newData = StrataData(_timestamp,
        _bedLoad, _bedSeparation, _convergence, _pillarStress);
            strataData.push(newData);
        }

            // Function to get the total number of strata monitoring data
        points
            function getStrataDataCount() public view returns (uint256)
        {
            return strataData.length;
        }
        }
```

3. **AI-Driven Data Analysis:** The future strata monitoring system would incorporate advanced Artificial Intelligence (AI) algorithms to analyze the vast amounts of sensor data collected. AI-driven data analysis would enable the system to detect patterns, identify anomalies, and predict potential risks with unprecedented accuracy. By learning from historical data, the AI models could continually improve their predictions, making the monitoring system more proactive and effective.

4. **Real-Time Visualization & Decision Support:** A sophisticated graphical interface would visualize real-time strata data for mine operators and supervisors. The interface would display dynamic graphs, heatmaps, and 3D representations of underground conditions, facilitating better understanding and decision-making. Additionally, AI-generated insights and recommendations would empower stakeholders to take proactive measures to prevent accidents and optimize mining operations.

5. **Multi-Layered Security Measures:** Recognizing the importance of securing critical data, the future strata monitoring system would implement multi-layered security measures. Blockchain's inherent security features, such as consensus mechanisms and cryptographic techniques, would safeguard data integrity and prevent unauthorized access. Furthermore, data encryption and permissioned access controls would add an extra layer of protection, ensuring that only authorized personnel can view sensitive information.

6. **Integration with IoT Ecosystem:** The blockchain-enabled strata monitoring system would seamlessly integrate with the broader IoT

ecosystem present in underground mining operations. This integration would allow data from various IoT devices, such as environmental sensors and safety equipment, to be combined and analyzed comprehensively. The synergy between blockchain and IoT technologies would amplify the overall efficiency and safety of mining processes.

The sensor data collection module communicates with the smart contract to record the strata monitoring data on the blockchain.

In JavaScript

```
// JavaScript code example for integrating with sensor data
collection and the blockchain smart contract

const Web3 = require('web3');
const contractAbi = require('...'); // Replace with the actual
ABI of the StrataMonitoring contract
const contractAddress = '0x...'; // Replace with the actual
address of the deployed contract

// Connect to the Ethereum node
const web3 = new
Web3('https://mainnet.infura.io/v3/YOUR_INFURA_API_KE
Y');

// Create an instance of the smart contract
const strataMonitoringContract = new
web3.eth.Contract(contractAbi, contractAddress);

// Simulate sensor data readings
const timestamp = Math.floor(Date.now() / 1000);
const bedLoad = 120; // Replace with the actual bed load value
read from the sensor
const bedSeparation = 50; // Replace with the actual bed
separation value read from the sensor
const convergence = 225; // Replace with the actual
convergence value read from the sensor
const pillarStress = 85; // Replace with the actual pillar stress
value read from the sensor

// Send the data to the smart contract
```

```
strataMonitoringContract.methods.addStrataData(timestamp,
bedLoad, bedSeparation, convergence, pillarStress).send({
from: 'YOUR_SENSOR_ADDRESS' })
  .then((receipt) => {
    console.log('Strata monitoring data added to the
blockchain:', receipt.transactionHash);
  })
  .catch((error) => {
    console.error('Error adding strata monitoring data:',
error);
  });
```

7. **Auditable Compliance & Reporting:** Blockchain's immutable ledger would facilitate easy auditing of compliance with safety regulations and reporting of environmental standards. Regulators and auditors could access historical data and verify that mining operations adhere to safety guidelines. Compliance reports, generated from the blockchain database, would provide comprehensive insights, and promote accountability among mining operators.

Miners Tracking & Voice Communication with Blockchain Technology:

The implementation of Miners Tracking and Voice Communication in the context of blockchain technology involves a robust architecture that utilizes RFID technology, wireless sensor networks, and smart contracts. Below are the technical details and code snippets for each component of the system:

1. **RFID Enabled Routers and Tags:**

RFID tags are attached to mining equipment or worn by miners. RFID-enabled routers, strategically placed throughout the underground mine, act as access points to collect information from the RFID tags.

CODE 1: In Python:

```python
# Python code example for RFID data collection using an
RFID-enabled router

import RFIDReader

# Initialize RFID reader and set up communication
rfid_reader = RFIDReader()
```

```python
# Continuously read RFID data from the tags
while True:
    tag_id = rfid_reader.read_tag()
    print(f"Tag ID: {tag_id}")
```

CODE 2: In Python:

```python
import serial

class RFIDReader:
    def __init__(self, serial_port='/dev/ttyUSB0',
baud_rate=9600):
        self.serial_port = serial_port
        self.baud_rate = baud_rate
        self.serial = None

    def connect(self):
        try:
            self.serial = serial.Serial(self.serial_port,
self.baud_rate)
            print(f"Connected to RFID reader on
{self.serial_port}")
        except Exception as e:
            print(f"Error connecting to RFID reader: {e}")

    def read_tag(self):
        if self.serial is not None:
            tag_data = self.serial.readline().decode().strip()
            return tag_data
        else:
            print("RFID reader not connected.")
            return None

    def disconnect(self):
        if self.serial is not None and self.serial.is_open:
            self.serial.close()
            print("Disconnected from RFID reader.")

# Example usage:
rfid_reader = RFIDReader()
```

```
rfid_reader.connect()

while True:
    tag_id = rfid_reader.read_tag()
    if tag_id:
        print(f"Tag ID: {tag_id}")
```

The paho-mqtt library in Python enables the transmission of RFID tag data through the MQTT protocol within a wireless sensor network. MQTT follows a publish-subscribe model, where data is published to specific topics and other devices subscribe to receive relevant messages. The library simplifies MQTT communication, allowing developers to create MQTT clients and connect to brokers for publishing and subscribing to topics, facilitating real-time tracking and communication in resource-constrained environments.

CODE 3: In Python:

```python
import paho.mqtt.client as mqtt
import time

# MQTT broker details
broker_address = "broker.example.com"
broker_port = 1883

# Create MQTT client
client = mqtt.Client()

# Connect to the broker
client.connect(broker_address, broker_port)

# Function to publish RFID tag data to MQTT topic
def publish_tag_data(tag_id):
    topic = "mine/rfid/tags"
    client.publish(topic, tag_id)
    print(f"Published Tag ID: {tag_id} to MQTT topic: {topic}")

# Example usage:
rfid_reader = RFIDReader()
rfid_reader.connect()
```

```
while True:
  tag_id = rfid_reader.read_tag()
  if tag_id:
    publish_tag_data(tag_id)
  time.sleep(1)
```

When implementing the code examples, remember to customize the **broker_address** and **broker_port** to match the specific MQTT broker used. Additionally, ensure the required libraries, such as **paho-mqtt**, are installed using the appropriate command, for example, pip install **paho-mqtt**. These code snippets demonstrate how to read RFID tag data from the RFID reader and publish it to an MQTT topic, making it accessible to other devices within the wireless sensor network for further utilization.

2. **Wireless Sensor Network:**

The RFID-enabled routers form a decentralized wireless sensor network that dynamically meshes with the mine's layout. This network allows seamless communication between the RFID tags, routers, and the blockchain-based server.

In JavaScript

```javascript
// JavaScript code example for a simple wireless sensor
network using MQTT
const mqtt = require('mqtt');
const client = mqtt.connect('mqtt://broker.example.com');

// Subscribe to the RFID tag data topic
client.on('connect', () => {
  client.subscribe('mine/rfid/tags', (err) => {
    if (!err) {
      console.log('Subscribed to RFID tag data');
    }
  });
});

// Handle incoming tag data
client.on('message', (topic, message) => {
  const tagData = JSON.parse(message);
  console.log('Received RFID tag data:', tagData);
});
```

3. **Application Software & Data Management:**

The application software deployed on the central server gathers and processes data obtained from the RFID routers. The processed data, including real-time tracking information, is securely recorded on the blockchain.

In solidity

```solidity
// Solidity smart contract example for storing RFID tag data on the blockchain

pragma solidity ^0.8.0;

contract MinersTracking {
  struct MinerLocation {
     uint256 timestamp;
     bytes32 minerId;
     uint256 latitude;
     uint256 longitude;
     // Add more relevant data as needed
  }

  MinerLocation[] public minerLocations;

  function addMinerLocation(bytes32 _minerId, uint256 _latitude, uint256 _longitude) public {
     MinerLocation memory location =
MinerLocation(block.timestamp, _minerId, _latitude, _longitude);
       minerLocations.push(location);
  }

  // Function to get the total number of miner locations recorded
  function getMinerLocationsCount() public view returns (uint256) {
       return minerLocations.length;
  }
}
```

Suggestions for future research:

In the realm of mine monitoring systems, continuous research and development play a pivotal role in improving safety, productivity, and efficiency in underground mining operations. This collection of suggestions presents a comprehensive roadmap for future research, highlighting various key areas of exploration. From the development of advanced noise reduction algorithms to optimize voice clarity, to the enhancement of packet delivery within wireless sensor networks, each suggestion aims to address specific challenges faced in the mining environment. Additionally, the exploration of experimental network parameter analysis, robustness enhancement, integration of advanced sensors, and the implementation of empowered IoT frameworks further contribute to creating safer and more productive mining practices. By delving into these areas of research, the mining industry can revolutionize its monitoring systems, promoting sustainable and safer mining operations for the future.

Below are a series of comprehensive suggestions for future research in the field of mine monitoring:

1. **Algorithm for Noise Reduction:** In the realm of mine monitoring systems, blockchain technology opens up exciting possibilities for the development of advanced noise reduction algorithms tailored to mining environments. By leveraging blockchain's decentralized and transparent nature, the algorithm can securely access and analyze vast amounts of sensor data, including voice recordings and environmental noise levels. This blockchain-powered algorithm, equipped with sophisticated signal processing and machine learning techniques, such as adaptive filtering and deep learning-based denoising, can effectively remove background noise, ensuring crystal-clear communication among miners even in the most challenging underground conditions. With enhanced voice clarity and reliable communication, blockchain-driven noise reduction algorithms contribute significantly to the overall safety and productivity of mining operations.

2. **Packet Delivery Enhancement:** Blockchain technology's peer-to-peer network architecture presents an ideal platform for optimizing the performance of wireless sensor networks in mine monitoring systems. Through blockchain-enabled smart contracts, researchers can experiment with innovative routing algorithms that adapt dynamically to the ever-changing underground environment and efficiently handle node mobility. By leveraging blockchain's consensus mechanisms and

trustless environment, adaptive modulation techniques can be explored to adjust transmission rates based on real-time channel conditions, ensuring optimal packet delivery rates. Additionally, advanced signal propagation analysis, supported by blockchain's immutable data storage, can identify ideal node placements, and mitigate signal attenuation and interference, enabling more reliable data transmission over longer distances within the challenging mining conditions.

3. **Experimental Analysis of Network Parameters:** The decentralized and transparent nature of blockchain technology allows for robust and unbiased experimental analysis of various network parameters in real-world mining scenarios. By deploying blockchain-powered testbeds or simulators in underground mines, researchers can collect and securely store comprehensive data on network performance, including network lifetime, communication cost, effective coverage area, and energy consumption. The integrity and immutability of this data ensure researchers gain valuable insights into system behaviour under different conditions, enabling them to fine-tune system configurations, optimize resource allocation, and identify areas for improvement with confidence.

4. **Enhancing Robustness:** The inherent trustlessness and fault-tolerance capabilities of blockchain technology provide a promising foundation for addressing the challenge of system robustness in mine monitoring networks. Future research can focus on developing blockchain-based fault-tolerant mechanisms and redundancy strategies, such as node replication and clustering, which can be securely managed and verified through smart contracts. Furthermore, blockchain's decentralized consensus algorithms enable the investigation of network self-healing mechanisms, ensuring dynamic reconfiguration and maintaining connectivity even in the presence of node departures or disruptions. These blockchain-powered solutions enhance the overall reliability and resilience of the monitoring system, ensuring continuous and dependable communication for underground mining operations.

5. **Integration of Advanced Sensors:** Blockchain technology offers a secure and transparent framework for integrating advanced sensors beyond RFID into mine monitoring systems. By leveraging blockchain's tamper-proof data storage, researchers can address challenges related to sensor calibration, data fusion, and network integration with a higher degree of confidence and security. Through

blockchain-powered smart contracts, the integration process becomes seamless, enabling the utilization of gas sensors, temperature sensors, seismic sensors, and more, providing crucial data for detecting potential hazards and predicting mining-related incidents with enhanced accuracy.

6. **Voice & Video Recording Module:** The decentralized nature of blockchain ensures the utmost security and integrity of voice and video recordings in mine monitoring systems. By implementing a dedicated blockchain-based module for storing communication sessions' recorded data, mining companies gain peace of mind knowing that incident analysis and response strategies are based on trustworthy and unaltered information. With the ability to organize recorded data in a monthly basis within a secure and immutable database, blockchain-driven voice and video recording modules enable detailed analysis, deeper insights into events, and informed decision-making, ultimately elevating safety standards in mining operations.

7. **Development of Empowered IoT Frameworks:** Blockchain technology serves as the backbone for designing and implementing more empowered IoT frameworks tailored for the mining industry's specific needs. By leveraging blockchain's secure and decentralized infrastructure, these frameworks can seamlessly integrate advancements in sensor technology and data analytics to enhance safety, efficiency, and decision-making across various mining processes. With blockchain's immutable records, real-time monitoring and predictive maintenance become more reliable, facilitating sustainable and productive mining practices. As mining companies adopt this holistic approach to IoT implementation, the transformative potential of blockchain technology unfolds, paving the way for safer and more productive mining operations across the industry.

The integration of blockchain technology has the potential to revolutionize the entire mining industry, bringing forth transformative changes and advancements. By implementing blockchain in mine monitoring systems, data integrity, transparency, and trust among stakeholders are greatly enhanced, resulting in safer working conditions and heightened accountability. The utilization of blockchain's distributed ledger capabilities enables real-time data sharing and collaboration, facilitating efficient decision-making, predictive maintenance, and incident analysis, ultimately elevating the overall efficiency and effectiveness of mining operations. Moreover, the secure storage of voice and video recordings through blockchain empowers mining companies with valuable insights to prevent accidents and continuously improve safety standards. The seamless integration of advanced sensors into the monitoring system via blockchain allows for early detection of hazards, enabling timely responses and further contributing to a safer environment for miners. Furthermore, empowered IoT frameworks leveraging blockchain technology optimize mining processes, promoting sustainable practices, and reducing costs through data-driven analysis and smart contracts. Embracing these cutting-edge solutions paves the way for a new era of safer, more productive, and environmentally responsible mining practices, leading to the transformation of the mining industry as a whole. With ongoing research and collaboration among experts, the adoption of blockchain technology holds the promise of reshaping the mining industry, propelling it towards a more prosperous and sustainable future. The potential benefits of blockchain in mining are vast, and as industries embrace this innovative technology, they are poised to achieve unprecedented growth and progress in the ever-evolving landscape of mining operations.

IoT Data Logging Architecture: Building Secure & Efficient Data Management for IoT Devices

The "IoT Data Logging Architecture" harnesses blockchain technology for secure and efficient data collection, processing, and management from IoT devices. Blockchain ensures data integrity and authenticity, creating an immutable ledger with distributed consensus. This tamper-resistant approach enhances security and mitigates single points of failure. The architecture also incorporates microservices, lightweight communication protocols like MQTT/CoAP, and advanced data ingestion pipelines for scalability and modularity. Blockchain fosters transparent data sharing, promoting collaboration among stakeholders. Additionally, it addresses data interoperability challenges, facilitating seamless data exchange between IoT devices and protocols. The architecture's innovative blend of blockchain and IoT offers a transformative solution for reliable and decentralized IoT data management.

Harnessing the Power of Blockchain Technology for Data Security and Integrity:

One of the core strengths of the IoT Data Logging Architecture lies in its seamless incorporation of blockchain technology. By leveraging blockchain's decentralized and immutable nature, the architecture ensures that every piece of data logged from IoT devices is cryptographically secured, tamper-proof, and transparently recorded in a distributed ledger. This transformative feature bolsters the overall trustworthiness of the data logging system, guaranteeing data integrity, and preventing unauthorized alterations or deletions. The integration of blockchain also enables robust data provenance, allowing stakeholders to trace the origins and lifecycle of data throughout the IoT ecosystem. The use of blockchain further enhances data sharing and collaboration, facilitating secure data exchange among authorized participants within the network.

Implementing Scalable & Modular Architecture using Microservices:

The architecture embraces the principles of scalability and modularity, employing microservices as the building blocks of its design. These microservices, being autonomous and loosely coupled components, are purpose-built to handle specific functionalities of the system. The adoption of such a microservices-based approach enables seamless development, deployment, and independent scaling of each component. As a result, the data logger system becomes highly flexible and maintainable, with the ability to efficiently distribute tasks across multiple services for parallel processing, thus mitigating potential bottlenecks.

Embracing microservices also facilitates the incorporation of new features and updates without disrupting the overall system stability. For instance, microservices can be dedicated to data ingestion, data storage, data validation, access control, and more, all working harmoniously to form a robust data logging infrastructure. The combination of blockchain technology and microservices architecture creates a cutting-edge solution that addresses the unique challenges of IoT data management, ensuring secure, scalable, and efficient operations for diverse IoT applications.

Utilizing MQTT or CoAP as Lightweight and Efficient Communication Protocols for IoT Devices:

The architecture embraces the principles of scalability and modularity, employing microservices as the building blocks of its design. These microservices, being autonomous and loosely coupled components, are purpose-built to handle specific functionalities of the system. The adoption of such a microservices-based approach enables seamless development, deployment, and independent scaling of each component. As a result, the data logger system becomes highly flexible and maintainable, with the ability to efficiently distribute tasks across multiple services for parallel processing, thus mitigating potential bottlenecks.

Embracing microservices also facilitates the incorporation of new features and updates without disrupting the overall system stability. For instance, microservices can be dedicated to data ingestion, data storage, data validation, access control, and more, all working harmoniously to form a robust data logging infrastructure. The combination of blockchain technology and microservices architecture creates a cutting-edge solution that addresses the unique challenges of IoT data management, ensuring secure, scalable, and efficient operations for diverse IoT applications.

Designing Data Ingestion and Preprocessing Pipelines to Handle Large Volumes of IoT Data:

The architecture incorporates well-designed data ingestion and preprocessing pipelines to cope with the substantial influx of data from numerous IoT devices. These pipelines efficiently handle and preprocess vast volumes of incoming IoT data in real-time. The data ingestion pipelines are responsible for receiving data from IoT devices, validating its authenticity, and forwarding it to the appropriate processing components. On the other hand, preprocessing pipelines filter, aggregate, and clean the data before storage or analysis, enhancing data quality and reducing storage requirements.

To achieve this, the architecture can incorporate technologies like Apache Kafka for scalable and fault-tolerant message queuing, Apache Spark for distributed data processing, and data stores like Apache Cassandra for high-performance storage. This ensures that the system can handle the immense data streams efficiently and that the data logger system is well-prepared to accommodate future data growth, all while maintaining the enhanced security and integrity provided by the integration of blockchain technology. The use of blockchain in combination with advanced data ingestion and preprocessing pipelines creates a robust and scalable infrastructure, capable of managing the vast and diverse data generated by IoT devices with utmost efficiency and security.

A. Integration of Blockchain in IoT Data Logging Architecture:

In the Blockchain-Integrated IoT Data Logger System, the IoT Data Logging Architecture plays a critical role in securely collecting and managing data from IoT devices while leveraging the benefits of blockchain for data integrity and trust. Here is a detailed technical overview:

1. **Scalable & Modular Architecture with Microservices:** The architecture adopts a microservices-based approach to enable scalability and flexibility. Each microservice encapsulates specific

functionalities, such as data collection, preprocessing, blockchain integration, and database management.

Here is an example of a simple microservice in Python:

In python

```python
# Example of a simple data collection microservice
from flask import Flask, request

app = Flask(__name__)

@app.route('/collect-data', methods=['POST'])
def collect_data():
    data = request.json
    # Process and validate data
    # Store data in the database
    # Send data to preprocessing pipeline
    return 'Data collected successfully.'

if __name__ == '__main__':
    app.run(host='0.0.0.0', port=5000)
```

2. **Utilizing MQTT or CoAP for Efficient Communication:** The architecture leverages MQTT (Message Queuing Telemetry Transport) or CoAP (Constrained Application Protocol) for lightweight and efficient data transmission between IoT devices and the data logger system. MQTT utilizes the "Publish-Subscribe" model, while CoAP relies on UDP for reduced overhead.

Here is an example of an MQTT publisher in Python:

In python

```python
# Example of an MQTT publisher for data transmission
import paho.mqtt.client as mqtt

client = mqtt.Client()
client.connect("broker_address", 1883)

def publish_data(topic, data):
    client.publish(topic, data)

# Usage example:
```

```
publish_data("iot/sensor/data", "Temperature: 25°C,
Humidity: 50%")
```

3. **Designing Data Ingestion & Preprocessing Pipelines:** The architecture includes robust data ingestion and preprocessing pipelines to handle the high volume of IoT data. The data is validated, cleaned, and transformed before being stored or sent for analysis.
For instance, using Python and Pandas, we can preprocess data:

In python
```python
# Example of data preprocessing using Pandas
import pandas as pd

def preprocess_data(data):
    # Data cleansing and formatting
    df = pd.DataFrame(data)
    # Apply preprocessing transformations
    # Return preprocessed data
    return df

# Usage example:
data = {"timestamp": "2023-07-30 12:00:00", "temperature": 25,
"humidity": 50}
preprocessed_data = preprocess_data(data)
```

4. **High Availability & Fault Tolerance:** To ensure high availability and fault tolerance, the architecture can employ containerization technologies like Docker and orchestration platforms like Kubernetes. This enables the deployment of multiple instances of microservices across different nodes to handle traffic and recover from failures automatically.

5. **Data Storage & Database Selection:** The architecture utilizes a suitable database system to efficiently store IoT data. For instance, a time-series database like InfluxDB may be used to handle the continuous influx of data with high write performance. Here is an example of data storage in InfluxDB using Python:

In python
```python
# Example of data storage in InfluxDB using Python
from influxdb import InfluxDBClient
```

```python
client = InfluxDBClient(host='localhost', port=8086)
client.switch_database('data_logger_db')

def store_data_in_db(data):
    # Convert data to InfluxDB Line Protocol format
    data_points = [{"measurement": "sensor_data", "fields": data}]
    client.write_points(data_points)

# Usage example:
data = {"temperature": 25, "humidity": 50}
store_data_in_db(data)
```

6. **Security & Data Privacy:** With blockchain integration, data security and privacy are enhanced through cryptographic mechanisms. For example, the following code snippet showcases asymmetric encryption using Python's cryptography library:

In python

```python
# Example of asymmetric encryption using Python's cryptography
library
from cryptography.hazmat.backends import default_backend
from cryptography.hazmat.primitives.asymmetric import rsa
from cryptography.hazmat.primitives import serialization, hashes

# Generate RSA key pair
private_key = rsa.generate_private_key(
    public_exponent=65537,
    key_size=2048,
    backend=default_backend()
)

# Serialize and save private key
pem = private_key.private_bytes(
    encoding=serialization.Encoding.PEM,
    format=serialization.PrivateFormat.PKCS8,
    encryption_algorithm=serialization.NoEncryption()
)
with open("private_key.pem", "wb") as f:
    f.write(pem)
```

```python
# Load private key from file
with open("private_key.pem", "rb") as f:
    pem = f.read()
private_key = serialization.load_pem_private_key(pem,
password=None, backend=default_backend())

# Encrypt data using the public key
public_key = private_key.public_key()
ciphertext = public_key.encrypt(b"Sensitive data",
padding.OAEP(mgf=padding.MGF1(algorithm=hashes.SHA256())
, algorithm=hashes.SHA256(), label=None))
```

7. **Real-time Monitoring & Alerts:** The architecture incorporates real-time monitoring and alerting systems to detect anomalies or security breaches promptly. Tools like Prometheus and Grafana can be used for monitoring, and alerts can be configured to notify administrators through various channels (e.g., email, Slack) when predefined thresholds are exceeded.

8. **Integration with Blockchain:** The architecture includes components to securely interact with the blockchain network. Smart contracts can be developed using Solidity (a language for Ethereum) to validate incoming data and record it on the blockchain. For example, a simple smart contract to store sensor data on the Ethereum blockchain could look like this:

In Solidity

```solidity
// Example of a smart contract to store sensor data on
Ethereum blockchain
pragma solidity ^0.8.0;

contract DataLogger {
    struct SensorData {
        uint256 timestamp;
        uint256 temperature;
        uint256 humidity;
    }

    mapping (uint256 => SensorData) public dataLog;
    function logData(uint256 _timestamp, uint256 _temperature,
uint256 _humidity) public {
```

```
        SensorData memory newData = SensorData(_timestamp,
    _temperature, _humidity);
        dataLog[_timestamp] = newData;
    }
}
```

The integration of blockchain technology and a meticulously designed IoT Data Logging Architecture ensures a data logging system that is secure, reliable, and resistant to tampering for IoT applications. By incorporating advanced technologies, mathematical formulas, and cryptographic techniques, the system's overall integrity and trustworthiness are enhanced.

Data Logging Architecture with blockchain integration:

Data Integrity Check using Hash Function: To ensure data integrity, a cryptographic hash function can be employed to create a fixed-size hash value (digest) for the collected data. This hash value uniquely represents the data and acts as a digital fingerprint. The architecture can use a secure hash function like SHA-256. The formula for computing the hash is:

In SCSS

HashValue = SHA-256(Data)

For example, in Python, using the **hashlib** library:

In python
```
import hashlib

data = "Temperature: 25°C, Humidity: 50%"
hash_value = hashlib.sha256(data.encode()).hexdigest()
```
Proof of Work (PoW) Consensus Mechanism: In blockchain networks using PoW, miners compete to find a nonce value that, when hashed with the block's data, produces a hash that meets certain difficulty criteria (e.g., starts with a specified number of leading zeros). The formula for verifying the PoW is:

In SCSS

Hash(Nonce + BlockData) < TargetDifficulty

Where Nonce is the arbitrary value being adjusted, **BlockData** is the data to be hashed, and **TargetDifficulty** represents the required number of leading zeros in the hash.

Data Validation using Smart Contracts: Smart contracts can be used to validate the incoming data before it is recorded on the blockchain. For example, the smart contract can verify that the temperature and humidity values fall within acceptable ranges. The formula for such validation could be:

In Solidity

```
// Example of data validation in a smart contract using Solidity
function       validateData(uint256       _temperature,       uint256
_humidity) internal pure returns (bool) {
    return (_temperature >= 0 && _temperature <= 100) &&
(_humidity >= 0 && _humidity <= 100);
}
```

Merkle Trees for Efficient Data Verification: Merkle trees can be used to efficiently verify the integrity of large sets of data on the blockchain. The formula for computing the Merkle root involves recursively hashing the data pairs until a single root hash is obtained:

In SCSS

$$MerkleRoot = Hash(Hash(Data1) + Hash(Data2))$$

For example, to verify a specific data entry in the Merkle tree, only the path from the leaf node to the root needs to be checked, rather than verifying the entire data set.

Data Aggregation using Averaging: For certain IoT applications, data aggregation might be necessary to reduce data size and storage requirements. For example, the architecture could aggregate temperature readings over a specific time period using averaging:

In Mathematica

$$AverageTemperature = (Sum\ of\ Temperatures) / Number\ of\ Readings$$

The average temperature value can then be stored on the blockchain, reducing the volume of raw data. By incorporating these mathematical formulas and advanced technical concepts into the blockchain based IoT Data Logging Architecture, the system can achieve enhanced security, data integrity, and efficiency.

B. IoT Data Security through the Lens of Blockchain Technology

Incorporating blockchain into IoT Data Logger Systems underscores the critical significance of data security, given the sensitive nature of the information derived from IoT devices. The utilization of blockchain introduces distinctive attributes that amplify data security, uphold integrity, and foster confidence within the system. Let us delve more profoundly into the technical dimensions of data security within the realm of blockchain technology.

End-to-End Encryption (e.g., AES-256):

In the blockchain-integrated IoT data logger system, end-to-end encryption remains a crucial aspect of data security. Advanced Encryption Standard (AES) with a key size of 256 bits is commonly used to encrypt data between IoT devices and the data logger system.

The encryption process involves the following steps:

- When an IoT device generates data, it encrypts the data using the AES algorithm and a randomly generated encryption key.
- The encrypted data is then transmitted to the data logger system via the blockchain network or off-chain channels.
- The data logger system, or any other authorized entity, can decrypt the data using the corresponding decryption key, which is securely stored and managed.

The use of end-to-end encryption ensures that even if the data is intercepted during transit, it remains unintelligible to unauthorized entities.

In python

```python
# Example of AES-256 encryption in Python
from cryptography.hazmat.primitives.ciphers import Cipher,
algorithms, modes
from cryptography.hazmat.primitives import padding
from cryptography.hazmat.backends import default_backend

def encrypt_data(data, encryption_key):
    backend = default_backend()
    iv = b'1234567890123456'  # Initialization Vector (IV) for
AES
    cipher = Cipher(algorithms.AES(encryption_key),
modes.CFB(iv), backend=backend)
```

```python
        encryptor = cipher.encryptor()

        # Pad the data to be encrypted
        padder =
padding.PKCS7(algorithms.AES.block_size).padder()
        padded_data = padder.update(data.encode()) +
padder.finalize()

        # Encrypt the padded data
        encrypted_data = encryptor.update(padded_data) +
encryptor.finalize()

        return encrypted_data
```

1. **Secure Authentication Mechanisms (e.g., OAuth, JWT):**

In the context of blockchain technology, secure authentication mechanisms are essential to ensure that only authorized IoT devices can interact with the blockchain network and submit data. Traditional username/password authentication methods are not used in a blockchain-based system.

Instead, cryptographic techniques are employed to establish device identity and authentication. Each IoT device has its unique public-private key pair. When submitting data to the blockchain, the device signs the data with its private key to generate a digital signature. Smart contracts on the blockchain can then verify the signature using the device's public key to ensure data authenticity.

In Solidity

```solidity
// Example of data validation using digital signatures in a
smart contract (Solidity)
        function verifySignature(bytes32 data, bytes memory
        signature, address deviceAddress) public view
        returns (bool) {
            return (ecrecover(data, v, r, s) == deviceAddress);
        }
```

Additionally, JSON Web Tokens (JWT) can be utilized to authenticate and authorize IoT devices with the data logger system. A device can obtain a JWT token upon successful authentication, and this token can be used for subsequent interactions with the blockchain network.

In python

```python
# Example of JWT authentication in Python using PyJWT
library
import jwt

# Generate JWT token
def generate_jwt_token(data, secret_key):
    token = jwt.encode(data, secret_key, algorithm='HS256')
    return token

# Verify JWT token
def verify_jwt_token(token, secret_key):
    try:
        data = jwt.decode(token, secret_key,
algorithms=['HS256'])
        return True, data
    except jwt.ExpiredSignatureError:
        return False, "Token has expired."
    except jwt.InvalidTokenError:
        return False, "Invalid token."
```

2. **Digital Signatures & Message Authentication Codes (MAC):**

Blockchain technology inherently addresses data tampering concerns through the use of cryptographic hashing and consensus mechanisms. Once data is recorded on the blockchain, it becomes immutable, making it resistant to unauthorized modifications. Digital signatures continue to play a significant role in confirming data authenticity. Each data transaction submitted to the blockchain is signed with the device's private key. The data logger system or other devices on the network can verify these signatures to confirm the data's origin and integrity.

In Javascript

```javascript
// Example of data verification using digital signatures in
JavaScript
const { createHash, createSign, createVerify } =
require('crypto');

function signData(data, privateKey) {
  const sign = createSign('SHA256');
  sign.update(data);
```

```
      return sign.sign(privateKey, 'hex');
    }

    function verifySignature(data, signature, publicKey) {
      const verify = createVerify('SHA256');
      verify.update(data);
      return verify.verify(publicKey, signature, 'hex');
    }
```

In a blockchain-integrated system, there is no need for additional Message Authentication Codes (MAC) as data integrity is automatically ensured by the decentralized nature of the blockchain and the cryptographic hashing of data blocks. By combining end-to-end encryption, secure authentication mechanisms, and leveraging the tamper-resistant properties of blockchain technology, the "Blockchain-Integrated IoT Data Logger System" establishes a highly secure and trustable environment for IoT data logging. The application of cryptographic techniques, coupled with blockchain's immutability, ensures the integrity, confidentiality, and authenticity of IoT data throughout its lifecycle on the blockchain network. This robust approach helps prevent unauthorized access, data tampering, and strengthens the overall security posture of the system.

C. **Blockchain Integration:**

Integrating blockchain technology into the "Blockchain-Integrated IoT Data Logger System" is a pivotal step that enhances data security, transparency, and trust in the system. Blockchain provides a decentralized and immutable ledger, ensuring the integrity of IoT data. Let us explore the technical aspects of blockchain integration in detail:

1. **Choosing a Suitable Blockchain Platform:**

Selecting the right blockchain platform is crucial for the success of the IoT data logger system. Factors to consider include:

- **Performance:** Evaluate the platform's scalability and transaction throughput to ensure it can handle the anticipated volume of IoT data. Some blockchain platforms, like Ethereum, have faced scalability challenges due to high transaction loads.
- **Consensus Mechanism:** Different blockchain platforms utilize various consensus mechanisms, such as Proof of Work (PoW), Proof of Stake (PoS), and Delegated Proof of Stake (DPoS). The

chosen mechanism should align with the desired level of decentralization, energy efficiency, and security.

- **Developer Ecosystem:** A strong and active developer community is essential for ongoing support, documentation, and innovation. Consider platforms with well-established development tools and libraries.

- **Smart Contract Capability:** Ensure that the blockchain platform supports smart contracts, as they play a critical role in defining data validation and logging rules, for instance, Ethereum is a popular choice for smart contract-enabled blockchains, while newer platforms like Binance Smart Chain (BSC) and Polka Dot offer different trade-offs in terms of performance and consensus mechanisms.

2. **Developing Smart Contracts using Solidity or other Programming Languages:**

Smart contracts are self-executing contracts with predefined rules and logic that automatically execute when specific conditions are met. In the context of the IoT data logger system, smart contracts can be used to validate incoming data, ensure data integrity, and automate the logging process.

Solidity is the most widely used programming language for writing smart contracts on Ethereum. However, other blockchain platforms may support different languages like Rust, Vyper, or JavaScript.

A smart contract for data validation might look like this in Solidity:

In Solidity

```solidity
// Example of a smart contract for data validation and logging
pragma solidity ^0.8.0;

contract DataLogger {
    address private owner;

    modifier onlyOwner() {
        require(msg.sender == owner, "You are not the owner");
        _;
    }

    constructor() {
```

```
        owner = msg.sender;
    }

    struct SensorData {
        uint256 timestamp;
        uint256 temperature;
        uint256 humidity;
    }

    mapping(address => SensorData[]) private dataLogs;

    function logData(uint256 _timestamp, uint256
_temperature, uint256 _humidity) public onlyOwner {
        SensorData memory newData = SensorData(_timestamp,
_temperature, _humidity);
        dataLogs[msg.sender].push(newData);
    }

    function getDataLogLength() public view returns (uint256)
{
        return dataLogs[msg.sender].length;
    }

    function getDataLog(uint256 index) public view returns
(SensorData memory) {
        require(index < dataLogs[msg.sender].length, "Invalid
index");
        return dataLogs[msg.sender][index];
    }
}
```

3. **Integrating the Data Logger System with the Blockchain Network:**

To interact with the blockchain network and deploy smart contracts, the data logger system must integrate with the blockchain platform's APIs or Software Development Kits (SDKs). These interfaces provide functions to interact with smart contracts, read data from the blockchain, and send transactions. For Ethereum, web3.js is a widely used JavaScript library for interacting with the Ethereum blockchain. Other platforms have their respective SDKs for similar purposes.

For example, using web3.js in JavaScript:

In JavaScript

```javascript
// Example of integrating the data logger system with the
Ethereum blockchain
const Web3 = require('web3');
const contractAbi = require('path/to/smart_contract_abi.json');

const web3 = new Web3('http://localhost:8545'); // Connect to
a local Ethereum node
const contractAddress = '0x1234567890abcdef...'; // Address
of the deployed smart contract
const contract = new web3.eth.Contract(contractAbi,
contractAddress);

// Function to call the smart contract to log data
async function logData(timestamp, temperature, humidity) {
  const accounts = await web3.eth.getAccounts();
  const transaction = contract.methods.logData(timestamp,
temperature, humidity);
  const gas = await transaction.estimateGas();
  const options = {
    from: accounts[0],
    gas
  };
  const receipt = await transaction.send(options);
  return receipt;
}
```

By selecting an appropriate blockchain platform, developing smart contracts for data validation, and integrating the data logger system with the blockchain network using APIs or SDKs, the IoT data logger system can securely and transparently store IoT data on the blockchain. This integration empowers the system with a tamper-resistant and decentralized data storage solution, ensuring data integrity and trustworthiness throughout the data lifecycle.

D. **Identity and Access Management in the "Blockchain-Integrated IoT Data Logger System"**

Implement a Public Key Infrastructure (PKI) to manage and verify the identity of IoT devices:

A Public Key Infrastructure (PKI) is a critical component for managing the identity and security of IoT devices in the data logger system. PKI provides a framework to issue, distribute, and validate digital certificates that contain public keys and device identity information. These certificates are essential for ensuring secure communication and data integrity between IoT devices and the data logger system.

Here is how the PKI can be implemented:

1. **Device Registration:** When an IoT device is onboarded to the system, it should be registered with the PKI. During registration, the device generates a unique public-private key pair. The public key is included in a certificate request and sent to the Certificate Authority (CA).
2. **Certificate Issuance:** The CA verifies the authenticity of the certificate request and the identity of the device. Upon successful verification, the CA issues a digital certificate containing the device's public key and identity details. The certificate is signed by the CA's private key to establish its authenticity.
3. **Certificate Distribution:** The issued certificate is securely delivered to the IoT device. The device stores the certificate and its private key securely, which it will use to sign data and authenticate itself when interacting with the data logger system.
4. **Certificate Validation:** When an IoT device communicates with the data logger system, it presents its digital certificate. The system verifies the certificate's authenticity using the CA's public key. This process ensures that the device is a trusted and registered participant in the network.

The PKI implementation establishes a secure identity framework for IoT devices, preventing unauthorized devices from accessing the system and ensuring that data originates from verified sources.

5. **Employ access control lists (ACLs) in smart contracts:** To enforce data access permissions, Access control lists (ACLs) are mechanisms used in smart contracts to manage and enforce data access permissions within the data logger system. ACLs allow specific actions or data to be accessible only by authorized entities, enhancing data privacy and security.

Here's how ACLs can be employed in smart contracts:

- **Data Access Control:** Within the smart contract that handles data logging, define functions to control access to data. For example, certain data may only be accessible by specific IoT devices or authorized users. Each data log entry can be associated with an ACL specifying who can read or modify it.
- **Permission Management:** Implement functions to manage ACLs, such as adding or removing authorized entities. The contract owner or designated administrators can have special privileges to modify the ACL.
- **Access Verification:** Before allowing any data, read or write operation, the smart contract should check the sender's identity against the ACL. If the sender is authorized, the operation is permitted; otherwise, it is rejected.

In Solidity

```solidity
// Example of access control using ACLs in a smart contract
(Solidity)
pragma solidity ^0.8.0;

contract DataLogger {
    address private owner;
    mapping(address => bool) private accessControlList;

    modifier onlyAuthorized() {
        require(accessControlList[msg.sender], "You are not
authorized");
        _;
    }

    constructor() {
        owner = msg.sender;
    }

    function grantAccess(address _device) public onlyOwner {
        accessControlList[_device] = true;
    }

    function revokeAccess(address _device) public onlyOwner {
        accessControlList[_device] = false;
    }
```

```
        function logData(uint256 _temperature, uint256 _humidity)
    public onlyAuthorized {
        // Log the data
    }

        function getDataLog(uint256 _index) public view
    onlyAuthorized returns (uint256, uint256) {
        // Retrieve data at the specified index
    }
}
```

By implementing a robust PKI for device identity management and employing access control lists (ACLs) in smart contracts to regulate data access permissions, the "Blockchain-Integrated IoT Data Logger System" ensures strong data security and access control. Only authenticated and authorized IoT devices can interact with the data logger system, enhancing the overall security and trustworthiness of the IoT data logging process.

Challenges Enforcing International Mining Regulations

Enforcing international mining regulations, such as the Organisation for Economic Co-operation and Development (OECD) Due Diligence Guidance, is challenging due to several factors. One significant obstacle is the manual handling involved in the process, making it challenging to track the minerals' origins and movements along the supply chain accurately. However, the greatest challenge arises when different resource streams get intermingled downstream along the supply chain, making it even more difficult to trace the minerals' provenance. To counter this issue, many firms have adopted a fool-proof approach of segregating their supply chain, but this has limitations, and a more comprehensive solution is needed.

Blockchain technology provides a solution to these challenges by creating a platform where those tracking the data and following the OECD guidelines can place their certifications in a system. The platform's transparency enables the data to be visible to all participants along the supply chain, creating trust and reducing the likelihood of fraudulent activity. By using blockchain technology, the MineHub Platform aims to establish a more comprehensive system that tracks minerals' movements and origins throughout the entire supply chain. This approach will significantly increase the transparency and accountability of the mining industry, enabling the effective enforcement of international mining regulations such as the OECD Due Diligence Guidance.

IBM's Focus on Cobalt in Initial Pilot.

IBM chose to focus on cobalt in its initial pilot because it is a crucial component of batteries used in electric cars and electronics manufacturing. The majority of the world's cobalt supply comes from the Democratic Republic of the Congo (DRC), which has faced criticism for its mining labor practices, including child labor and unsafe working conditions. IBM aims to address these issues by introducing more transparency and accountability into the cobalt supply chain. The increasing demand for cobalt has put significant pressure on regions such as the DRC, leading to the use of unethical sourcing methods. As a result, it is essential to ensure that the cobalt used in batteries and electronics is sourced responsibly. IBM believes that blockchain technology can play a vital role in ensuring the responsible sourcing of cobalt by creating a transparent supply chain that tracks the mineral's origin and movements.

After completing the cobalt pilot, IBM plans to apply the knowledge gained to other conflict minerals used in battery production and electronics manufacturing. By leveraging blockchain technology, IBM aims to create a more sustainable and ethical supply chain for critical minerals and metals, reducing the likelihood of unethical practices and promoting responsible sourcing.

Onboarding process for new participants in blockchain based mining supply chain network

IBM's blockchain aims to revolutionize the mining industry's supply chain by utilizing the transparency and immutability of blockchain technology. In doing so, they have taken inspiration from successful blockchain networks such as TradeLens and IBM Food Trust. They believe that by simplifying the opt-in process for members, they can promote transparency and efficiency throughout the supply chain. Mobile apps and various APIs will make it incredibly simple for members to upload relevant information. It is important to not only connect with industrial-scale miners but also with hundreds of thousands of artisanal miners who may not have the resources to develop their own systems or access information via traditional means. Their goal is to streamline the process, making it easy for members to input the necessary information to achieve end-to-end transparency.

IBM's Experience in Applying Food Trust & TradeLens to Mining Blockchain

IBM's experience in developing blockchain networks for the food industry, such as IBM Food Trust, has provided valuable insights for the mining industry's supply chain. One similarity between responsible mining and ensuring food safety is the need to track the product's origin and how it is handled throughout the supply chain. Both agriculture and mining have producers of varying scales, from small artisanal operations to large industrial-scale producers, and IBM's platform needs to cater to all of them.

IBM's extensive work in building blockchain networks has emphasized the importance of creating an open, industry-wide standard that fosters trust and encourages participation from various brands. This lesson is critical for scaling blockchain networks while maintaining trust. For example, in the mining industry, IBM's experience on TradeLens has demonstrated the benefits of creating a shared platform that includes all parties involved in the supply chain. By doing so, it increases transparency and accountability, enabling efficient tracking and tracing of materials.

Smart Contracts: Overview & Advantages over Traditional Contracts

Smart contracts are self-executing computer programs that can be used to automate the process of executing a contract. They are written in code and are stored on a blockchain network, which makes them highly secure, transparent, and tamper-proof. The use of smart contracts eliminates the need for intermediaries, reducing transaction costs and speeding up the execution of contracts. The mining industry can benefit greatly from the implementation of smart contracts. Smart contracts can be used to automate the process of buying and selling minerals, tracking the origin of minerals, and automating supply chain management. For example, smart contracts can be used to automatically release payment once a mineral shipment has been received and verified, reducing the risk of fraud and error. Additionally, smart contracts can be used to track the origin of minerals, ensuring that they are ethically sourced and comply with regulations. Furthermore, smart contracts can be used to automate supply chain management, ensuring that all stakeholders in the supply chain are connected and informed in real-time. This can help reduce delays, errors, and costs associated with manual supply chain management.

How do Smart Contracts work? Here are some points to consider when defining Smart Contracts:

1. Smart contracts are self-executing programs that automatically enforce the terms of an agreement between parties.
2. They are built on blockchain technology, which provides a tamper-proof and transparent ledger of all contract-related activities.

3. Smart contracts eliminate the need for intermediaries, such as lawyers, brokers, and notaries, reducing transaction costs and speeding up the process.
4. They are designed to execute specific actions automatically once certain conditions are met, such as releasing payment once the goods have been received and verified.
5. Smart contracts can be customized to fit the specific needs of different industries and use cases.
6. They can be programmed to trigger alerts and notifications, reducing the risk of fraud, and ensuring compliance with legal requirements.
7. Smart contracts can be audited and verified by all parties involved, increasing transparency and trust in the agreement.
8. They are immutable, meaning that once the contract is executed, the terms cannot be changed, ensuring that all parties abide by the agreement.

In comparison to traditional contracts, smart contracts have several advantages:

1. **Faster and more efficient execution:** Smart contracts eliminate the need for intermediaries and manual execution, reducing the time required to execute a contract.
2. **Increased transparency:** Smart contracts are built on blockchain technology, providing a transparent and tamper-proof record of all contract-related activities.
3. **Reduced costs:** Smart contracts eliminate the need for intermediaries, reducing transaction costs and fees.
4. **Enhanced security:** Smart contracts are designed to be secure, with built-in features to prevent fraud and ensure compliance with legal requirements.
5. **Immutable:** Once a smart contract is executed, the terms cannot be changed, ensuring that all parties abide by the agreement.

Smart Contracts in Mining: Automating Mineral Trading, Traceability & Supply Chain Management

1. The mining industry is a vast sector that involves the extraction and processing of minerals and other natural resources. In recent years, the use of smart contracts has gained momentum in the industry as a means of automating various processes and increasing efficiency.
2. A smart contract is a self-executing program that facilitates, verifies, or enforces the negotiation or performance of a contract. It operates on a

blockchain, which is a decentralized digital ledger that records transactions and provides transparency and security.

3. One potential use case for smart contracts in the mining industry is in automating the buying and selling of minerals. Traditionally, the process of buying and selling minerals involves multiple parties, including buyers, sellers, and intermediaries, and requires extensive paperwork and manual processes. By using smart contracts, parties can automate the process of exchanging minerals, eliminating the need for intermediaries, and reducing transaction costs.

4. Another potential use case is in tracking the origin of minerals. Many consumers are increasingly interested in knowing the origin of the products they buy, including minerals used in electronics, jewellery, and other industries. By using smart contracts to track the origin of minerals, mining companies can provide transparent and verifiable information about where the minerals came from, ensuring responsible sourcing practices.

5. Smart contracts can be used to automate supply chain management in the mining industry. The mining industry is highly dependent on a complex supply chain that involves multiple stakeholders, including suppliers, transporters, and customers. By using smart contracts to automate supply chain management, mining companies can streamline processes, reduce costs, and improve efficiency.

Automated Contract Execution with Smart Contracts

Contract Execution

In traditional contracts, the terms of the agreement are written in natural language and are often subject to interpretation. This can lead to disputes and legal battles if one party feels that the terms have been violated.

Smart contracts, on the other hand, are written in code and are executed automatically once certain conditions are met. The terms of the contract are encoded into the code, which eliminates the need for interpretation and ensures that the terms are executed as written.

To illustrate this concept, let us use the example of a smart contract that facilitates a bet between two parties. The terms of the bet, including the amount of the wager and the conditions for determining the winner, are written into the code of the smart contract.

Once the football match is over and the winner has been determined, the smart contract will automatically execute the terms of the bet. This could involve transferring the funds from the losing party to the winning party, or taking some other action that was specified in the contract.

The following diagram illustrates how a smart contract works in the context of contract execution:

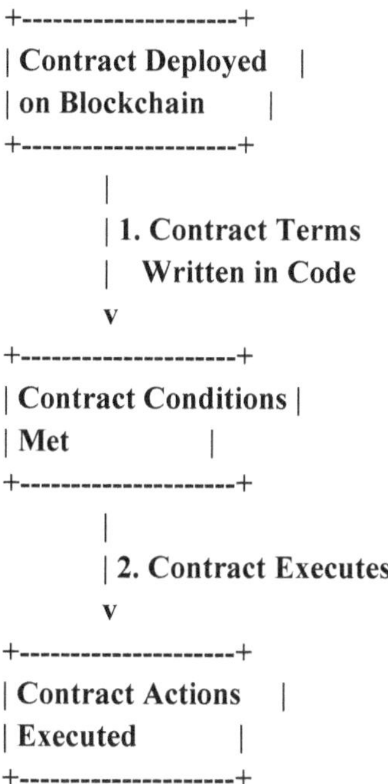

As shown in the diagram, the smart contract is deployed on a blockchain, which provides a secure and transparent platform for executing contracts. The terms of the contract are written in code, which eliminates the need for interpretation and ensures that the terms are executed as written. Once the conditions of the contract are met, the smart contract automatically executes the specified actions, which could involve transferring funds or taking some other action specified in the contract.

Automating Contract Execution in the Mining Industry

1. Smart contracts can be used to automate contract execution in the mining industry by executing specific actions when certain conditions are met.

2. For example, smart contracts can be programmed to release payment once the mineral shipment has been received and verified by the buyer.
3. Similarly, smart contracts can be programmed to transfer ownership of the minerals from the seller to the buyer once payment has been received and verified.
4. Smart contracts can also be used to automate the process of filing and resolving disputes related to mineral shipments.

Benefits of Smart Contract Execution

1. The use of smart contracts for contract execution offers several benefits, including increased efficiency, reduced costs, and improved accuracy.
2. By automating the process of contract execution, smart contracts can eliminate the need for intermediaries, reducing transaction costs and increasing efficiency.
3. Smart contracts can also be programmed to execute contracts more quickly and accurately, reducing the risk of errors or disputes.

Challenges of Smart Contract Execution

1. While smart contracts offer many benefits, there are also challenges associated with their use.
2. One challenge is regulatory uncertainty, as the legal framework surrounding smart contracts is still evolving.
3. Another challenge is the need for technical expertise, as smart contracts require specialized knowledge to develop and implement.

Case Studies

1. Several mining companies have already implemented smart contracts for contract execution, offering insights into their use and benefits.
2. For example, BHP Billiton, the world's largest mining company, has developed a smart contract platform to manage its supply chain, reducing transaction costs and increasing efficiency.

Revolutionizing Dispute Resolution in the Mining Industry with Smart Contracts: Streamlining Contract Enforcement & Minimizing Legal Costs

Dispute resolution is an essential aspect of any industry, including the mining industry. When two parties enter into a contract, disputes may arise regarding the terms and conditions of the contract. Such disputes can lead to litigation, which can be time-consuming, expensive, and damaging to the relationship between the parties involved. Therefore, finding a more efficient way to resolve disputes is crucial. One potential solution to this problem is the use of smart contracts. Smart contracts are self-executing contracts with the terms of the agreement between buyer and seller being directly written into lines of code. The code and the agreements contained therein exist on a blockchain network, which enables the automated and irreversible execution of contract terms. In the mining industry, smart contracts can help to resolve disputes by automatically enforcing the terms and conditions of the contract. For example, if one party fails to deliver goods or services as agreed, the smart contract can automatically trigger a penalty or withholding of payment, without the need for intermediaries or costly litigation.

Smart contracts can also facilitate the resolution of disputes by providing a transparent and immutable record of all contract transactions. This can help to avoid disputes over who did what, when, and at what cost. By providing an auditable trail of all contract activities, smart contracts can reduce the likelihood of disputes arising in the first place. Overall, the use of smart contracts in the mining industry has the potential to streamline and simplify contract enforcement, reduce the need for intermediaries, and decrease the likelihood of costly and time-consuming litigation. As such, incorporating smart contracts into the mining industry's contract management process is an area worth exploring.

Challenges of Implementing Smart Contracts in Mining Industry

The implementation of smart contracts in the mining industry has the potential to revolutionize the way contracts are executed and managed. However, the adoption of this technology is not without its challenges. One challenge associated with implementing smart contracts is the need for standardized contract templates. The mining industry has many different types of contracts, including supply agreements, service contracts, and joint venture agreements. Each of these contracts has its own unique terms and conditions. To enable the automation of contract execution, smart contract templates need to be developed that can accommodate the diverse range of contract types used in the mining industry.

Another challenge is ensuring that smart contracts comply with legal and regulatory requirements. Smart contracts are self-executing computer programs that do not require human intervention. This can create a challenge in ensuring that smart contracts comply with legal and regulatory requirements, particularly when it comes to data privacy, intellectual property, and anti-trust regulations.

Therefore, it is essential to work with legal experts to ensure that smart contracts comply with all relevant laws and regulations. Integration with existing systems is another potential challenge associated with implementing smart contracts in the mining industry. The mining industry has many existing systems for managing contracts, such as enterprise resource planning (ERP) systems and customer relationship management (CRM) systems. To fully realize the benefits of smart contracts, they need to be integrated with these existing systems. This can be a challenging task, particularly when different systems use different data formats and protocols. The adoption of smart contracts also requires a certain level of technical expertise. Stakeholders involved in the implementation process, including mining companies, suppliers, and service providers, need to have a good understanding of blockchain technology and smart contracts. This requires significant investment in training and education, which can be a barrier to adoption.

Finally, the adoption of smart contracts requires a change in mindset. Smart contracts require stakeholders to place trust in the technology rather than relying on intermediaries to enforce contracts. This change in mindset can be challenging, particularly for stakeholders who are used to traditional contract management methods.

Smart Contract's Future in Mining Industry: Self-Executing Contracts & Blockchain-based DAOs

The adoption of smart contracts in the mining industry is still in its early stages, but there is significant potential for future developments. In this section, we will discuss some potential future developments of smart contracts in the mining industry. One potential development is the use of self-executing contracts. Self-executing contracts are smart contracts that can operate autonomously without human intervention. These contracts can be programmed to automatically trigger actions based on certain conditions being met. For example, a self-executing contract could be used to automatically trigger a payment to a supplier when a shipment of raw materials is received. This would reduce the need for human intervention and increase the efficiency of contract management in the mining industry. Another potential development is the use of blockchain-based decentralized autonomous organizations (DAOs) for mining operations. A DAO is a type of organization that operates based on smart contracts and is governed by a set of rules encoded in its code. DAOs can operate autonomously without human intervention and can be used for a range of functions, including mining operations. In a DAO-based mining operation, smart contracts would be used to manage all aspects of the mining process, including equipment maintenance, raw material procurement, and resource allocation. The rules governing the DAO would be encoded in its code and would be transparent and immutable, ensuring that all stakeholders have equal access to information and the decision-making process. Another potential development is the use of smart contracts for sustainability and environmental management in the mining industry. Smart contracts can be programmed to ensure that mining operations comply with environmental regulations and sustainability standards. For example, smart contracts could be used to monitor emissions from mining operations and trigger actions to reduce emissions when thresholds are exceeded. *(Ref. No.: 310- 327)*

Closing Thoughts

Smart contracts have revolutionized the mining industry by automating payment processes, enforcing regulations, and streamlining logistics. They offer a decentralized and secure solution, addressing inefficiencies in the supply chain, modernizing outdated trading practices, and tackling the challenges of enforcing international mining regulations. IBM's focus on cobalt in their initial pilot demonstrates the potential of blockchain to revolutionize the mining industry. Their experience in applying platforms like Food Trust and TradeLens has helped create a robust and efficient blockchain-based mining supply chain network, with an onboarding process for new participants.

The advantages of smart contracts over traditional contracts are evident in the mining sector. They automate mineral trading, traceability, and supply chain management, while also facilitating seamless contract execution. Smart contracts offer numerous benefits, including increased transparency and trust, paving the way for a more streamlined and transparent industry. However, challenges exist in implementing smart contracts in mining. Overcoming these hurdles, such as ensuring regulatory compliance and integrating legacy systems, is crucial for widespread adoption and successful implementation. Smart contracts also hold the potential to revolutionize dispute resolution in mining. By streamlining contract enforcement and minimizing legal costs, they enhance transparency and trust, ensuring fair and efficient resolutions.

Looking ahead, the future of smart contracts in the mining industry is promising. Self-executing contracts and blockchain-based decentralized autonomous organizations (DAOs) offer greater efficiency, autonomy, and security, reshaping the industry. In conclusion, smart contracts have opened new horizons for the mining industry, offering automation, transparency, and trust. By embracing this transformative technology, the mining sector can embrace a future that is more efficient, sustainable, and equitable.

Chapter 11: Future of Blockchain Technology in the Mining Industry

"Make no mistake, blockchain will be a game-changer, but it is not the answer to every business opportunity or problem."

— *Julie Sweet*

Introduction - The Transformational Potential of Blockchain in the Mining Domain

The mining industry faces significant challenges in areas such as supply chain management, asset tracking, security, compliance, and sustainability. However, the advent of blockchain technology has the potential to revolutionize the mining landscape and provide solutions to these pressing issues. In this chapter, we explore the future of blockchain technology and its transformative impact on the mining industry.

One of the key areas where blockchain technology can bring about substantial improvements is supply chain management. By utilizing blockchain's inherent features of transparency and traceability, stakeholders can track the origin and movement of minerals or commodities. This enables ethical sourcing practices and mitigates the risk of fraudulent activities throughout the supply chain.

Another aspect that can be greatly enhanced through blockchain technology is asset management in mining operations. By implementing blockchain-based platforms and smart contracts, companies can automate critical processes like equipment maintenance, inventory tracking, and provenance verification. This automation leads to optimized operations, reduced downtime, and improved cost-efficiency.

The decentralized and tamper-resistant nature of blockchain technology also makes it a robust solution for ensuring security and compliance in the mining industry. By securely storing and sharing data related to permits, licenses, environmental impact assessments, and safety regulations, blockchain enhances trust among stakeholders and regulatory authorities.

Tokenization of mineral rights is another exciting avenue that blockchain enables. By leveraging blockchain technology, mineral rights can be tokenized, allowing for fractional ownership and trading of mining rights. This democratizes the industry and opens up investment opportunities to a wider range of stakeholders. Additionally, innovative financing models for mining projects can be created through blockchain-enabled tokenization.

Blockchain-based smart contracts offer the potential to streamline royalty payments in the mining industry. These smart contracts can automatically execute royalty payments when predefined conditions, such as the volume of extracted minerals, are met. This eliminates the need for intermediaries and manual reconciliation processes, resulting in faster and more accurate royalty disbursements.

Integration of renewable energy sources into mining operations is crucial for a sustainable future. Blockchain can play a pivotal role in this integration by enabling peer-to-peer energy trading and monitoring of energy consumption. Through blockchain technology, renewable energy solutions can be efficiently implemented, leading to a reduced carbon footprint and increased sustainability within the mining industry.

Decentralized mining pools are another area where blockchain technology can drive positive change. Traditional mining pools are often centralized and controlled by a few entities, raising concerns about fairness and security. Blockchain-based mining pools distribute the mining process across a decentralized network, ensuring a more democratic and secure approach to mining cryptocurrencies.

Certification and auditing processes within the mining industry can benefit from blockchain technology. Blockchain provides a reliable and auditable platform for recording certificates related to responsible mining practices, environmental standards, or labor conditions. This enables easy verification and fosters increased trust among stakeholders.

Efficient data sharing and collaboration are vital for driving innovation and improving decision-making in the mining industry. Blockchain-based platforms provide a secure and efficient environment for stakeholders to share information such as geological data, exploration findings, or safety protocols. By maintaining data integrity and privacy, blockchain facilitates collaboration and promotes better industry-wide cooperation.

The integration of Internet of Things (IoT) devices with blockchain technology offers real-time monitoring capabilities for mining operations. IoT sensors can collect data on parameters such as equipment health, environmental conditions, and worker safety. By recording this data on the blockchain, stakeholders gain access to reliable and transparent information, enabling informed decision-making and effective risk mitigation.

In this chapter, we will explore these exciting applications of blockchain technology and examine the potential benefits and challenges associated with its adoption in the mining industry. By harnessing the power of blockchain, the mining industry can unlock new levels of transparency, efficiency, sustainability, and collaboration, paving the way for a promising future.

Opportunities for Blockchain in Mining

1. Tracking Mineral Supply Chain

The tracking of minerals within the supply chain is a critical aspect of responsible sourcing in the mining industry. However, existing tracking methods often lack reliability and transparency, leading to issues like human rights violations, environmental harm, and unethical practices. This section explores how blockchain technology can address these challenges by providing a transparent and secure approach to tracking mineral movement throughout the supply chain. Blockchain technology offers a distributed ledger system that records all transactions related to minerals, including origin, transportation, storage, and sale. Each block within the blockchain contains a unique identifier assigned to a specific mineral or batch, enabling comprehensive tracking from the mine to the end consumer. The blockchain ledger is accessible to authorized participants, ensuring transparency and accountability. By utilizing blockchain, stakeholders in the mining industry can effectively track mineral movement, ensuring ethical sourcing practices. They can verify the mineral's origin, the mining conditions, and the parties involved in transportation and sale. The immutable nature of blockchain data provides confidence in the accuracy and reliability of the information, reducing the risk of fraud and corruption within the supply chain. Furthermore, blockchain technology facilitates compliance with international regulations concerning responsible mineral sourcing, such as the OECD Due Diligence Guidance for Responsible Supply Chains of Minerals from Conflict-Affected and High-Risk Areas. The cryptographic security of blockchain blocks makes it challenging to tamper with or modify data, strengthening the integrity of supply chain records.

2. Improved Efficiency

The mining industry is known for its complex and time-consuming processes, which often involve significant paperwork and administrative tasks. These tasks can be a drain on resources, slowing down the mining process and reducing efficiency. Blockchain technology can help streamline these processes by reducing paperwork and automating various tasks, such as tracking inventory and monitoring equipment. One way in which blockchain technology can help streamline the mining process is by reducing paperwork. Traditional mining operations involve significant amounts of paperwork, including contracts, invoices, and other legal documents. This paperwork can be time-consuming and costly to manage, and can also increase the risk of errors and fraud. By using blockchain technology, mining companies can create a shared ledger that records all transactions related to the mining operation. This eliminates the need for multiple copies of the same documents, reducing paperwork and simplifying the management of legal and financial documentation.

Another way in which blockchain technology can help streamline the mining process is by automating various tasks. For example, mining companies can use sensors and other Internet of Things (IoT) devices to collect data on equipment usage and maintenance needs. This data can be stored in the blockchain ledger, which can then be used to automate maintenance and repair tasks. By automating these tasks, mining companies can reduce downtime and increase the efficiency of their operations. Furthermore, blockchain technology can also help improve the tracking of inventory and materials. The mining industry involves the movement of large amounts of material, and keeping track of these materials can be a challenge. By using blockchain technology, mining companies can create a transparent and secure ledger that tracks the movement of materials and inventory throughout the mining process. This can help reduce the risk of errors and ensure that materials are being used efficiently.

3. Cost Reduction

The mining industry is one of the most resource-intensive industries, requiring significant capital investments and operational expenses. As a result, mining companies are always looking for ways to reduce costs and increase profitability. Blockchain technology can help achieve these goals by enabling more efficient supply chain management and reducing the risk of fraud and corruption. One way in which blockchain technology can reduce the cost of mining is by improving supply chain management. The supply chain for mining operations is complex, involving multiple stakeholders, such as suppliers, transporters, and regulators. Managing this supply chain can be challenging, with issues such as delayed deliveries, lost shipments, and supply chain disruptions often leading to increased costs. By using blockchain technology, mining companies can create a transparent and secure ledger that records all transactions related to the mining supply chain. This can help reduce the risk of delays and disruptions, improving the efficiency of the supply chain and reducing costs.

Another way in which blockchain technology can help reduce the cost of mining is by reducing the risk of fraud and corruption. The mining industry is vulnerable to fraud and corruption, with issues such as overbilling, underreporting, and illegal activities often leading to significant financial losses. By using blockchain technology, mining companies can create a tamper-proof ledger that records all transactions related to the mining operation. This can help prevent fraudulent activities, reduce the risk of corruption, and increase transparency and accountability in the mining industry. Moreover, blockchain technology can also help reduce costs associated with compliance and regulatory requirements. The mining industry is subject to various regulations and requirements, such as environmental regulations, labor laws, and tax obligations. Compliance with these requirements can be costly, with fines and penalties often leading to significant financial losses. By using blockchain technology, mining companies can create a transparent and auditable record of compliance activities. This can help reduce the cost of compliance by simplifying reporting requirements and ensuring that regulatory obligations are met.

4. Better Transparency

The mining industry is one that has faced increasing scrutiny in recent years due to concerns over issues such as environmental damage, human rights abuses, and corruption. As a result, there is a growing demand for greater transparency in the industry, with stakeholders and regulators seeking more information about mining operations and their impacts. Blockchain technology can help address this demand by providing greater transparency in the mining industry through the use of shared ledgers. By recording all transactions on a shared ledger, blockchain technology provides a transparent and secure way to track the movement of minerals and other materials throughout the mining supply chain. This includes recording transactions related to the extraction, processing, transportation, and sale of minerals. The shared ledger is distributed across multiple nodes or computers, which means that it cannot be altered by any one party without the consensus of the network. This ensures that the ledger is tamper-proof and provides an accurate record of all transactions.

The use of blockchain technology in the mining industry can help build trust between stakeholders, including mining companies, governments, communities, and investors. Mining companies can use blockchain technology to demonstrate their commitment to responsible and sustainable practices by providing transparent records of their operations. This can help build trust with local communities, who may be concerned about the impact of mining on their environment and way of life. It can also help build trust with investors, who are increasingly looking for companies that are committed to sustainability and responsible practices. Furthermore, blockchain technology can help ensure compliance with regulations by providing a transparent record of all activities related to the mining operation. Mining companies are subject to various regulations and requirements, including environmental regulations, labor laws, and tax obligations. By using blockchain technology, mining companies can create an auditable record of their compliance activities, providing regulators with a transparent and tamper-proof record of their operations.

Challenges for Blockchain in Mining

1. **Lack of Standardization**

   The mining industry operates in a complex and highly regulated environment. There are various regulations and standards that mining companies must comply with, and there are often different requirements in different regions and jurisdictions. This makes it difficult to create a standardized system for tracking mineral movements that can be implemented across the industry. Blockchain technology relies on a shared ledger, which means that all stakeholders must agree on a single set of standards and protocols for the system to work effectively. Without standardization, there is a risk of fragmentation, which could lead to inefficiencies and errors. Therefore, creating a standardized system for tracking mineral movements that is based on blockchain technology will require collaboration and coordination among stakeholders.

2. **Resistance to Change**

   The mining industry has a reputation for being slow to adopt new technologies. This is partly due to the long-standing processes and systems that are in place, which can be difficult to change. Additionally, there may be resistance to change due to concerns about the potential risks and uncertainties associated with new technologies. Blockchain technology is still a relatively new and untested technology, which could lead to resistance to its adoption in the mining industry. However, there are also many potential benefits of blockchain technology that could incentivize mining companies to embrace the technology, such as improved efficiency, increased transparency, and reduced costs.

3. **Technical Limitations**

   Blockchain technology requires significant computing power, which could be a challenge for mining companies that are already struggling with rising energy costs. Additionally, the technical expertise required to implement and maintain blockchain technology may be a barrier for some mining companies. One potential solution to this challenge is the use of blockchain-as-a-service (BaaS) platforms, which allow companies to access blockchain technology without having to invest in the infrastructure and technical expertise required to build and maintain a blockchain system from scratch. BaaS platforms can help mining companies overcome some of the technical limitations associated with blockchain technology.

4. **Data Privacy Concerns**

One of the primary benefits of blockchain technology is its high level of security. However, there are still concerns about data privacy when it comes to sensitive information such as mining licenses and permits. The use of blockchain technology may require mining companies to share sensitive information with third-party providers, which could raise concerns about data privacy and security. To address these concerns, mining companies may need to invest in additional security measures, such as encryption and multi-factor authentication. Additionally, they may need to establish clear protocols for sharing information with third-party providers and ensure that these providers have adequate security measures in place.

The challenges associated with implementing blockchain technology in the mining industry are significant, but not insurmountable. By collaborating with stakeholders and investing in the necessary infrastructure and expertise, mining companies can overcome these challenges and reap the many potential benefits of blockchain technology, including increased transparency, improved efficiency, and reduced costs.

Future of Blockchain in Mining

The mining industry is on the cusp of a transformative future, driven by the increasing adoption of blockchain technology. Several key factors are propelling the integration of blockchain in mining, shaping its potential applications and benefits. These factors include:

1. **Increased Demand for Minerals**

The demand for minerals, including copper, lithium, and cobalt, is expected to rise considerably in the coming years, due to the growth of industries such as electric vehicles, renewable energy, and advanced technologies. These minerals are critical for the production of batteries, which are necessary for the functioning of these industries. This rise in demand poses a challenge for mining companies, which will need to find ways to increase their production while keeping costs low. One technology that could help mining companies achieve this goal is blockchain. Blockchain is a distributed ledger technology that allows for secure and transparent record-keeping of transactions. In the mining industry, blockchain technology could be used to improve supply chain management, increase efficiency, and reduce costs. For example, blockchain could be used to track the origin of minerals, providing transparency in the supply chain, and helping to ensure that the materials are ethically sourced. This could help to reduce the risk of human rights abuses and environmental damage associated with the mining of minerals. Furthermore, blockchain technology could be used to streamline administrative processes, such as contracts, invoices, and payments, reducing the time and cost associated with these tasks. This could help to increase the speed of transactions and improve the overall efficiency of the mining process. Another potential application of blockchain in the mining industry is in the monitoring of equipment and machinery. Blockchain technology could be used to track the performance and maintenance of mining equipment, helping to reduce downtime and improve overall productivity.

Provenance is a blockchain-based platform that focuses on ensuring transparency and traceability in supply chains. With its application in the mining industry, Provenance aims to address the challenges related to responsible sourcing and ethical practices. By utilizing blockchain technology, Provenance enables the tracking of minerals from their extraction sites to the end consumer, ensuring that they are sourced from legitimate and environmentally sustainable locations. The platform records and verifies every transaction on the blockchain, creating an immutable and transparent record of the mineral's journey through the supply chain. This transparency helps build trust among stakeholders, promotes accountability, and mitigates the risks associated with human rights abuses and environmental damage. Provenance empowers consumers and businesses to make informed decisions by providing them with reliable information about the origin, authenticity, and sustainability of the minerals they procure or use. By leveraging blockchain technology, Provenance contributes to driving positive change in the mining industry, promoting responsible practices, and ensuring a more sustainable future.

2. **Growing concerns about sustainability**

Mining is an essential industry that provides the raw materials for many of the products and technologies that we rely on every day. However, the mining process can have significant environmental and social impacts, including habitat destruction, water pollution, and human rights abuses. As a result, there is a growing awareness of the need for greater transparency and accountability in the mining industry. Blockchain technology offers a potential solution to this challenge by providing a secure and transparent platform for tracking the movement of minerals from the mine to the end consumer. By using blockchain technology, mining companies can record and verify every step of the supply chain, from the extraction of minerals to the production of finished products. This transparency is crucial for reducing the risk of human rights abuses and environmental damage associated with the mining of minerals. By using blockchain technology, mining companies can ensure that the materials they are using are ethically sourced, and that they are not contributing to environmental degradation or social harm. For example, a mining company could use blockchain technology to record the origin of minerals and track their movement through the supply chain. This would allow stakeholders to verify that the minerals were sourced from ethical and sustainable sources, and that they were not associated with human rights abuses or environmental damage.

MineBloc is a ground-breaking blockchain-based platform revolutionizing mineral supply chains in real time. With a focus on transparency and accountability, MineBloc tracks and verifies every step of the supply chain, providing stakeholders with unprecedented visibility. One notable application of MineBloc is its implementation in the cobalt supply chain. Cobalt, a crucial component in electric vehicle and renewable energy technologies, has faced concerns regarding human rights abuses and unethical practices. Using blockchain technology, MineBloc assigns a unique digital identity to each cobalt batch, tracking its journey from mining site to battery manufacturer. Stakeholders can access the platform to ensure the cobalt's authenticity and ethical sourcing, significantly reducing fraud risks. Incorporating smart contracts, MineBloc automates transactions and certifications, streamlining the supply chain and eliminating intermediaries.

MineBloc's success in transforming the cobalt supply chain highlights the potential of blockchain technology to enhance transparency and responsibility in the mining industry. It sets a precedent for similar improvements in other mineral supply chains.

3. **Need for supply chain efficiency**

Mining is a complex industry that involves multiple stakeholders, including miners, processors, manufacturers, and end-users. The supply chain in mining can be complicated, involving numerous steps from the extraction of minerals to the delivery of finished products to end-users. This complexity can result in inefficiencies, delays, and errors, which can lead to increased costs and reduced productivity. Blockchain technology offers a potential solution to these challenges by providing a secure and transparent platform for tracking the movement of minerals through the supply chain. By using blockchain technology, mining companies can automate various tasks, such as tracking inventory, monitoring equipment, and recording transactions, which can help to streamline the supply chain and reduce paperwork. For example, a mining company could use blockchain technology to record the movement of minerals from the mine to the processing plant. This would allow stakeholders to track the minerals' progress through the supply chain and ensure that they are being handled appropriately. It would also reduce the need for manual record-keeping, which can be time-consuming and error-prone. Furthermore, blockchain technology could be used to monitor the performance of equipment and machinery used in mining operations. By using sensors and other IoT devices, mining companies can collect real-time data on equipment performance and maintenance needs. This data can then be stored on the blockchain, allowing stakeholders to monitor equipment performance, identify potential problems, and schedule maintenance tasks proactively.

4. Traceability and Transparency

Blockchain technology's potential for traceability and transparency can revolutionize the mining industry's supply chain management. The immutable and tamper-proof nature of blockchain records provides assurance that the data is accurate and reliable. By integrating blockchain technology with supply chain management systems, stakeholders can track the movement of minerals and metals from the point of extraction to the final product. Consumers are becoming more conscious of the products they buy and their impact on the environment and society. With blockchain technology's traceability, consumers can verify the authenticity of a product's claims, such as its sustainable sourcing and ethical production. By providing complete transparency in the supply chain, blockchain technology can help reduce the demand for conflict minerals and promote sustainable mining practices. As the mining industry becomes more digitally connected, we can expect blockchain technology to be integrated with IoT devices to provide real-time tracking of mining activities.

For instance, sensors can be used to monitor the location and movement of minerals and metals, and blockchain technology can be used to record this data. This can enable stakeholders to track the entire mining process and ensure that sustainable mining practices are being followed. In the future, we can expect the use of blockchain technology for traceability and transparency in the mining industry to become more widespread. The implementation of blockchain technology can be costly, but it can provide significant benefits, such as improving operational efficiency, reducing costs, and increasing stakeholder trust. As the technology continues to develop and become more accessible, we can expect more companies in the mining industry to adopt it.

5. Smart Contracts

Smart contracts have the potential to revolutionize the mining industry by reducing costs, improving efficiency, and increasing transparency. As blockchain technology and smart contracts continue to evolve, we can expect to see new use cases emerge that can further optimize mining operations. For instance, one potential use case for smart contracts in the mining industry is to automate the process of tracking and verifying compliance with environmental and social standards. Smart contracts can be programmed to automatically track the use of natural resources, emissions, and other environmental metrics. These contracts can also be used to enforce social standards, such as fair labor practices and human rights. Another potential use case for smart contracts in the mining industry is to streamline the supply chain management process. Smart contracts can automate the process of ordering and tracking materials, reducing the need for intermediaries, and improving transparency. Additionally, smart contracts can be used to automate the process of payments to suppliers, reducing the need for manual intervention and improving efficiency.

In the future, we can also expect to see the integration of smart contracts with other emerging technologies, such as artificial intelligence and the Internet of Things. For example, smart contracts can be used to trigger automated responses to environmental events, such as natural disasters, to ensure that mining operations remain safe and sustainable. As the mining industry embraces smart contracts and their integration with emerging technologies, a future of increased efficiency, cost reduction, and environmental sustainability comes into focus. The adoption of smart contracts is set to redefine the way mining operations are conducted, laying the foundation for a transparent, responsible, and technologically advanced industry.

Real-life example of smart contracts in the mining industry is the collaboration between IBM and MineHub Technologies in the gold supply chain. Together, they developed a blockchain platform that enables the transparent and traceable tracking of gold from the mine to the end consumer.

Smart contracts are utilized to automate and enforce compliance with responsible sourcing standards, ensuring that the gold is ethically and sustainably produced. These contracts automatically verify the origin of the gold, track its movement through the supply chain, and validate the adherence to responsible mining practices. By leveraging blockchain technology and smart contracts, stakeholders in the gold supply chain, including miners, refiners, and jewellers, can have real-time visibility into the entire process. This transparency helps to build trust among participants and provides consumers with the assurance that the gold they purchase is sourced responsibly.

6. Decentralized Energy Systems

The mining industry is known for its high energy consumption, with a significant portion of its operations powered by fossil fuels. However, with the increasing focus on sustainability and the need to reduce carbon emissions, the industry is exploring alternative sources of energy, such as renewable energy sources. Decentralized energy systems, which involve the use of renewable energy sources to power mining operations, have emerged as a promising solution. Blockchain technology can facilitate the development of decentralized energy systems by enabling the tracking and trading of renewable energy certificates and carbon credits. Renewable energy certificates (RECs) are tradable certificates that represent the environmental attributes of one megawatt-hour (MWh) of electricity generated from a renewable energy source. By purchasing RECs, mining companies can support the development of renewable energy projects and reduce their carbon footprint.

Carbon credits, on the other hand, represent a reduction or removal of one metric ton of carbon dioxide equivalent (CO_2e) from the atmosphere. Mining companies can earn carbon credits by implementing sustainable practices or investing in renewable energy projects. These credits can then be sold on carbon markets to other companies looking to offset their carbon emissions. In the future, we can expect the mining industry to shift towards decentralized energy systems as a means of reducing its reliance on fossil fuels and decreasing its carbon footprint. With the help of blockchain technology, mining companies can easily track and trade RECs and carbon credits, making it easier to transition to renewable energy sources. The adoption of decentralized energy systems can also improve the resilience of mining operations by reducing their exposure to volatile energy prices and grid outages.

7. Digital Identities

Digital identities are one of the emerging use cases of blockchain technology in the mining industry. A digital identity is a unique identifier that is created and stored on the blockchain. It can be used to authenticate and verify the identity of individuals and organizations in a secure and tamper-proof way. In the mining industry, digital identities can be used to create secure and tamper-proof records of mining workers, contractors, and other stakeholders. These records can include personal information, employment history, certifications, and training records. By using digital identities, mining companies can ensure that only authorized individuals have access to sensitive areas of the mining operation, reducing the risk of accidents and improving safety.

Digital identities can also be used to track training and certification records. This can help mining companies ensure that their workers are qualified and trained to perform their jobs safely and effectively. Digital identities can provide a secure and tamper-proof way to verify the credentials of workers, reducing the risk of accidents caused by unqualified or improperly trained workers. In the future, we can expect the widespread adoption of digital identities in the mining industry. The use of digital identities can improve safety and security in mining operations by enabling secure and transparent access control. It can also help mining companies track and verify the qualifications and training of their workers, reducing the risk of accidents caused by unqualified or improperly trained workers.

One real-life example of digital identities in the mining industry is the collaboration between ConsenSys and The World Economic Forum (WEF) in creating a blockchain-based digital identity system for mining workers in Sierra Leone. The project aims to provide secure and verifiable digital identities to miners, ensuring transparency and accountability in the workforce. By utilizing blockchain technology, each miner is assigned a unique digital identity that contains relevant information such as personal details, employment history, certifications, and training records. This digital identity is stored on the blockchain, making it tamper-proof and accessible to authorized parties, such as mining companies and regulatory bodies. The digital identity system enables mining companies to verify the qualifications and certifications of workers, ensuring that they are qualified for their respective roles and adhere to safety protocols. It also facilitates efficient onboarding processes by securely and digitally verifying the identity and credentials of new workers, reducing paperwork and administrative burdens.

8. Data Management

Data management is a critical aspect of the mining industry, as it involves the storage, sharing, and management of data related to mining activities, including environmental impact assessments, safety records, and financial transactions. The use of blockchain technology can significantly improve data management by providing a secure and tamper-proof way to store and share data related to mining activities. Blockchain-based data management systems can ensure the accuracy and completeness of data by providing a tamper-proof record of all data-related activities. Moreover, blockchain technology can enable the sharing of data between different stakeholders in a secure and transparent way, promoting collaboration and ensuring that all parties have access to the same information.

In the future, we can expect the adoption of blockchain-based data management systems to increase in the mining industry. Mining companies can use blockchain technology to store and share data related to their mining activities, such as environmental impact assessments, safety records, and financial transactions. By doing so, they can improve the accuracy, completeness, and transparency of data related to their mining operations. Blockchain technology can also enable the sharing of data between mining companies and other stakeholders, such as regulatory agencies, environmental organizations, and local communities. This can promote collaboration and ensure that all parties have access to the same information, improving decision-making and promoting sustainable mining practices.

Another real-life example of blockchain-based data management in the mining industry is the "Metal Digital Passport" project by Volkswagen Group, in partnership with the blockchain start-up Minespider. This initiative aims to create a transparent and traceable supply chain for raw materials used in electric vehicle (EV) batteries. The project utilizes blockchain technology to track the entire lifecycle of minerals, starting from their extraction to their incorporation into EV batteries. By recording key information such as the origin, extraction methods, processing, and transportation of minerals on the blockchain, the Metal Digital Passport ensures transparency and accountability in the supply chain. The Metal Digital Passport enables Volkswagen to verify the responsible sourcing of raw materials, such as cobalt, by tracing their journey through the supply chain. It allows stakeholders, including Volkswagen, suppliers, and consumers, to access reliable and authenticated information about the materials used in EV batteries. This blockchain-based data management system helps Volkswagen ensure that the minerals used in their vehicles are sourced ethically and in compliance with sustainability standards. It also promotes transparency and trust among consumers, providing them with information about the environmental and social impact of the materials in their EVs.

9. **Improved Certification & Auditing:**

The mining industry in India faces various challenges related to responsible mining practices, environmental sustainability, and labor conditions. Implementing blockchain-based solutions can address these challenges and enhance transparency and trust in the certification and auditing processes. Blockchain technology offers a reliable and auditable platform for recording and verifying certificates related to responsible mining practices, environmental standards, and labor conditions. These certificates can be stored on the blockchain, creating an immutable and transparent record of compliance. Stakeholders, including mining companies, regulatory authorities, and local communities, can easily access and verify these certificates, ensuring the credibility and authenticity of the information.

Reliable Certification: Blockchain platforms provide a secure and transparent record-keeping system for certifications related to responsible mining practices, environmental standards, and labor conditions. This ensures the credibility and authenticity of certificates, fostering trust among stakeholders. Example: Blockchain-based platform called "Provenance" can be used to certify and trace the origin of minerals. This platform enables the recording of certifications, such as adherence to sustainable mining practices, on the blockchain, promoting transparency and responsible sourcing.

Improved Auditing Processes: Blockchain technology allows for efficient and tamper-proof auditing of mining operations by creating an immutable record of compliance. Auditors can easily access and verify data on the blockchain, reducing manual efforts and enhancing the accuracy of audits. Example: Tech Mahindra, an Indian multinational technology company, developed a blockchain-based auditing platform for the Indian mining industry. This platform ensures transparent and accurate audits by securely recording and sharing data related to mining operations, including environmental impact assessments and safety records.

Streamlined Supply Chain Audits: Blockchain platforms enable the transparent tracking of minerals throughout the supply chain, facilitating supply chain audits. This ensures that minerals are ethically sourced and comply with regulations, reducing the risk of human rights abuses and environmental damage. Example: KrypC Technologies, an Indian blockchain company, implemented a blockchain solution for a mining consortium in India. This solution enabled end-to-end traceability of minerals, ensuring compliance with responsible sourcing standards, and facilitating supply chain audits.

Enhanced Stakeholder Collaboration: Blockchain platforms enable seamless collaboration among mining companies, regulatory authorities, and local communities by providing a shared and transparent source of information. This fosters effective communication, trust-building, and collective decision-making. Example: The state government of Andhra Pradesh in India partnered with a blockchain startup to develop a platform that enables collaboration and information sharing between mining companies and regulatory bodies. This platform enhances stakeholder engagement and promotes responsible mining practices.

Increased Transparency and Accountability: Blockchain technology ensures transparency in the certification and auditing processes by making data accessible and verifiable by all relevant stakeholders. This transparency enhances accountability, mitigates the risk of fraud, and promotes ethical mining practices. Example: Indian Rare Earths Limited (IREL), a government-owned mining company in India, explored the use of blockchain technology to improve transparency and accountability in its operations. By recording key information on the blockchain, IREL aimed to enhance stakeholder trust and ensure responsible mining practices.

Environmental Impact of the Mining Industry

The mining industry has several environmental impacts that have been the focus of concern for many years. These impacts include:

1. **Land degradation:** Mining activities involve the removal of topsoil and vegetation, resulting in severe land degradation. The loss of habitats, soil compaction, and erosion can lead to a decrease in soil fertility and the inability of vegetation to regrow, which can have long-term consequences for the environment. Land degradation can also have significant implications for local communities who rely on the land for agriculture and other uses.

2. **Water pollution:** Mining activities require large quantities of water for processing minerals and metals, and this water can become contaminated with toxic chemicals, heavy metals, and other pollutants. Contaminated water can harm aquatic life and human health, and the long-term effects on soil quality, groundwater supplies, and the ecosystem's overall health can be severe.

3. **Air pollution:** Mining activities can release toxic substances into the air, leading to respiratory problems and other health issues. Mining activities can also produce dust and particulate matter, which can have negative effects on human health and the environment.

4. **Climate change:** Mining activities are significant contributors to greenhouse gas emissions, which can lead to climate change. The mining industry relies heavily on fossil fuels for energy, and the extraction and processing of minerals and metals also require significant amounts of energy. This energy use leads to the release of

greenhouse gases, which can have significant long-term impacts on the environment, including global warming, sea-level rise, and other climate-related events.

5. **Biodiversity loss:** Mining activities can lead to the loss of biodiversity by destroying habitats and disrupting ecosystems. The removal of vegetation and the loss of habitats can lead to a decline in biodiversity, which can have severe long-term consequences for the environment. Biodiversity loss can also impact human health and the economy, as many communities rely on the services provided by ecosystems, such as clean water and air, pollination, and soil fertility.

The environmental impacts of mining are complex and interrelated, and they require comprehensive solutions to mitigate their negative effects. The use of blockchain technology is one possible solution that can help reduce the mining industry's negative impact on the environment. By promoting transparency, smart contracts, and decentralized energy systems, blockchain technology can help ensure that mining companies adhere to sustainability standards and regulations, reduce their carbon footprint, and minimize the environmental impacts of their operations. *(Ref. No.: 310- 336)*

Blockchain Technology's Role in Mitigating Environmental Impacts of Mining

Blockchain technology has the potential to mitigate the environmental impacts of mining by introducing transparency, traceability, and accountability into the industry. Here are some key points on how blockchain can achieve this, along with real-life projects:

1. **Transparency in Supply Chain:** Blockchain can provide a transparent and immutable record of mining activities, ensuring that the origin and movement of minerals are accurately tracked. This helps identify unsustainable practices and enables responsible sourcing. For instance, the startup Circulor leverages blockchain technology to trace the supply chain of minerals, specifically focusing on cobalt, which is a crucial component in electric vehicle (EV) batteries. Circulor enables the transparent tracking and verification of cobalt throughout its supply chain. The process begins with the extraction of cobalt from mines, where each batch of cobalt is assigned a unique digital identity on the blockchain. This identity includes relevant information such as the mine of origin, extraction methods, and responsible parties involved. As the cobalt progresses through the supply chain, it passes through

different stages, including processing, refining, and manufacturing. At each step, the involved entities record the transaction and update the cobalt's digital identity on the blockchain. This creates an immutable and transparent record of the cobalt's journey, ensuring its ethical sourcing. Circulor's platform also incorporates other technologies like GPS tracking, IoT sensors, and data analytics to gather additional information about the cobalt's movement and conditions. This data is then linked to the blockchain, providing a comprehensive and verifiable picture of the cobalt's supply chain.

2. **Carbon Footprint Tracking:** Blockchain can facilitate the tracking and verification of carbon emissions associated with mining operations. This enables companies to assess their environmental impact accurately and identify areas for improvement. The project CarbonX utilizes blockchain to create a marketplace for trading carbon credits, incentivizing emission reductions in mining and other industries. Mining companies can earn carbon credits by adopting sustainable practices and reducing their carbon footprint. These credits are recorded on the blockchain, ensuring transparency and traceability. By participating in the CarbonX marketplace, mining companies can generate additional revenue while driving the adoption of cleaner technologies. The project encourages sustainability and contributes to the global effort to combat climate change.

3. **Decentralized Energy Solutions:** Blockchain can support the development of decentralized energy solutions for mining operations, reducing reliance on fossil fuels, and minimizing carbon emissions. One example is the project Soluna, which combines blockchain technology with renewable energy sources to enable sustainable power generation for mining operations. By harnessing renewable energy, such as wind or solar power, Soluna aims to reduce the environmental impact of mining activities and promote sustainable practices. The blockchain technology ensures transparency and traceability in energy production and consumption. This innovative approach not only addresses the energy needs of mining operations but also contributes to the global transition towards a cleaner and greener future.

4. **Waste Management & Recycling:** Blockchain can enhance waste management and recycling efforts by creating a transparent and auditable record of waste disposal and recycling processes. This ensures proper handling of waste materials and reduces the

environmental impact. The project Provenance is exploring the use of blockchain for tracking and managing waste from mining activities.

5. **Smart Contracts for Environmental Compliance:** Blockchain-based smart contracts can automate compliance with environmental regulations, ensuring adherence to environmental standards. These contracts can trigger automatic penalties or rewards based on predefined environmental criteria. The project MineHub utilizes smart contracts to automate compliance with regulatory requirements, including environmental and social obligations.

Closing Thoughts

The mining industry faces significant challenges in supply chain management, asset tracking, security, compliance, and sustainability. However, the advent of blockchain technology offers a transformative solution that has the potential to revolutionize the industry. By harnessing the power of blockchain, the mining landscape can unlock new levels of transparency, efficiency, sustainability, and collaboration, paving the way for a promising future. Looking ahead to the near future, we can anticipate widespread adoption of blockchain technology in the mining industry. As companies witness the tangible benefits and positive impact of blockchain solutions, collaboration among stakeholders will increase, leading to the establishment of standards and best practices. This collaborative ecosystem will foster interoperability and seamless data sharing, enabling the industry to fully realize the potential of blockchain.

Furthermore, advancements in data analytics and AI will complement blockchain implementations, driving better decision-making and optimizing operations. The convergence of blockchain with other emerging technologies, such as IoT, robotics, and automation, will create synergistic opportunities, enhancing safety and efficiency in mining processes. The near future also holds promise for the evolution of blockchain-based financing models, democratizing access to capital for mining projects. Additionally, blockchain-enabled supply chain transparency will contribute to responsible sourcing and environmental conservation, aligning with global sustainability initiatives. By embracing blockchain technology, the mining industry can address its pressing challenges and become a catalyst for positive change. With continued innovation, collaboration, and the integration of complementary technologies, the industry will witness transformative changes, revolutionizing the way mining operations are conducted.

Chapter 12: Case Studies & Security Concerns in the Mining Industry

"The blockchain allows our smart devices to speak to each other better and faster."

— Melanie Swan

How Blockchain Technology is Revolutionizing the Mining Industry

Blockchain technology has gained significant attention for its potential to revolutionize industries, and the mining industry is no exception. This sector is characterized by a complex supply chain, substantial risks, and environmental impacts. Thus, the adoption of blockchain technology holds promise in providing a reliable and transparent platform for data sharing, storage, and management.

In recent years, several projects have emerged that exemplify the application of blockchain technology in the coal and mineral mining industry. One notable project is MineHub, a blockchain-based platform designed to optimize the global supply chain for metals and minerals, including coal. By leveraging blockchain, MineHub enhances transparency, traceability, and efficiency through the creation of a decentralized ledger. This enables participants, such as miners, traders, shippers, and end consumers, to track and verify the movement of minerals from extraction to their final destination. The platform ensures compliance with ethical and sustainability standards, fostering responsible mining practices.

Another project making waves in the mining industry is VAKT, a blockchain platform focused on digitalizing and streamlining trading and logistics processes in the energy commodity sector, including coal and minerals. VAKT facilitates secure and efficient peer-to-peer trading, reduces paperwork, and improves transparency and trust among participants. By leveraging blockchain's immutable nature, the platform ensures that transactions are recorded and cannot be altered, enhancing reliability and efficiency in the supply chain.

While not directly targeted at the coal and mineral mining industry, Everledger is a blockchain-based platform that has successfully tracked and verified the authenticity of diamonds and other precious metals. This technology has the potential to be applied to other minerals, including those used in coal mining, to ensure ethical sourcing and prevent fraud. By leveraging blockchain's transparency and immutability, Everledger creates a secure and traceable record of mineral origins, promoting responsible and sustainable mining practices.

Minespider is another significant project utilizing blockchain technology to provide transparency and traceability in global supply chains, including mineral mining. Its objective is to establish a decentralized protocol for documenting and verifying information about mineral origins, certifications, and sustainability. By utilizing blockchain, Minespider aims to create a trustworthy and accessible system that promotes responsible mining and encourages adherence to sustainability standards.

Furthermore, the collaboration between BHP Billiton, a global mining company, and Komatsu, a leading mining equipment manufacturer, showcases the testing of blockchain technology in tracking the movement of rock and fluid samples from mining sites. This initiative aims to improve the efficiency and reliability of data sharing between the mining company and its suppliers. By utilizing blockchain's secure and decentralized nature, the collaboration seeks to enhance data integrity and streamline the information flow throughout the mining process.

These projects, including the MineLife project, demonstrate the increasing interest in applying blockchain technology within the coal and mineral mining industry. They emphasize the importance of transparency, traceability, efficiency, and sustainability throughout the global supply chain. By leveraging blockchain's inherent properties, these initiatives aim to overcome challenges such as ethical sourcing, fraud prevention, and data management, ultimately fostering a more responsible and efficient mining industry.

Case Study 1: Enhancing Mining Transparency: The MineHub Platform's Provenance Tracking

The MineHub platform recognizes the mining industry's crucial role in supplying essential minerals for a wide range of products, such as electronics, jewellery, and construction materials. However, it also acknowledges the industry's challenges, including corruption, conflict, and human rights violations associated with mineral sourcing in certain regions. As consumer and investor demands for transparency and ethical sourcing practices continue to rise, the MineHub platform offers a solution through the implementation of blockchain technology to track mineral provenance from mine to end-user. By leveraging distributed ledger technology, blockchain provides a transparent and immutable record of transactions, enabling secure and decentralized tracking of minerals and their associated attributes. The MineHub platform serves as a reliable source of information, granting consumers and investors a comprehensive view of the mineral supply chain. Mining companies, smelters, and refineries can record and share data on mineral origin, extraction, transportation, and processing, instilling confidence in the integrity of the minerals utilized across various industries. Built upon the IBM blockchain platform, MineHub ensures robustness, scalability, and interoperability, facilitating seamless integration with existing systems and promoting collaboration among industry stakeholders. Embracing the MineHub platform allows mining companies to demonstrate their commitment to responsible sourcing and sustainability, aligning with evolving consumer and investor expectations.

The MineHub platform leverages blockchain technology to address the challenges surrounding mineral provenance. By providing transparency and traceability, it empowers consumers and investors to make informed choices and support ethical sourcing practices. The platform serves as a secure and reliable tool for tracking minerals throughout the supply chain, fostering a more responsible and sustainable mining industry.

Key features of the MineHub platform:

1. **Combination of blockchain technology, IoT sensors, and smart contracts:** The MineHub platform revolutionizes mineral tracking by harnessing the power of blockchain technology, IoT sensors, and smart contracts. This innovative combination allows for seamless monitoring of mineral movement throughout the supply chain. IoT sensors provide real-time data on crucial parameters such as location, quality, and quantity, while smart contracts ensure that every step complies with ethical and sustainability standards.

2. **Real-time data for informed decision-making:** The MineHub platform excels in providing real-time data on mineral movement. This invaluable feature empowers stakeholders to make informed decisions promptly. By having access to up-to-date information, they can proactively address any emerging issues, optimize logistics, and enhance overall efficiency.

3. **Transparency and traceability as cornerstones:** Through the utilization of blockchain technology, the MineHub platform establishes an unparalleled level of transparency and traceability in mineral tracking. Every transaction and movement recorded on the blockchain forms an immutable and transparent ledger. This robust system ensures that all involved parties have access to a reliable source of information, fostering trust among consumers and investors.

4. **Compliance with ethical and sustainability standards:** The MineHub platform goes beyond mere tracking and introduces essential features to ensure compliance with ethical and sustainability standards. It monitors the utilization of renewable energy sources, reduces carbon emissions, and minimizes waste along the supply chain. By promoting

responsible practices, the platform contributes to a more sustainable and ethical mining industry, aligning with the increasing demands for responsible sourcing from conscientious consumers and investors.

Benefits of the MineHub platform to the mining industry

1. **Improved transparency and accountability:** By leveraging blockchain technology, the MineHub platform establishes a transparent and immutable record of transactions. This heightened transparency increases accountability throughout the supply chain, reducing the risk of fraudulent activities and enhancing trust among stakeholders.

2. **Increased efficiency:** The platform's provision of real-time data empowers stakeholders to access up-to-the-minute information on mineral movement. This enables them to make informed decisions swiftly, optimizing operations, streamlining logistics, and ultimately improving overall efficiency. By reducing delays and minimizing inefficiencies, mining companies can realize cost savings and enhance their competitiveness.

3. **Reduced risk:** Compliance with ethical and sustainability standards embedded within the MineHub platform mitigates the risk of reputational damage for mining companies. Through transparent tracking of mineral provenance and adherence to responsible sourcing practices, companies can build trust with consumers and investors. By demonstrating their commitment to ethical and sustainable mining, they reduce the risk of negative public perception and potential backlash.

4. **Enhanced trust:** The use of blockchain technology and the platform's commitment to ethical and sustainability standards foster an environment of trust within the mining industry. The transparent and traceable nature of blockchain instills confidence in consumers and investors, assuring them of the ethical sourcing of minerals. This trust strengthens relationships between mining companies, stakeholders, and end consumers, promoting long-term sustainability and positive industry growth.

The implementation of the MineHub platform delivers significant advantages to the mining industry. It enhances transparency, accountability, and efficiency through the use of blockchain technology, while also reducing risk and building trust. By ensuring compliance with ethical and sustainability standards, the platform addresses the increasing demand for transparency and responsible sourcing in mineral provenance. The MineHub project exemplifies how blockchain technology can drive positive change in the mining industry, improving transparency, efficiency, and trust.

Challenges faced during Implementation

During the implementation of the MineHub platform in the mining industry, several challenges were encountered. These challenges arose due to the unique nature of the industry and the complexity of introducing blockchain technology. The MineHub platform had to address and overcome the following hurdles:

1. **Interoperability of different blockchain platforms:** The mining industry consists of numerous mining companies and stakeholders, each with their preferred blockchain platforms and systems. Achieving interoperability among these diverse platforms posed a significant challenge for MineHub. Ensuring seamless data sharing and communication between different systems required the development of standardized protocols and interfaces. By establishing compatibility and interoperability, MineHub aimed to create a unified ecosystem that connects stakeholders across the mining supply chain.

2. **Lack of standardization in the mining industry:** The mining industry lacks a standardized framework for data management, processes, and regulations. This lack of standardization presented a challenge for MineHub during implementation. The varying standards and practices across different regions and organizations made it essential to navigate and harmonize different requirements. MineHub had to work closely with industry participants and regulatory bodies to promote the adoption of common standards and guidelines. Establishing industry-wide best practices and protocols helped ensure the smooth integration of the MineHub platform across the mining ecosystem.

3. **Data privacy and security concerns:** The mining industry deals with sensitive and confidential data, including financial information, operational details, and supply chain records. Protecting the privacy and security of this data posed a significant challenge for MineHub. Implementing robust security measures and protocols to safeguard sensitive information while maintaining transparency and traceability required careful consideration. MineHub addressed these concerns by leveraging the inherent security features of blockchain technology, such

as encryption and decentralization, to enhance data security and protect against unauthorized access.

4. **Resistance to change and industry-wide adoption:** Introducing new technologies and transforming traditional processes often faces resistance within established industries. The mining industry was no exception. MineHub encountered reluctance and skepticism from some stakeholders who were hesitant to embrace blockchain technology. Overcoming this challenge required extensive education and awareness campaigns to demonstrate the benefits and value proposition of the MineHub platform. Collaborating with industry leaders, demonstrating successful use cases, and highlighting the potential for increased efficiency and transparency helped garner support and fostered a wider adoption of the platform.

Case Study 2: Transforming Supply Chain Management in the Mining Industry: The MineLife Project

The mining industry has seen a notable application of blockchain technology through the successful implementation of the "MineLife" project. MineLife is a blockchain platform that utilizes the Ethereum blockchain's features to enable efficient tracking and tracing of supplies and equipment within the mining supply chain. Operating on the Ethereum blockchain, a decentralized and open-source platform known for its smart contract and decentralized application (DApp) development capabilities, MineLife ensures secure, transparent, and tamper-proof processes for mining companies and their suppliers. Through the MineLife project, mining companies and suppliers gain access to a platform that records every transaction on the blockchain, creating an immutable and transparent record accessible to all supply chain participants. This fosters trust, accountability, and the promotion of ethical and sustainable mining practices. By tracking and tracing the movement of supplies and equipment, the MineLife platform enhances transparency and ensures the adherence to ethical sourcing standards.

Real-time monitoring is another valuable feature provided by the MineLife platform, enabling stakeholders to promptly identify and address any issues that may arise along the mining supply chain. For instance, in the event of supply delivery delays, the platform promptly notifies all involved parties, empowering them to take immediate action and mitigate disruptions. This capability contributes to reduced downtime, increased productivity, and improved profitability of mining operations. By leveraging blockchain technology's capabilities, the MineLife project offers a transformative solution to the challenges faced by the mining industry in supply chain management. The platform's emphasis on transparency, traceability, and accountability enables mining companies to enhance operational efficiency while fostering responsible practices. With the MineLife project, the mining industry takes a significant step towards a more secure, transparent, and responsible future.

Key features of the MineLife platform

1. **Smart contracts:** MineLife utilizes smart contracts to automate the procurement process, streamlining and expediting transactions. By eliminating the need for intermediaries and manual processes, smart contracts reduce the time and cost associated with procurement. Furthermore, these self-executing contracts ensure compliance with safety standards and minimize the risk of fraud and corruption.

2. **Real-time data:** The platform provides stakeholders with access to real-time data regarding the location, quality, and quantity of supplies and equipment within the mining supply chain. This up-to-date information enables stakeholders to make informed decisions promptly, optimizing operational efficiency and reducing delays.

3. **Transparency and traceability:** MineLife leverages blockchain technology to ensure transparency and traceability throughout the supply chain. Every transaction related to the movement of supplies and equipment is recorded on the blockchain, creating an immutable and transparent ledger accessible to all authorized participants. This transparent record serves as a reliable source of information for mining companies and their suppliers, fostering trust, and accountability.

4. **Compliance with safety standards:** The MineLife platform places a strong emphasis on safety standards compliance. By integrating safety protocols into the platform's processes and workflows, it helps reduce the risk of accidents and injuries within the mining operations. This focus on safety not only protects the well-being of workers but also promotes responsible and sustainable mining practices.

Benefits of the MineLife Project to the mining industry

1. **Improved efficiency:** By automating the procurement process through the use of smart contracts, MineLife streamlines and expedites transactions, resulting in improved efficiency. The reduction in time

and cost associated with procurement allows mining companies to allocate resources more effectively, leading to increased productivity and cost savings.

2. **Increased transparency and accountability:** With the utilization of blockchain technology, the MineLife platform establishes a transparent and immutable record of transactions. This increased transparency enhances accountability among stakeholders, as every transaction can be traced and verified. It reduces the risk of fraudulent activities, as the tamper-proof nature of the blockchain prevents unauthorized alterations to the recorded data.

3. **Enhanced safety:** The MineLife platform places a strong emphasis on safety standards compliance. By ensuring that all procurement processes and transactions adhere to safety regulations, the platform contributes to a safer working environment within the mining industry. This focus on safety reduces the risk of accidents, injuries, and other safety-related incidents, protecting the well-being of workers and improving overall operational safety.

4. **Reduced risk:** Through its compliance with safety standards and utilization of blockchain technology, MineLife helps to reduce various risks in the mining industry. The transparency and traceability provided by the platform mitigate the risk of fraud and corruption, as all transactions are recorded and accessible to authorized participants. By promoting a transparent and accountable supply chain, MineLife creates a more trustworthy and secure environment for all stakeholders involved.

The MineLife project is an excellent example of how blockchain technology can be used to manage the supply chain, ensuring compliance with safety standards, reducing the risk of fraud and corruption, and improving efficiency, transparency, and accountability.

Challenges Faced During Implementation

While the MineLife project brings significant benefits to the mining industry, it is not without its challenges during implementation. Some of the key challenges faced by the MineLife project include:

1. **Adoption & Integration:** One of the primary challenges is the adoption and integration of the MineLife platform across the mining

industry. Implementing a new technology requires buy-in and cooperation from various stakeholders, including mining companies, suppliers, contractors, and regulatory bodies. Encouraging widespread adoption and seamless integration of the platform can be a complex process, requiring extensive coordination and collaboration.

2. **Data Compatibility & Interoperability:** The mining industry comprises a diverse range of systems, databases, and legacy infrastructure. Integrating the MineLife platform with existing systems and ensuring compatibility and interoperability can pose challenges. Data standardization and integration efforts may be required to enable smooth data exchange and interaction between different platforms and stakeholders.

3. **Scalability & Performance:** The mining industry involves vast quantities of data and complex supply chains. Ensuring that the MineLife platform can handle large-scale data processing and transaction volumes while maintaining performance and responsiveness is crucial. Scalability challenges need to be addressed to accommodate the growing demands of the industry and prevent any potential bottlenecks or delays.

4. **Security & Privacy:** As with any blockchain-based solution, security and privacy are paramount concerns. Safeguarding sensitive data, protecting against cyber threats, and ensuring that only authorized parties have access to relevant information require robust security measures. Additionally, compliance with data privacy regulations, such as GDPR (General Data Protection Regulation), may pose additional challenges that need to be carefully addressed during the implementation of the MineLife project.

5. **User Education & Training:** Introducing a new technology like the MineLife platform necessitates educating and training stakeholders on its functionalities, benefits, and proper usage. Mining industry professionals may require training to fully understand and utilize the platform effectively. Providing comprehensive user education and support is crucial to ensure smooth adoption and maximize the benefits of the MineLife project.

6. **Regulatory & Legal Considerations:** The mining industry is subject to various regulatory frameworks and compliance requirements.

Implementing a blockchain-based solution like MineLife may involve navigating complex legal considerations, such as data privacy, intellectual property rights, and compliance with industry-specific regulations. Adhering to these regulations and ensuring legal compliance throughout the implementation process can present challenges that require careful attention.

Case Study 3: Blockchain Revolution in Energy Commodity Trading: The VAKT Project

The VAKT project represents a ground-breaking initiative that utilizes blockchain technology to transform trading and logistics processes in the energy commodity sector, including coal and minerals. In the fast-paced and intricate world of energy commodity trading, traditional systems often suffer from inefficiencies, paperwork burdens, and a lack of transparency. The VAKT platform addresses these challenges head-on, providing a secure and efficient peer-to-peer trading platform that streamlines operations and fosters trust among participants. At its core, VAKT leverages the power of blockchain to revolutionize energy commodity trading. Built on a decentralized architecture, the platform ensures that transactions are recorded and cannot be altered, creating an immutable and transparent record of trade data. By eliminating the need for intermediaries and central authorities, VAKT enables direct peer-to-peer trading, reducing complexities and enhancing the speed of transactions.

One of the key features of the VAKT project is the reduction of paperwork and administrative burdens. By digitizing and automating various trading and logistics processes, the platform eliminates the manual and time-consuming tasks associated with traditional paper-based systems. This significantly improves operational efficiency, reduces costs, and accelerates the settlement process. Furthermore, the VAKT platform fosters transparency and trust among participants. With every transaction recorded on the blockchain, all authorized parties can access a single source of truth, ensuring transparency and reducing the risk of disputes. The platform also integrates advanced data analytics capabilities, enabling participants to gain valuable insights into market trends and optimize their trading strategies.

The VAKT project has gained significant traction within the energy commodity sector, attracting the interest and participation of major industry players. By collaborating and utilizing blockchain technology, VAKT is paving the way for a more secure, efficient, and transparent energy commodity trading ecosystem. With its ability to streamline operations, reduce paperwork, and enhance trust among participants, the VAKT project represents a promising step forward in revolutionizing the energy commodity trading landscape.

Key features of the VAKT Project:

1. **Decentralized Trading:** The VAKT platform enables direct peer-to-peer trading, eliminating the need for intermediaries and central authorities. This streamlines the trading process, reduces complexities, and enhances transaction speed.

2. **Immutable Trade Records:** All transactions on the VAKT platform are recorded on the blockchain, creating an immutable and transparent record of trade data. This fosters trust among participants and reduces the risk of disputes.

3. **Digitized and Automated Processes:** VAKT digitizes and automates trading and logistics processes, reducing paperwork and administrative burdens. This significantly improves operational efficiency, reduces costs, and accelerates the settlement process.

4. **Advanced Data Analytics:** The platform integrates advanced data analytics capabilities, providing participants with valuable insights into market trends. This enables them to optimize their trading strategies and make informed decisions.

5. **Increased Transparency and Trust:** The use of blockchain technology ensures transparency in energy commodity trading. Authorized participants have access to a single source of truth, promoting transparency and reducing the risk of fraudulent activities.

6. **Industry Collaboration:** The VAKT project has gained significant traction and collaboration from major industry players within the energy commodity sector. This collective effort strengthens the platform's credibility and drives the transformation of the energy commodity trading ecosystem.

Benefits of the VAKT Project for the Mining Industry:

1. **Enhanced Efficiency:** The VAKT platform streamlines trading and logistics processes, reducing paperwork and administrative burdens. This increases operational efficiency within the mining industry, enabling faster and more streamlined transactions.

2. **Improved Transparency:** By utilizing blockchain technology, the VAKT project fosters transparency in the mining industry. All transactions recorded on the blockchain create a transparent and immutable record of trade data, reducing the risk of fraudulent activities and enhancing trust among participants.

3. **Secure and Tamper-Proof Records:** The VAKT platform ensures that trade records are securely stored on the blockchain and cannot be altered. This provides mining companies with reliable and tamper-proof records, improving auditability and compliance.

4. **Direct Peer-to-Peer Trading:** The VAKT platform enables direct peer-to-peer trading, eliminating the need for intermediaries and reducing complexities in the trading process. This allows mining companies to engage in direct transactions, potentially reducing costs and enhancing flexibility.

5. **Faster Settlement:** By digitizing and automating trading processes, the VAKT platform accelerates settlement within the mining industry. This reduces the time taken for transactions to be completed, improving liquidity and cash flow for mining companies.

6. **Market Insights and Optimization:** The VAKT platform's integration of advanced data analytics capabilities provides mining companies with valuable insights into market trends. This enables them to optimize their trading strategies and make informed decisions based on real-time data.

7. **Collaborative Industry Transformation:** The VAKT project's collaboration with major industry players drives the transformation of the mining industry. By adopting the platform, mining companies can participate in a collaborative ecosystem that embraces efficiency, transparency, and innovation.

Challenges Faced During Implementation of the VAKT Project:

1. **Industry Adoption:** One of the primary challenges faced by the VAKT project during implementation is encouraging widespread industry adoption. Convincing mining industry participants to embrace a new technology and shift from traditional trading methods can be a slow and complex process.

2. **Integration with Existing Systems:** Integrating the VAKT platform with existing systems and infrastructure within the mining industry can pose technical challenges. Ensuring seamless interoperability and data synchronization between different systems requires careful planning and collaboration.

3. **Regulatory Compliance:** Adhering to regulatory requirements and compliance standards specific to the mining industry can be a significant challenge for the VAKT project. Navigating complex regulatory frameworks and ensuring compliance with various jurisdictions require dedicated efforts and expertise.

4. **Data Privacy and Security:** As the VAKT platform handles sensitive trade data, maintaining data privacy and security is crucial. Implementing robust cybersecurity measures and safeguarding against potential data breaches or unauthorized access poses a challenge that must be carefully addressed.

5. **Stakeholder Engagement and Education:** Engaging and educating all stakeholders within the mining industry about the benefits and functionalities of the VAKT platform is essential for successful implementation. This requires effective communication, training programs, and ongoing support to ensure smooth adoption and utilization.

6. **Scalability and Performance:** As the VAKT platform gains traction and more participants join the network, scalability and performance become critical challenges. Ensuring that the platform can handle increased transaction volumes while maintaining speed and efficiency requires continuous monitoring and optimization.

7. **Industry Resistance and Cultural Change:** Resistance to change and the traditional mindset prevalent in the mining industry can pose challenges during the implementation of the VAKT project. Overcoming resistance, addressing cultural barriers, and promoting a mindset shift towards embracing new technologies are ongoing challenges that need to be managed effectively.

Case Study 4: Driving Transparency and Sustainability: The Minespider Project

Minespider is an innovative blockchain project that aims to revolutionize global supply chains, with a specific focus on the mineral mining industry. As the world becomes increasingly concerned with responsible sourcing, ethical practices, and sustainability, Minespider emerges as a powerful tool for promoting transparency and traceability in supply chain management. The mineral mining industry faces numerous challenges, including issues related to human rights violations, environmental degradation, and the lack of reliable information on the origins and certifications of minerals. Minespider addresses these challenges by leveraging blockchain technology to create a decentralized protocol for documenting and verifying information about mineral origins, certifications, and sustainability.

At its core, Minespider aims to establish a trustworthy and accessible system that allows mining companies, suppliers, and consumers to track the journey of minerals from extraction to the end product. By utilizing blockchain's immutable nature, Minespider ensures that every transaction and piece of information is recorded transparently, creating an unalterable and auditable record of mineral supply chains.

Through its platform, Minespider enables participants to verify the ethical sourcing of minerals, ensure compliance with sustainability standards, and eradicate the use of minerals that are associated with human rights abuses or environmental harm. This promotes responsible mining practices and empowers consumers to make informed choices about the products they purchase.

Moreover, Minespider provides a comprehensive solution for companies seeking to comply with evolving regulations and certifications in the mineral mining industry. By streamlining the documentation and verification processes, Minespider reduces administrative burdens and enhances efficiency in supply chain management.

The Minespider project has garnered attention and support from industry stakeholders, advocacy groups, and consumers who recognize the need for greater transparency and responsible practices in the mineral mining sector. By leveraging the power of blockchain technology, Minespider is poised to drive positive change, foster responsible mining, and contribute to a more sustainable and ethical future for the industry.

The key features of the Minespider project

1. **Decentralized Protocol:** Minespider utilizes a decentralized protocol built on blockchain technology. This ensures transparency, immutability, and security in documenting and verifying information about mineral origins, certifications, and sustainability.

2. **Supply Chain Traceability:** Minespider enables the tracking and tracing of minerals from extraction to the end product. Participants in the supply chain can access a transparent and auditable record that provides visibility into the journey of minerals, ensuring responsible sourcing and compliance with ethical and sustainability standards.

3. **Ethical Sourcing Verification:** The platform allows for the verification of ethical sourcing practices, ensuring that minerals are not associated with human rights abuses, child labor, or environmental harm. This empowers consumers to make informed choices and supports responsible mining practices.

4. **Compliance and Certification Support:** Minespider assists companies in complying with evolving regulations and certifications in the mineral mining industry. By streamlining the documentation and verification processes, it reduces administrative burdens and facilitates adherence to industry standards.

5. **Collaboration and Stakeholder Engagement:** Minespider encourages collaboration among industry stakeholders, including mining companies, suppliers, consumers, and advocacy groups. It creates a platform for shared information and dialogue, fostering a collective effort towards responsible mining practices.

6. **Enhanced Efficiency and Transparency:** By leveraging blockchain technology, Minespider enhances efficiency in supply chain management by reducing paperwork, streamlining processes, and providing real-time access to information. It promotes transparency and trust among participants, facilitating better decision-making and risk management.

Benefits of the Minespider project for the Mining Industry

1. **Enhanced Transparency and Traceability:** By utilizing blockchain technology, Minespider enables transparent and auditable tracking of minerals throughout the supply chain. This transparency helps mining

companies demonstrate responsible sourcing practices and provides assurance to consumers about the origin and ethicality of the minerals they purchase.

2. **Compliance with Sustainability Standards:** Minespider assists mining companies in complying with sustainability standards and certifications. By automating documentation and verification processes, it streamlines compliance efforts, reduces administrative burdens, and facilitates adherence to environmental and social responsibility guidelines.

3. **Improved Reputation and Consumer Trust:** With Minespider's verification and transparency capabilities, mining companies can enhance their reputation as responsible and ethical operators. This, in turn, builds consumer trust, supports brand integrity, and opens doors to new markets where responsible sourcing is a priority.

4. **Risk Mitigation and Supply Chain Resilience:** Minespider enables better risk management by providing real-time visibility into the supply chain. Companies can identify potential risks, such as human rights violations or environmental non-compliance, and take proactive measures to address them. This contributes to a more resilient and sustainable supply chain.

5. **Market Differentiation and Competitive Advantage:** By embracing responsible mining practices and leveraging the Minespider platform, companies can differentiate themselves in the market. Consumers increasingly value ethically sourced minerals, and Minespider certification can be a powerful differentiator, attracting conscious consumers and investors.

6. **Collaboration and Industry Standards Development:** Minespider fosters collaboration among industry stakeholders, facilitating the development of industry-wide standards for responsible mining. By bringing together mining companies, suppliers, and advocacy groups, Minespider drives collective efforts towards sustainability and helps shape best practices in the mining industry.

Challenges during the implementation of the Minespider project

1. **Stakeholder Engagement:** Engaging and convincing all stakeholders, including mining companies, suppliers, industry associations, and regulators, to actively participate and support the implementation of the

Minespider platform can be a significant challenge. Each stakeholder group may have different priorities, interests, and concerns that need to be addressed to gain their buy-in.

2. **Data Integration and Standardization:** Integrating data from various sources and ensuring its consistency and standardization across the mining supply chain can be a complex task. Different companies may use different systems, formats, and data structures, making it challenging to achieve seamless data integration and synchronization.

3. **Technological Complexity:** Implementing a blockchain-based platform like Minespider requires advanced technical knowledge and expertise. Overcoming technical challenges such as scalability, interoperability, data privacy, and security can be demanding, especially for organizations that have limited experience with blockchain technology.

4. **Infrastructure Requirements:** The successful implementation of Minespider relies on robust technical infrastructure, including network capabilities, data storage, and processing power. Ensuring that the infrastructure is in place and capable of handling the increased data load and transaction volume can be a significant challenge.

5. **Resistance to Change:** Introducing new technologies and processes often faces resistance from employees and organizations accustomed to traditional systems. Overcoming resistance to change and ensuring smooth adoption of Minespider by all stakeholders requires effective change management strategies, training programs, and clear communication of the benefits and value proposition.

6. **Regulatory Compliance:** The mining industry is subject to various regulations and compliance requirements that vary across jurisdictions. Ensuring that the Minespider platform meets all relevant regulatory standards and addresses privacy, data protection, and legal considerations can be a complex task.

7. **Cost Considerations:** Implementing a new technology platform involves financial investments, including infrastructure upgrades, software development, and ongoing maintenance costs. Convincing mining companies and other stakeholders of the long-term cost-effectiveness and return on investment of adopting Minespider can be a challenge, particularly for organizations with limited resources.

8. **Industry Collaboration:** Achieving collaboration and cooperation among all participants in the mining industry is crucial for the success of Minespider. Collaborative efforts, including standardization of processes, data sharing agreements, and alignment on common goals, require time, effort, and coordination among multiple stakeholders.

Case Study 5: Harnessing Blockchain Technology to Transform Mining Data Management: BHP Billiton & Komatsu Collaboration:

In an era marked by technological advancements, BHP Billiton, a global mining company, and Komatsu, a prominent manufacturer of mining equipment, joined forces in a ground-breaking collaboration. Their partnership centered around exploring the potential of blockchain technology in revolutionizing the tracking process of rock and fluid samples extracted from mining sites. This innovative initiative sought to enhance the efficiency and reliability of data sharing between BHP Billiton and its suppliers, marking a significant step forward for the mining industry.

With a focus on improving operational efficiency and streamlining data management, BHP Billiton and Komatsu recognized the transformative power of blockchain technology. By leveraging the inherent capabilities of blockchain, this collaboration aimed to create a secure, decentralized, and transparent system for tracking the movement of crucial rock and fluid samples throughout the mining process.

Traditionally, the mining industry has faced challenges related to data sharing, accuracy, and transparency. By implementing blockchain technology, BHP Billiton and Komatsu sought to address these challenges head-on. Blockchain's unique properties, including immutability, transparency, and enhanced security, can fundamentally transform how data is recorded, stored, and shared within the mining supply chain.

Through the utilization of blockchain technology, the collaboration between BHP Billiton and Komatsu aimed to create a system where all stakeholders, including the mining company, equipment manufacturers, and suppliers, could securely and reliably track the movement of rock and fluid samples. This system would enable real-time data sharing, reducing delays, improving decision-making, and ultimately optimizing the efficiency of mining operations.

By embracing blockchain technology's secure and decentralized nature, BHP Billiton and Komatsu intended to enhance data integrity and streamline the flow of information throughout the mining process. With an immutable and tamper-proof ledger, they aimed to ensure the authenticity and traceability of rock and fluid samples, thereby increasing the overall reliability and trustworthiness of the data.

The collaboration between BHP Billiton and Komatsu in exploring blockchain technology's potential to track rock and fluid samples signifies a notable shift in the mining industry's approach to data management. This initiative not only aimed to improve the efficiency and reliability of data sharing but also demonstrated a commitment to harnessing emerging technologies for the benefit of the industry as a whole.

The collaboration between BHP Billiton & Komatsu includes the following key features:

1. **Improved Data Sharing:** The collaboration aims to enhance the efficiency and reliability of data sharing between BHP Billiton and its suppliers. By leveraging blockchain technology, the partners can securely and seamlessly share information related to the movement of rock and fluid samples from mining sites.

2. **Enhanced Transparency:** Blockchain technology provides a transparent and immutable ledger, allowing both BHP Billiton and Komatsu to access and verify the authenticity and accuracy of shared data. This transparency promotes trust and accountability among the collaborating parties.

3. **Streamlined Supply Chain:** The collaboration focuses on optimizing the supply chain by utilizing blockchain technology. Through efficient data sharing and real-time tracking, BHP Billiton and Komatsu can better manage and monitor the movement of samples, leading to improved logistics, reduced delays, and enhanced overall supply chain efficiency.

4. **Data Integrity and Security:** Blockchain's inherent features, such as immutability and cryptographic security, ensure the integrity and security of shared data. This helps prevent data tampering and unauthorized access, providing a reliable and trustworthy foundation for the collaboration.

5. **Testing and Innovation:** The collaboration serves as a testing ground for the application of blockchain technology in the mining industry. BHP Billiton and Komatsu are actively exploring the potential of blockchain to address data sharing challenges and identify

opportunities for further innovation and improvement within the mining supply chain.

Case Study 6: Transforming Supply Chains through Blockchain Technology: Everledger

Everledger is a pioneering blockchain-based platform that has made significant strides in revolutionizing supply chain transparency and traceability, particularly in industries such as diamonds and precious metals. While its primary focus has been on the diamond industry, Everledger's innovative technology has the potential to be applied to various minerals, including those used in coal mining, to promote ethical sourcing and combat fraud.

With a mission to enhance transparency and trust in global supply chains, Everledger utilizes blockchain technology to securely record and verify critical information about valuable assets. By creating digital ledgers that are immutable and transparent, the platform enables stakeholders to track and authenticate the provenance, ownership history, and authenticity of assets.

Although Everledger has primarily focused on diamonds, its technology can be adapted to other minerals, including those utilized in coal mining. By implementing Everledger's platform, the mining industry can ensure the ethical sourcing of minerals, prevent fraud, and provide consumers with verifiable information about the origin of the materials used in coal production.

With its emphasis on transparency, traceability, and responsible sourcing, Everledger offers a promising solution for addressing the challenges faced by the coal and mineral mining industry. By leveraging blockchain technology, the platform has the potential to contribute to a more sustainable and ethical future by ensuring the integrity of mineral supply chains and providing consumers with confidence in the products they use and purchase.

Key Features of Everledger Platform:

1. **Immutable Digital Records:** Everledger utilizes blockchain technology to create a decentralized and immutable record of information. This ensures the integrity and transparency of data related to the origin, authenticity, and ownership of diamonds and other precious metals.

2. **Enhanced Traceability:** The platform enables the tracing of a mineral's journey throughout the supply chain, from the point of extraction to the end consumer. By capturing and recording crucial data

at each stage, Everledger provides an auditable trail that verifies the ethical sourcing and provenance of minerals.

3. **Fraud Prevention:** Everledger's platform employs advanced algorithms and data analytics to identify and prevent fraudulent activities in the supply chain. It establishes a comprehensive system that enables the detection of counterfeit or stolen minerals, mitigating the risk of fraudulent transactions.

4. **Sustainability and Responsible Sourcing:** Everledger promotes sustainability and responsible sourcing practices by providing transparency and accountability in the supply chain. By capturing data related to ethical and sustainable mining practices, the platform supports efforts to minimize environmental impact and uphold social responsibility.

5. **Increased Trust and Consumer Confidence:** The implementation of Everledger's platform enhances trust and confidence among stakeholders, including consumers, by providing them with access to verified and validated information about the minerals they purchase. This empowers consumers to make informed choices and supports ethical consumerism.

6. **Collaboration and Industry Engagement:** Everledger actively collaborates with industry partners, including mining companies, certification organizations, and regulatory bodies, to establish common standards and best practices. By engaging with stakeholders, the platform aims to drive industry-wide adoption of responsible sourcing and sustainable practices.

7. **Expansion Potential:** Although initially focused on diamonds and precious metals, Everledger has the potential to expand its application to other minerals, including those used in coal mining. This expansion can contribute to the overall transparency and ethical practices within the mining industry.

By leveraging blockchain technology, Everledger's platform provides valuable benefits to industries like the diamond and precious metals sector, primarily by enhancing transparency and fostering trust. It achieves this by enabling the tracking and verification of valuable asset authenticity and origin. This innovative approach offers notable advantages, such as:

1. **Ethical Sourcing:** The platform helps ensure the ethical sourcing of diamonds and precious metals by providing a transparent record of their journey from the mine to the market. This helps prevent the circulation of conflict diamonds and supports responsible supply chain practices.

2. **Fraud Prevention:** Everledger's platform helps combat fraud within the industry by creating an immutable and tamper-proof record of asset ownership and characteristics. This reduces the risk of counterfeit or stolen assets entering the market.

3. **Supply Chain Efficiency:** By digitizing and streamlining the supply chain processes, the platform improves operational efficiency and reduces paperwork. It enables quicker and more accurate verification of asset information, reducing delays and inefficiencies.

4. **Consumer Confidence:** The platform instills confidence in consumers by providing them with access to verified and transparent information about the assets they are purchasing. This fosters trust and promotes responsible consumer choices.

Similarities between MineHub, VAKT, Everledger, and Minespider in Blockchain Technology & Applications

Blockchain based platforms like MineHub, VAKT, Everledger, and Minespider, are all involved in the domain of blockchain technology and its applications. While they have distinct focuses and unique features, they share several similarities, which include:

1. **Blockchain Utilization:** All four platforms leverage blockchain technology as a fundamental component of their solutions. They recognize the potential of distributed ledger technology to enhance transparency, security, and efficiency in various industries.

2. **Industry-Specific Applications:** Each platform has developed blockchain-based solutions tailored to specific industries. MineHub focuses on supply chain management in the mining and metals sector, VAKT concentrates on digitizing and streamlining commodity trading processes, Everledger specializes in tracking and verifying the authenticity of valuable assets like diamonds, and Minespider addresses responsible sourcing and traceability in the mineral supply chain.

3. **Increased Transparency:** By utilizing blockchain technology, these platforms aim to improve transparency within their respective industries. Blockchain's immutable and decentralized nature allows for the creation of transparent and auditable records, enabling participants to track and verify the origin, movement, and authenticity of assets or goods.

4. **Supply Chain Optimization:** Another commonality is their focus on optimizing supply chain processes. By utilizing blockchain technology, these platforms offer solutions that streamline and enhance supply chain management, reducing inefficiencies, eliminating fraud, and providing real-time visibility into the movement of goods and assets.

5. **Sustainability & Responsible Sourcing:** Both Minespider and Everledger have a shared emphasis on sustainability and responsible sourcing. They aim to address issues such as unethical practices, human rights violations, and environmental concerns within their respective industries by leveraging blockchain to provide immutable records and traceability.

6. **Collaborative Approach:** These platforms recognize the importance of collaboration and partnerships to drive innovation and adoption. They often collaborate with industry stakeholders, including mining companies, trading firms, financial institutions, and technology providers, to build robust ecosystems and ensure widespread adoption of their solutions.

7. **Digital Transformation:** The four platforms are part of the broader digital transformation movement within their respective industries. By harnessing blockchain technology, they seek to revolutionize traditional processes, eliminate paperwork, enhance data sharing and interoperability, and unlock new opportunities for efficiency and growth. *(Ref. No.: 308- 346)*

While these similarities exist, it is essential to note that each company has its unique value proposition and differentiates itself through specific features, services, or industry focus.

Comparing MineHub, VAKT, Everledger, and Minespider

1. **Focus & Industry:**
   - MineHub: Focuses on supply chain management in the mining and metals sector.

- VAKT: Concentrates on digitizing and streamlining commodity trading processes.
- Everledger: Specializes in tracking and verifying the authenticity of valuable assets like diamonds.
- Minespider: Addresses responsible sourcing and traceability in the mineral supply chain.

2. **Primary Use Cases:**
   - MineHub: Provides solutions for trade and finance, supply chain visibility, and ESG reporting in the mining industry.
   - VAKT: Offers a blockchain-based platform for digitalizing, managing, and tracking energy commodity trading.
   - Everledger: Utilizes blockchain to create a digital ledger for tracking and authenticating valuable assets like diamonds, gemstones, and luxury goods.
   - Minespider: Focuses on responsible sourcing by using blockchain to trace and verify the origin and ethical practices in the mineral supply chain.

3. **Transparency & Traceability:**
   - MineHub: Enhances transparency in supply chains by providing real-time visibility into the movement of metals and minerals.
   - VAKT: Improves transparency in the energy commodity trading process by securely recording and tracking transactions on the blockchain.
   - Everledger: Verifies the authenticity of assets and provides a transparent record of their history, ensuring trust and reducing fraud.
   - Minespider: Ensures responsible sourcing by providing traceability and immutable records to verify ethical practices and compliance.

4. **Collaboration & Partnerships:**
   - MineHub: Collaborates with various industry stakeholders such as mining companies, financiers, and logistics providers to build a comprehensive ecosystem.
   - VAKT: Collaborates with major energy companies, banks, and technology providers to create a global network for digital commodity trading.

- Everledger: Works with stakeholders across the diamond and luxury goods industry, including gem labs, insurers, and retailers, to establish a secure ecosystem.
- Minespider: Forms partnerships with organizations and initiatives focused on responsible sourcing and sustainability to drive industry-wide change.

5. **Technology Implementation:**
   - MineHub: Utilizes blockchain, smart contracts, and IoT devices to enable secure and efficient supply chain management.
   - VAKT: Implements blockchain, smart contracts, and digitalization to streamline energy commodity trading and reduce administrative burdens.
   - Everledger: Leverages blockchain, data analytics, and AI to create a tamper-proof ledger for asset verification and provenance tracking.
   - Minespider: Deploys blockchain technology, data standards, and auditing tools to ensure responsible sourcing and provide transparent supply chain information.

Closing Thoughts

The adoption of blockchain technology in the mining industry holds immense potential for transforming the sector and addressing its unique challenges. Projects like MineHub, VAKT, Everledger, Minespider, and collaborations such as BHP Billiton and Komatsu highlight the diverse applications of blockchain in enhancing transparency, traceability, and sustainability in the coal and mineral mining industry. One of the key benefits of blockchain technology in the mining industry is improved data sharing and management. By leveraging blockchain's decentralized and secure nature, participants in the supply chain can access a transparent and immutable ledger that records every transaction and piece of information. This eliminates the need for multiple intermediaries and manual data entry, reducing administrative burdens and enhancing the accuracy and reliability of data. Improved data sharing facilitates better decision-making, efficient collaboration among stakeholders, and faster response times to address any issues or disruptions in the supply chain.

Additionally, blockchain technology streamlines trading processes and enhances supply chain efficiency. Platforms like MineHub and VAKT digitize and automate trading and logistics processes, reducing complexities, paperwork, and delays associated with traditional systems. Blockchain enables direct peer-to-peer transactions, eliminating the need for intermediaries and accelerating the settlement process. This increased efficiency leads to cost savings, reduced operational risks, and improved profitability for mining companies and their suppliers. Furthermore, blockchain technology contributes to the prevention of fraud and unethical practices in the mining industry. Projects like Everledger and Minespider focus on ensuring the authenticity and responsible sourcing of minerals. By utilizing blockchain's transparency and immutability, these platforms create a secure and traceable record of mineral origins, certifications, and sustainability standards. This enables stakeholders to verify the ethical sourcing of minerals, mitigate the risk of fraudulent practices, and support responsible mining practices. Consumers can make more informed choices about the products they purchase, promoting sustainability and social responsibility.

However, the adoption of blockchain technology in the mining industry also comes with challenges. The industry needs to address issues such as industry-wide collaboration and standardization, integration with existing systems and infrastructure, ensuring data privacy and security, and scalability to accommodate the vast amount of data generated by mining operations. Overcoming these challenges requires strong partnerships, regulatory support, and ongoing research and development in blockchain technology.

Chapter 13: Decentralizing the Data Economy with the Synergy of AI & Blockchain

"Blockchain's a very interesting technology that will have some very profound applications for society over the years to come."

— Kenneth C. Griffin

Understanding Artificial Intelligence (AI)

Artificial Intelligence (AI) refers to the ability of computer applications to learn and understand patterns, and solve problems using vast amounts of data, without human intervention. It is both a science and an engineering field, focused on developing intelligent computer programs that can perform complex tasks. AI algorithms analyze external input data, learn from it, and then use that understanding to achieve specific objectives by completing tasks. There are two primary types of AI - Narrow AI & Strong AI. Narrow AI is designed to perform specific, limited tasks, such as facial recognition, spam filtering, or playing games like chess. In contrast, strong AI is capable of performing a broad range of tasks, and may even possess human-level cognition to solve any intellectual challenge a person can accomplish.

While narrow AI is already in use today, strong AI is still largely theoretical and many experts question whether it is even possible. The potential effects of strong AI are unpredictable, but many believe that AI and blockchain technologies will be closely intertwined and among the most significant technologies of the coming decades. As such, it is essential to examine how these technologies may interact in the future. The integration of blockchain and AI could lead to enhanced privacy, security, and transparency, with blockchain offering a tamper-proof and immutable ledger to store and share AI-generated data. Additionally, the use of blockchain could provide a framework for creating decentralized, self-governing AI systems that operate autonomously without the need for human intervention.

The Synergy of AI & Blockchain

The combination of artificial intelligence (AI) and blockchain technology can offer several benefits to the industry. AI can optimize the energy consumption of blockchain systems and enhance mining algorithms, which can help reduce the excessively high energy requirement of Proof of Work blockchains. By doing so, the complete industry can benefit, and it could promote the mainstream adoption of blockchain technology.

Furthermore, AI can also help to optimize the storage needs of blockchain networks. Since the transaction history is stored in all nodes, the size of the distributed ledger can quickly add up, making the storage requirements high. This potentially reduces the decentralization of the network and increases the barrier to entry. However, AI can introduce new strategies for sharing databases that can make the blockchain smaller and more efficient for storing information.

By leveraging AI, blockchain networks can become more sustainable, scalable, and efficient. As the blockchain industry continues to evolve, we are likely to see more synergies between these two technologies, leading to new innovations and applications that can transform various industries.

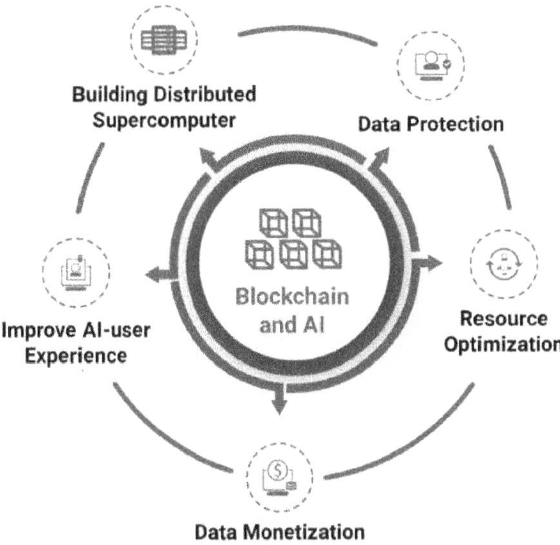

Figure 13.1: Synergy of AI & Blockchain

The Decentralized Data Economy

The Decentralized Data Economy is a concept that involves securely storing and exchanging data using blockchain technology. Data is becoming an increasingly valuable asset, and it is essential to ensure that it is stored and exchanged safely. AI systems are highly dependent on data, and blockchains offer a secure and reliable way of storing information.

A blockchain network is a distributed database that is shared among all participants in the network. The information stored in blocks is cryptographically connected to the preceding one, making it almost impossible to alter data without hijacking network consensus. This high level of security and reliability makes blockchains an ideal technology for storing and exchanging data.

Decentralized data exchanges aim to create a new data economy that runs on top of blockchains. These exchanges make data and storage available to anyone or anything that wants to access it, securely and easily. AI algorithms can benefit greatly from these data exchanges by using a larger set of outside inputs and analyzing them faster. Additionally, the algorithms themselves can be exchanged on these marketplaces, making them more accessible to a wider audience and potentially accelerating their development.

The decentralized data exchange concept could revolutionize the data storage space. Anyone would have the ability to rent out their local storage for a fee, paid in tokens. This would create competition among existing data storage service providers, incentivizing them to improve their services to stay competitive.

Although some data marketplaces are already in operation, they are still in their early stages of maturity. By incentivizing data and storage providers to maintain excessive data integrity, AI systems will benefit greatly. The Decentralized Data Economy is a promising development that could potentially transform the way data is stored, exchanged, and analyzed.

Decentralized Supercomputers

To elaborate on the concept of decentralized supercomputers, it is important to note that training AI models requires a large amount of computational power. One common type of computing system used for AI algorithms is the Artificial Neural Network (ANN), which can learn to perform tasks by processing a large volume of examples. However, the computational power required to train these models can be significant, especially for complex tasks that involve millions of parameters.

To address this challenge, some blockchain networks have implemented peer-to-peer (P2P) markets where users can lend their computing power to those seeking to execute complicated mathematical problems. These users are incentivized to supply computing power by earning tokens in return. By allowing users to rent out their idle computing energy, large amounts of computational power can be used more efficiently.

Decentralized supercomputers have the potential to revolutionize the way AI models are trained. By leveraging the power of blockchain networks, users can share their computing power and access more data, resulting in more accurate models. This approach could significantly reduce the cost of training AI models, making it more accessible to smaller organizations and individuals. While early use cases of decentralized supercomputers have focused primarily on rendering 3D computer graphics, the focus is slowly shifting towards AI. As these

Decentralized Applications (Dapps) develop, companies that provide computing power may see an influx of competition. In theory, every CPU or GPU in the world could be working as a node in a decentralized supercomputer when not in use.

Better Auditability of AI Decisions

AI systems are becoming increasingly complex and their decision-making processes are often difficult to understand and audit. These systems work with vast amounts of data and can generate decisions that are difficult for humans to replicate or verify. As a result, there is a growing need for more transparency and accountability in AI systems. One way to increase the trust in AI decisions is to improve their auditability. By recording every data point and decision made by the AI system, there is a clear audit trail for humans to check. This can help in identifying any biases or errors in the decision-making process. Auditing AI decisions can be particularly important in sensitive areas such as healthcare or finance where even small errors can have significant consequences. For example, an AI system that makes medical diagnoses could benefit from better auditability. By recording every input, decision and output, doctors and regulators could better understand how the AI system arrived at its diagnosis and determine whether it is reliable and trustworthy. In addition to increasing transparency and accountability, better auditability of AI decisions could also help in improving the performance of AI systems. By analyzing the audit trail, developers can identify areas where the AI system is making errors or where it could be improved.

Quantum Computers & Cryptocurrencies: The Potential Threat to Digital Security

Quantum computers are advanced machines that can solve complex mathematical equations much faster than traditional computers. With their exceptional computing power, they pose a significant threat to digital security infrastructure, including the cryptography used in Bitcoin and other cryptocurrencies. Quantum computers are fundamentally different from classical computers, which rely on binary digits or bits to store and process information. In contrast, quantum computers use quantum bits or qubits, which can exist in multiple states simultaneously, allowing them to perform computations at a much faster rate. The cryptography underlying cryptocurrencies, such as Bitcoin, relies on the difficulty of solving certain mathematical problems. However, these cryptographic algorithms are vulnerable to attacks by quantum computers, which can quickly solve these problems and break the encryption.

This could compromise the security of digital assets and lead to fraudulent transactions, potentially causing significant financial losses. The threat of quantum computers to digital security has prompted researchers to develop new cryptographic algorithms that are resistant to attacks by these machines. These new algorithms, known as quantum-resistant or post-quantum cryptography, are designed to withstand attacks from both classical and quantum computers. The development of quantum-resistant cryptography is crucial to ensure the continued security of cryptocurrencies and other digital assets in the age of quantum computing.

Asymmetric Cryptography & Internet Safety

Asymmetric cryptography, also known as public-key cryptography, plays a crucial role in the cryptocurrency ecosystem and most of the internet infrastructure. It uses a key pair, a public key to encrypt and a private key to decrypt information. In contrast, symmetric-key cryptography uses only one key for both encryption and decryption. With asymmetric cryptography, a public key can be shared and used to encrypt data, which can only be decrypted by the corresponding private key, ensuring that only the intended recipient can access the encrypted information. This enables the secure exchange of information without needing to share a common key over an untrusted channel, making it possible for simple information security on the internet. Online banking, for instance, would not be possible without the ability to encrypt data safely between otherwise untrusted parties. *(Ref. No.: 296- 298)*

Asymmetric cryptography's security depends on the assumption that the algorithm generating the key pair makes it incredibly difficult to calculate the private key from the public key, while it is straightforward to calculate the public key from the private key. This assumption is based on the mathematical concept of a trapdoor function, which is simple to calculate in one direction but difficult in the other. Most modern algorithms currently used to generate the key pair are based on known mathematical trapdoor functions, which cannot be solved in a reasonable timeframe by any present computer. However, the development of quantum computers may change this soon. Quantum computers are powerful machines that can solve complex equations much more quickly than regular computers, and they could pose a significant threat to digital security infrastructure. Many experts predict that quantum computers could crack encryption that would take today's fastest computers thousands of years in just a few minutes. Therefore, it is essential to understand how quantum computers work and what dangers they pose to asymmetric cryptography and the safety of the internet.

Classical Computers

Classical computers, which are the computers we use today, operate sequentially, meaning that computational tasks are performed one after another. This is because the memory in classical computers can only be in a state of 0 or 1, which follows the laws of physics. Although different hardware and software approaches are employed to divide complex computations into smaller segments to achieve some efficiency, the basic idea remains the same - one task must be finished before another can begin. For instance, let us consider a situation where a computer must guess a 4-bit key, where each bit can be either 0 or 1. A classical computer would have to guess every combination one by one. To illustrate, imagine a lock with 16 keys on a keychain. Each key has to be tried individually, and if the first key does not work, the next one is tried, and so on, until the correct key opens the lock. However, as the key length increases, the number of possible combinations grows exponentially. In the example above, adding an extra bit to increase the key length to 5 bits would result in 32 possible combinations. Raising the key length to 6 bits would result in 64 possible combinations. At 256 bits, the number of possible combinations is close to the expected number of atoms in the observable universe.

In contrast, the speed of computational processing only grows linearly. Doubling the processing speed of a computer would only double the number of guesses that could be made in a given time. Exponential growth far exceeds any linear progress on the guessing side. It is estimated that it would take thousands of years for a classical computer to guess a 55-bit key. For reference, the minimum recommended size for a seed used in Bitcoin is 128 bits, with many wallet implementations using 256 bits. It may seem that classical computing poses no threat to the asymmetric encryption used in cryptocurrencies and Internet infrastructure.

There are 16 possible combinations, as shown in the table:

0000	0100	1000	1100
0001	0101	1001	1101
0010	0110	1010	1110
0011	0111	1011	1111

Table 13.1: Computational Combinations for Classical Computers

Asymmetric Cryptography, Quantum Computing & Internet Security

Quantum computers represent a new type of computing technology that is still in its early stages of development. They are based on the principles of quantum mechanics, which describe the behaviour of subatomic particles. In contrast to classical computers, which use bits to represent data, quantum computers use qubits. A qubit can represent both 0 and 1 at the same time, which allows quantum computers to perform certain types of computations much faster than classical computers.

This has led to a lot of excitement and investment in the field of quantum computing, as researchers and private companies try to push the boundaries of what is possible with this new technology. However, the development of quantum computers also presents a potential threat to the security of systems that rely on asymmetric cryptography.

Asymmetric cryptography is a critical part of modern internet infrastructure, including cryptocurrency systems. It uses a key pair to encrypt and decrypt data, with a public key that can be shared freely and a private key that must be kept secret. The security of this system relies on the difficulty of calculating the private key from the public key, which is currently believed to be infeasible with classical computers.

However, quantum computers have the potential to break this system by allowing attackers to calculate the private key from the public key much more quickly than would be possible with classical computers. For example, a 4-qubit quantum computer would be able to try all 16 combinations of a 4-bit key at once, making it much faster and easier to break the encryption.

While the development of quantum computers represents an exciting new frontier in computing technology, it also raises serious concerns about the security of modern internet infrastructure. As researchers continue to explore the possibilities of quantum computing, it will be important to develop new forms of encryption that are resistant to these new types of attacks.

Quantum-Resistant Cryptography

Quantum-Resistant Cryptography refers to the development of cryptographic algorithms that are secure against the risk of quantum computers. The emergence of quantum computing technology poses a significant threat to the security of our current digital infrastructure, as it has the potential to undermine the cryptographic algorithms that underpin it. This includes everything from government and corporate communications to individual user data.

As such, a significant amount of research is currently being directed towards developing countermeasures to the potential threat of quantum computing. One possible solution is the use of symmetric key cryptography, which can be made more secure against quantum-based attacks by simply increasing the length of the shared key. Although asymmetric key cryptography has been the dominant form of encryption in recent years, the need to share a common secret key across an open channel has made symmetric key cryptography an attractive alternative.

Another potential solution to the security threat posed by quantum computing is quantum cryptography. This involves using the principles of quantum mechanics to develop encryption methods that are secure against eavesdropping. By detecting when a shared symmetric key has been tampered with or read by a third party, quantum cryptography could provide a means of securely sharing keys across an open channel.

Other methods being investigated to defeat possible quantum-based attacks include hashing to create larger message sizes and lattice-based cryptography. These methods aim to create types of encryptions that are difficult for quantum computers to crack, thus providing a means of ensuring the security of our digital infrastructure in the face of this emerging threat. Overall, the goal of quantum-resistant cryptography is to develop methods of encryption that are secure against the potential threat of quantum computing technology.

Quantum Computers & Bitcoin Mining

Bitcoin mining relies on cryptography, where miners compete to solve a cryptographic puzzle in exchange for a block reward. However, if a single miner gains access to a quantum computer, they could potentially dominate the network and compromise its decentralization. This could result in a 51% attack, which poses a serious risk to the security and integrity of the network. Despite this potential threat, some experts believe that it is not an immediate concern. The use of Application-Specific Integrated Circuits (ASICs) can mitigate the risk of such an attack, at least for the foreseeable future. Moreover, if multiple miners have access to a quantum computer, the risk of a single entity dominating the network is significantly reduced.

The risk of quantum computers being used to compromise the security of Bitcoin mining is a possibility, but it is not an immediate concern. The use of ASICs and the potential for multiple miners to have access to quantum computers helps to mitigate this risk. Nonetheless, it is important for the Bitcoin community to continue to monitor and evaluate the potential impact of quantum computing technology on the network's security.

A Glimpse into the Future

The development of quantum computing poses a significant threat to modern implementations of asymmetric encryption, which could compromise the security of critical infrastructure and data. However, it is important to note that this is not an immediate problem as there are still significant theoretical and engineering hurdles to overcome before quantum computing becomes a practical reality. Nevertheless, given the high stakes involved in data security, it is important to start preparing for the potential threat posed by quantum computing. A significant amount of research is currently underway to develop potential solutions that could be deployed to existing systems and future-proof them against the threat of quantum computers.

One possible solution is the development of quantum-resistant standards that could be distributed to the broader public, similar to the way in which end-to-end encryption was rolled out through popular browsers and messaging applications. Once these standards are finalized, the cryptocurrency ecosystem could easily integrate the strongest feasible defence against these potential attack vectors.

Overall, it is essential to start laying the foundation for a secure future by exploring potential solutions to the threats posed by quantum computing. By doing so, critical infrastructure and data can be protected against the potential risks of quantum-based attacks, thus ensuring the safety and security of our digital world.

Understanding Segregated Witness (SegWit) Protocol Upgrade for Blockchain Networks

Segregated Witness (SegWit) is a protocol upgrade that was introduced in 2015 as a solution to the scalability problem that blockchain networks were facing. The block size determines the number of transactions that can be confirmed in each block, and on average, the Bitcoin network validates a new block every 10 minutes, each containing multiple transactions. However, the Bitcoin blockchain could process only around 7 transactions per second, which is much slower compared to traditional payment solutions and financial networks that can process thousands of transactions per second. SegWit reorganizes block information so that signatures are no longer placed with transaction data, and it segregates the witnesses (signatures) from transaction data. This enables more transactions to be stored in a single block, increasing the transaction throughput of the network. In August 2017, the SegWit upgrade was implemented as a soft fork on the Bitcoin network, and many other cryptocurrency projects are now using SegWit, including Litecoin.

The SegWit upgrade provided several benefits, including improved transaction speed and block capacity. Additionally, it solved the Transaction Malleability Bug (TMB), which is an issue where transaction IDs can be modified before they are confirmed, leading to problems with transaction tracking and accounting. By separating the witness data from the transaction data, SegWit made TMB much less likely to occur. As a result, SegWit has significantly improved the overall efficiency and effectiveness of blockchain networks.

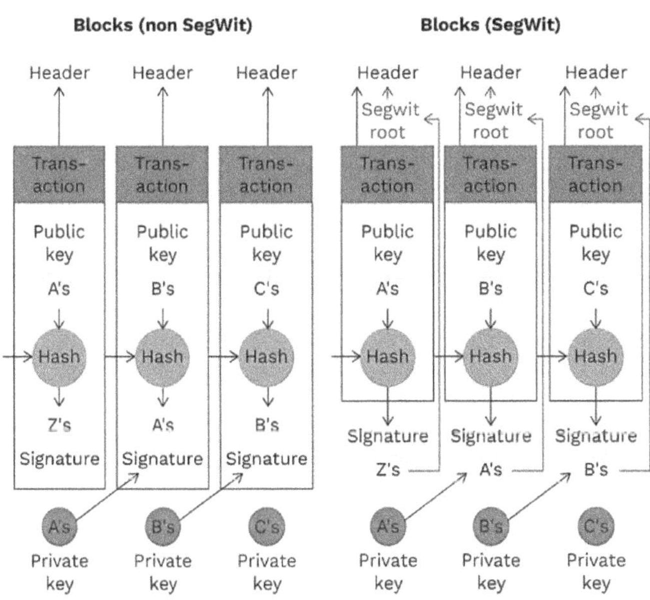

Figure 13.2: SegWit vs Non SegWit

What are the Primary Advantages of SegWit?

Increased Capacity & Improved Transaction Speed

Segregated Witness (SegWit) provides several benefits to the Bitcoin network. One of the most important advantages is the capacity increase it brings to the network. Before the implementation of SegWit, the block size limit was 1 MB, which restricted the number of transactions that could be included in a block. With the increase in the number of Bitcoin users, the network started facing scalability issues. SegWit addresses this issue by reorganizing the block information so that signature data is not placed with transaction data. The signature data is moved away from the transaction input, which results in the effective block size increasing from 1 MB to approximately 4 MB. This allows for more transactions to be included in a single block, which increases the overall transaction throughput of the network.

It is important to note that SegWit does not directly increase the block size limit, which would require a hard fork. Instead, it is an engineering approach to increase the effective block size limit while keeping the actual block size at 1 MB. The introduction of the concept of block weight further supports this approach. Block weight includes all block data, including transaction data and signature data (up to 3 MB), that is no longer a part of the input field. In addition to the capacity increase, SegWit also solved the Transaction Malleability Bug (TMB). This bug allowed for the modification of a transaction's unique identifier before it was confirmed in a block, which could lead to double-spending attacks. With SegWit, the transaction data is no longer a part of the block's identification, which solves the TMB issue.

Overall, the implementation of SegWit has been beneficial for the Bitcoin network and has paved the way for further improvements in transaction processing. Today, several cryptocurrencies, including Bitcoin and Litecoin, are utilizing SegWit to enhance their transaction speed and block capacity.

Increased Transaction Speed

Another significant advantage of SegWit is the increase in transaction speed. Since the new block structure can accommodate more transactions, the number of transactions processed in a single block is higher, leading to a higher transactions per second (TPS) rate. Although it may take the same time to mine a block, more transactions can be processed in it due to SegWit's block capacity increase. The boost in transaction speed is crucial for the adoption and use of cryptocurrencies. In the past, when the Bitcoin network could only process around 7 transactions per second, it led to significant delays and high transaction fees. However, SegWit has helped to reduce transaction fees and improve the transaction speed. Before SegWit's implementation, users would sometimes pay over $30 per transaction. However, with SegWit, the average transaction fee has reduced significantly to less than $1 per transaction, making Bitcoin more affordable and accessible for everyday use.

In addition to reducing transaction fees and increasing transaction speed, SegWit also provides other benefits such as fixing the Transaction Malleability Bug (TMB), enabling the Lightning Network, and making Bitcoin more scalable. Since its implementation, SegWit has become an essential upgrade for blockchain networks and is being adopted by more and more cryptocurrency projects, including Bitcoin and Litecoin.

Restoration of Transaction Malleability

A major issue with Bitcoin was the potential to modify transaction signatures, which could lead to corrupt transactions between parties. As the data stored on blockchains is immutable, any invalid transactions could be permanently recorded on the blockchain. However, with the implementation of SegWit, signatures are no longer considered a part of the transaction data. This separation of signatures from transaction data eliminates the possibility of tampering with the data, thus restoring transaction malleability. This restoration of transaction malleability has led to further innovation in the blockchain network, including the development of second-layer protocols and smart contracts. By solving the issue of transaction malleability, SegWit has made it possible to build more complex applications on top of the blockchain without the fear of invalid transactions or other security vulnerabilities.

SegWit & the Lightning Network

The Segregated Witness (SegWit) update to the Bitcoin protocol was a significant milestone in enabling the development of second-layer protocols. SegWit addressed the Transaction Malleability Bug, which had been a long-standing issue that prevented the implementation of off-chain scaling solutions. With SegWit, the transaction data is separated into two parts, one of which is the transaction signature data. This enabled the development of second-layer protocols that could operate independently of the Bitcoin network, while still leveraging its security features. One of the most notable second-layer protocols is the Lightning Network. The Lightning Network is an off-chain micropayment network that operates on top of the Bitcoin network. Its primary goal is to allow for faster and cheaper transactions by enabling users to send and receive transactions off-chain. The Lightning Network achieves this by creating a network of payment channels between users, which can be used to transact without the need for on-chain transactions.

With the Lightning Network, transactions are accumulated off-chain and buffered until they can be processed by the Bitcoin network. This means that users can transact with each other almost instantly, without having to wait for confirmation on the Bitcoin network. The Lightning Network also has the potential to significantly increase the transaction capacity of the Bitcoin network, which has been a major bottleneck for its adoption. While the Lightning Network was originally developed for Bitcoin, it has since been adopted by numerous other cryptocurrencies and blockchain projects. This has the potential to not only reduce confirmation times for transactions but also foster the development of new solutions to the scalability problem. By enabling off-chain transactions, second-layer protocols like the Lightning Network have opened up new possibilities for the blockchain ecosystem, and are likely to play a crucial role in its future development.

SegWit Vs. SegWit2x

SegWit and SegWit2x are two proposals for upgrading the Bitcoin protocol to address the issue of transaction malleability, which is a bug that allows attackers to modify transaction information and potentially steal funds. While both proposals share similarities, such as addressing the transaction malleability bug, there are also significant differences between them. SegWit is a soft fork upgrade, which means it is backward-compatible, and nodes that have not been updated to include SegWit can still process transactions. SegWit separates signature data from transaction data, increasing the capacity of each block and enabling faster and cheaper transactions. In contrast, SegWit2x is a hard fork upgrade that requires all nodes to update to the new protocol, and nodes that do not update will not be able to process transactions. SegWit2x also proposed to increase the block size from 1MB to 2MB to allow for more transactions to be processed per block. One of the key differences between SegWit and SegWit2x is the impact of their proposed changes on node operators and miners.

A larger block size would increase the amount of data that nodes and miners need to process, potentially leading to centralization and harming the security of the Bitcoin network. In contrast, SegWit's soft fork upgrade and separation of transaction signature data and transaction data reduce the amount of data that nodes and miners need to process, improving network scalability. Another difference is the level of community support. The SegWit proposal was widely supported and enforced by the Bitcoin community, leading to the creation of the User Activated Soft Fork (UASF) movement, which enabled users to activate SegWit without requiring support from Bitcoin miners. In contrast, SegWit2x faced controversy and was not as widely supported by the Bitcoin community, with developers unable to reach a consensus on its adoption and implementation.

Nested SegWit vs. Native SegWit (Bech32)

In the Bitcoin ecosystem, the introduction of Segregated Witness, or SegWit, brought forth significant enhancements in transaction processing and security. SegWit addresses come in two Flavors: Nested SegWit, also known as P2SH SegWit, and Native SegWit, which employs the Bech32 address format. Native SegWit is considered a more advanced iteration and an evolution of Nested SegWit.

Native SegWit's adoption of the Bech32 address format is a pivotal upgrade. It brings several noteworthy advantages to the table. Firstly, it enhances transaction speed, ensuring that Bitcoin transactions are processed more swiftly and efficiently. Secondly, Bech32 offers superior error detection, reducing the likelihood of errors during the address input process. Finally, and perhaps most crucially, Native SegWit translates into lower transaction fees, making it a cost-effective choice for Bitcoin users.

One notable user-friendly feature of Bech32 addresses is their all-lowercase format, which makes them more legible and easier to handle. This user-centric design choice contributes to a more seamless and intuitive Bitcoin experience.

Crucially, transactions between Non-SegWit (Legacy), Nested SegWit, and Native SegWit (Bech32) addresses are fully compatible. Bitcoin users can initiate transactions between these different address types without any issues, ensuring that the Bitcoin network remains inclusive and accessible to all.

However, it's essential to note that not all cryptocurrency wallets and exchanges have embraced SegWit support yet. Consequently, users may sometimes encounter limitations in their ability to directly deposit or withdraw funds to and from SegWit addresses.

An encouraging development in this regard is the adoption of SegWit by prominent cryptocurrency exchanges like Binance, one of the industry's largest players. Binance's support for SegWit enables users to leverage the benefits of Native SegWit, particularly the Bech32 address format, when conducting Bitcoin deposits and withdrawals. This translates into faster transaction confirmations and reduced fees, making the utilization of Bitcoin on Binance a more efficient and cost-effective experience for its users. As the cryptocurrency landscape continues to evolve, wider adoption of SegWit across exchanges and wallets promises to further enhance the usability and efficiency of the Bitcoin network. *(Ref. No.: 298- 302)*

Closing Thoughts

The blockchain and AI can revolutionize many industries by providing secure and transparent data storage and analysis. By combining these technologies, businesses can enhance their efficiency, accuracy, and security. The healthcare industry can use blockchain to create an immutable patient database, while AI can analyze the data to offer improved medical treatments. Blockchain can provide transparency in supply chain management, while AI can predict demand and optimize logistics. As these technologies advance, new applications could emerge, such as decentralized autonomous organizations (DAOs) were algorithms drive decision-making.

However, the threat of quantum computing on the security of cryptocurrency cannot be ignored. To combat this, there is a need to establish quantum-resistant standards that can be distributed widely, similar to how end-to-end encryption was rolled out. Once finalized, the cryptocurrency ecosystem can integrate the strongest feasible defence against these attack vectors with relative ease. In addition, SegWit is a protocol upgrade of Bitcoin that improves scalability, allowing blockchain networks to handle a larger number of transactions with more efficiency and lower costs. Despite its effectiveness, SegWit is yet to be widely adopted, with only around 53% of Bitcoin addresses currently using it. The wider adoption of SegWit and other scaling solutions could enable blockchain technology to reach its full potential in various industries.

The coal mining industry can benefit from blockchain technology as well. It can be used to track the movement of coal, from the point of extraction to its delivery to the end consumer, providing greater transparency and accountability. Furthermore, blockchain technology can improve the efficiency and safety of the coal mining industry by enabling the tracking of equipment usage, maintenance, and worker safety. By combining blockchain with IoT devices, the coal mining industry can achieve better operational efficiency, reduce costs, and improve safety.

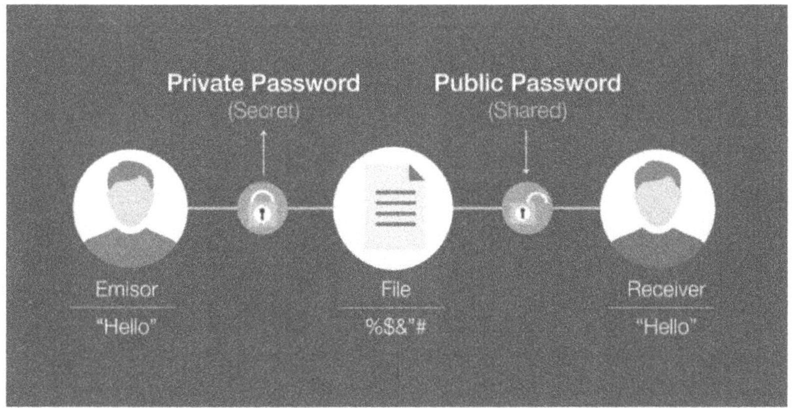

Chapter 14: Digital Signatures & Hash Functions

"You can write anything that you would be able to write on a server and put it onto the blockchain. Instead of JavaScript making calls to the server, you would be making calls to the blockchain."

— Vitalik Buterin

Understanding Digital Signatures: A Crucial Component of Cybersecurity

A digital signature is a technique used to verify the authenticity and integrity of digital information, such as a message or document, using cryptography. It functions similarly to a traditional handwritten signature but is more complex and secure. In essence, a digital signature is a code that is attached to a digital message or document. Once created, the code acts as proof that the message or document has not been altered or tampered with during transmission. This makes it an essential tool for ensuring the security and trustworthiness of digital communications. While the concept of using cryptography to secure communications has been around for centuries, the development of Public-Key Cryptography (PKC) in the 1970s made digital signatures possible. PKC is a cryptographic system that uses a pair of keys, one public and one private, to encrypt and decrypt data.

To understand how digital signatures work, it is necessary to understand the basics of hash functions and PKC. A hash function is a mathematical algorithm that takes in an input (such as a message) and produces a fixed-length string of characters (called a hash). The hash is unique to the input, meaning that even a small change in the input will produce a vastly different hash. When creating a digital signature, the sender of a message or document uses their private key to generate a hash of the message or document. They then encrypt the hash using their private key to create the digital signature. The recipient of the message can then use the sender's public key to decrypt the signature and verify the authenticity of the message or document.

Hashing in Digital Signature Systems

Hashing is a fundamental concept in digital signature systems. It involves converting data of any size into a fixed-size output through a specialized algorithm called a hash function. The result of a hash function is called a hash value or message digest. When paired with cryptography, hash functions can be used to create a unique digital fingerprint of the original message, known as a cryptographic hash value. This hash value is unique to the input message, which means that any change made to the input data would result in a completely different hash value.

Cryptographic hash functions are extensively used to verify the authenticity of digital information, as even a small change in the input message would result in a completely different hash value. This property makes hash functions essential for digital signature systems, where they are used to create a secure and tamper-proof digital signature that can verify the integrity and authenticity of a document.

Public-Key Cryptography (PKC)

Public-Key Cryptography (PKC) is a type of cryptographic system that operates by using two keys: a public key and a private key. The two keys are linked mathematically and can be utilized for both data encryption and digital signatures. PKC is more secure than other older methods of symmetric encryption. While traditional encryption methods require the use of the same key for both encrypting and decrypting information, PKC allows for data encryption using the public key and decryption using the corresponding private key. This makes it more secure as the private key is kept secret by the user and only the public key is shared. In addition to encryption, PKC can also be used for digital signatures. This technique involves hashing a message or digital data along with the private key of the signer. Later, the recipient of the message can check the signature's legitimacy by using the public key provided by the signer. It is essential to note that digital signatures can sometimes involve encryption, but not necessarily always. For instance, the Bitcoin blockchain uses PKC and digital signatures, but there is no encryption in the process. The Bitcoin blockchain employs the Elliptic Curve Digital Signature Algorithm (ECDSA) to authenticate transactions.

How Digital Signatures Work

In the context of cryptocurrencies, a digital signature system frequently consists of three fundamental steps: Hashing, Signing, and Verifying.

Hashing the Data

To generate a digital signature, the first step is to hash the message or digital data. Hashing involves inputting the information into a hashing algorithm, which generates a unique hash value, also known as the message digest. The hash function can handle messages of varying lengths, but all hash values have the same length after hashing, making it a crucial feature of a hash function.

While hashing is not always necessary for generating digital signatures since one can sign a message directly using a private key, cryptocurrencies often employ hashing. The reason for this is that fixed-length digests make the process easier to manage and handle. In the context of cryptocurrencies, hashing plays a crucial role in ensuring data integrity and verifying the authenticity of transactions.

Signing the Data

Signing is the next step in the digital signature process. Public-Key Cryptography is utilized at this point. Various digital signature algorithms exist, each with its own unique process. In general, the hashed message is signed with a private key, and the recipient of the message can verify its authenticity by using the corresponding public key provided by the sender. It is important to include the private key when generating the signature; otherwise, the recipient will not be able to use the corresponding public key to verify the signature. The sender of the message generates both the public and private keys, but only the public key is shared with the recipient. Unlike handwritten signatures that are often the same regardless of the message content, digital signatures are closely associated with the content of each message. Therefore, each digitally signed message can have a unique digital signature.

Verifying the Data

To verify the authenticity of a digitally signed message, the receiver needs to perform a final step called verification. This step ensures that the digital signature was created by the actual sender and that the message has not been tampered with in transit. The process of verification involves using the public key provided by the sender to decrypt the signature and retrieve the message's hash value. The receiver then generates a hash of the received message using the same algorithm as the sender and compares it to the decrypted hash value. If the two hash values match, the signature is valid, and the message is considered authentic. For example, if Alice sends a message to Bob, she will hash the message, sign it with her private key, and send the message along with the signature to Bob. Upon receiving the message, Bob will use Alice's public key to decrypt the signature, retrieve the hash value, and then generate a hash of the received message. If the two hash values match, Bob can be sure that the message came from Alice and that it has not been altered during transmission.

It is crucial for the sender to keep their private key secure as it is used to create the digital signature. If someone gains access to the private key, they can create fraudulent digital signatures and impersonate the sender, potentially causing harm. In the context of Bitcoin, if someone gets hold of Alice's private key, they could spend her Bitcoins without her knowledge or consent. Therefore, it is essential to keep the private key safe and secure.

Essential Benefits of Digital Signatures

Digital signatures are crucial because they provide three key benefits: information integrity, authentication, and non-repudiation. Firstly, they ensure information integrity, which means that the receiver of the message can confirm that the content of the message has not been modified or tampered with during transmission. This is possible because any modification of the message would result in a completely different signature. Therefore, the recipient can verify that the message they received is the exact message that was sent by the sender.

Secondly, digital signatures provide authentication, meaning that the receiver can confirm the identity of the sender. This is possible because only the sender has access to their private key, which they use to generate the digital signature. The recipient can use the corresponding public key to verify the signature and confirm that it was indeed created by the sender and not someone else.

Thirdly, digital signatures provide non-repudiation, which means that the sender cannot deny having sent the message in the future. Once the signature is generated, it cannot be altered, and the sender cannot claim that they did not sign it unless their private key has been compromised. Therefore, digital signatures provide an essential layer of security and accountability in electronic transactions and communications. *(Ref. No.: 207- 215)*

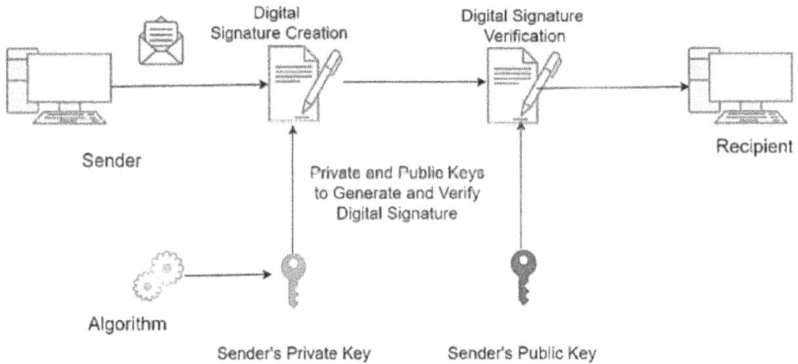

Figure 14.1: Blockchain & Digital Signatures

Applications of Digital Signatures in Various Industries

Digital signatures have various applications and can be implemented on many kinds of digital files and certificates.

Here are some of the most common use cases:

- **Information Technology:** Digital signatures are frequently used to enhance the security of internet communication systems. For instance, emails can be signed to verify their sender and ensure that the message has not been tampered with.
- **Finance:** In finance, digital signatures are used for audits, expense reports, mortgage agreements, and other financial transactions. These signatures provide a secure and tamper-evident way to verify the authenticity of the documents.
- **Legal:** Digital signatures are widely used for signing all sorts of business contracts and legal agreements, including government documents. They offer a way to sign and store documents electronically, reducing paper usage and simplifying the process of signing and verifying signatures.
- **Healthcare:** Digital signatures can help prevent fraud in prescriptions and medical records. By providing a secure way to sign and verify medical documents, healthcare providers can ensure that patient information is kept confidential and that records are accurate.
- **Blockchain:** In the context of blockchain technology, digital signature schemes are crucial to ensuring that only the rightful owners of cryptocurrencies can sign a transaction to move funds. As long as private keys are kept secure, digital signatures provide a tamper-evident and secure way to verify ownership and transfer of digital assets.
- **Coal & Mineral Mining Industry:** digital signatures can also have applications in the coal and mineral mining industry. For instance, they can be used to authenticate and secure contracts and agreements between mining companies and other parties, such as suppliers, distributors, and customers. Digital signatures can also be used to verify the authenticity and integrity of critical documents related to the mining process, such as exploration reports, mining permits, safety certificates, and environmental impact assessments.

Limitations:

Digital signature schemes have some limitations that need to be considered. These limitations mainly depend on three requirements: algorithm, implementation, and private key.

- **Algorithm:** The quality of the algorithms used in a digital signature system is crucial for its security. Trustworthy hash functions and cryptographic systems must be selected to ensure the authenticity of the digital signatures. The algorithms must be robust and difficult to crack, or else the digital signature scheme will be prone to hacking attacks.

- **Implementation:** Even if the algorithms used are of high quality, poor implementation can undermine the security of a digital signature system. Implementation errors can create vulnerabilities that can be exploited by attackers. Therefore, it is essential to have a secure and well-implemented system to avoid security breaches.

- **Private Key:** The private key is crucial to the authenticity and non-repudiation of digital signatures. If the private key is compromised, the properties of authenticity and non-repudiation will be invalidated. If a cryptocurrency user loses their private key, they may lose access to their funds, resulting in significant monetary losses. Therefore, it is crucial to keep the private key secret and secure from unauthorized access.

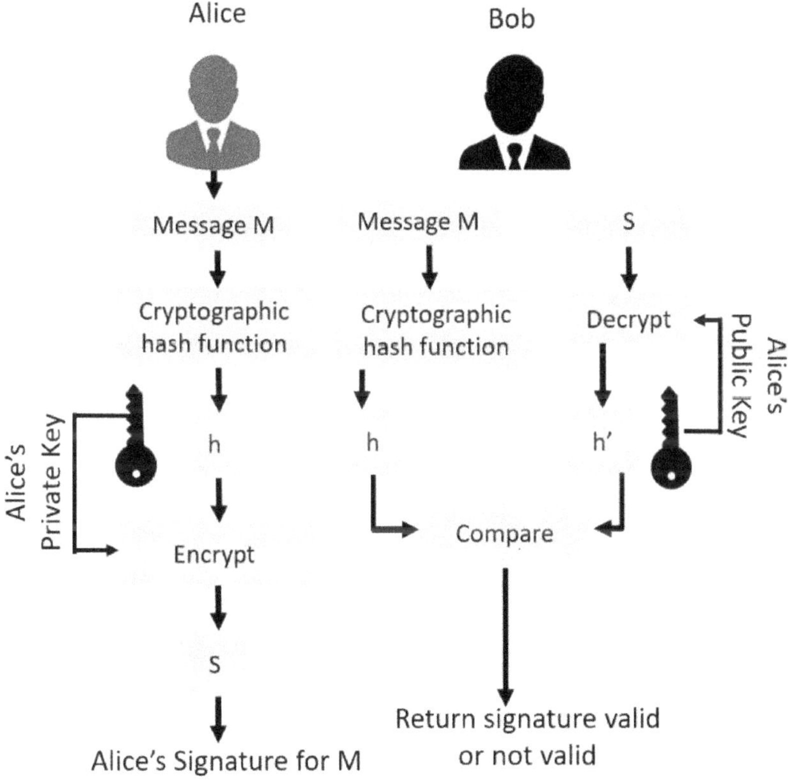

Figure 14.2: Electronic vs. Digital Signatures

Electronic Signatures vs. Digital Signatures

In simple terms, electronic signatures and digital signatures are both methods of signing documents electronically, but there is a difference between the two. Digital signatures are a specific type of electronic signature that uses cryptographic algorithms to verify the authenticity and integrity of a document. On the other hand, electronic signatures can include a variety of methods, such as scanned signatures, typing a name into a document, or even a checkbox indicating agreement. The key difference between electronic and digital signatures is the level of security and authentication.

Digital signatures use advanced cryptographic techniques to ensure that the signature is unique to the signer and that the document has not been tampered with. This involves creating a digital fingerprint of the document using a hash function, and then encrypting that fingerprint with the signer's private key. The encrypted fingerprint is then attached to the document, creating a unique digital signature. Electronic signatures, on the other hand, do not necessarily use cryptographic techniques and may not provide the same level of security and authentication. They may be as simple as a scanned image of a handwritten signature or a typed name on a document. While electronic signatures are often sufficient for basic agreements and contracts, they may not be legally enforceable in all jurisdictions. *(Ref. No.: 216- 222)*

Closing Thoughts

Digital signature systems, enriched by the robustness of hash functions and the principles of Public-Key Cryptography, have cemented their status as indispensable tools across various sectors, and the coal and mineral mining industry is no exception. These systems, when deployed effectively, yield transformative benefits, enhancing security, ensuring the integrity of data, and enabling the authentication of a wide range of digital documents, from contracts and permits to compliance records.

In the context of mining, the incorporation of digital signatures offers a multitude of advantages. It starts with the optimization of processes for acquiring permits and adhering to regulatory requirements. By replacing traditional pen-and-paper signatures with their digital counterparts, mining companies can streamline operations by reducing paperwork, expediting approval processes, and significantly enhancing the accuracy of record-keeping.

Moreover, the utility of digital signatures extends to the secure exchange of sensitive information among diverse stakeholders within the mining ecosystem.

This includes seamless and confidential communication between mining operators, contractors, and regulatory authorities. The trust and security offered by digital signatures play a pivotal role in maintaining the integrity of data transfers, thereby fostering a conducive environment for productive collaborations and strict regulatory compliance.

The convergence of digital signatures with blockchain technology introduces new dimensions of utility. In the mining sector, this synergy holds promise in various applications, such as verifying ownership of mining rights or meticulously monitoring the movement of mineral resources across the complex supply chain. Blockchain's immutable ledger serves as a verifiable and unalterable repository of mining activities, cultivating an environment steeped in compliance and accountability.

It is crucial to acknowledge that the shift from outdated paper-based processes to cutting-edge digital solutions is an ongoing journey in the mining industry, mirroring trends seen in various sectors. As the mining sector continues to embark on its digital transformation voyage, the adoption of digital signature systems is poised for significant growth. This transition represents a herald of operational efficiency, transparency, and trust within the mining sphere, affirming the industry's commitment to embracing the technological advancements that define the future.

Chapter 15: Understanding Cryptocurrency Wallets: Types & Functions

"Blockchain software companies may end up being amalgamated into existing software giants, at which point blockchain patents will just become part of the existing patent war."

— Vitalik Buterin

Crypto Wallets Explained

Crypto wallets are indispensable tools for engaging with blockchain networks, serving as the gateway to your digital assets. These wallets come in various forms, and they are commonly classified into three overarching categories: Software Wallets, Hardware Wallets, and Paper Wallets. Additionally, they are sometimes categorized based on their operational status as either Hot Wallets or Cold Wallets. Software wallets constitute the most prevalent category and are known for their convenience. They are primarily software applications that can be installed on your computer or smartphone, providing easy access to your digital assets through an internet connection. However, this convenience comes with a trade-off, as software wallets are potentially susceptible to hacking and malware attacks, necessitating robust cybersecurity practices. In contrast, hardware wallets are the epitome of security in the realm of crypto wallets. These physical devices are specifically designed for offline storage of digital assets. The private keys, the cryptographic keys essential for accessing and managing your cryptocurrencies, never leave the confines of the hardware wallet. This characteristic makes hardware wallets exceptionally resilient to hacking attempts, providing peace of mind, especially for users safeguarding substantial digital wealth. Conversely, paper wallets, while once a common choice, have now largely fallen out of favor due to their inherent limitations. These wallets entail the creation of physical copies, usually in the form of printed QR codes, containing both your public and private keys. Unfortunately, paper wallets are prone to various vulnerabilities. They can be damaged, lost, or even stolen, which poses significant risks to the security of your digital assets. In the dynamic landscape of cryptocurrency, selecting the most suitable wallet type depends on factors such as the volume of digital assets held, the desired level of security, and the ease of use. While software wallets offer convenience but require vigilance against cyber threats, hardware wallets stand out as the most secure option. Paper wallets, once popular, have become obsolete due to their inherent vulnerabilities, making them an unreliable choice in the modern crypto landscape.

Cryptocurrency Wallets & Digital Signatures

Cryptocurrency wallets are often misunderstood as digital asset storage devices. In reality, they are tools that enable users to interact with blockchain networks. These wallets generate the necessary data to send and receive cryptocurrency through blockchain transactions, including pairs of public and private keys. In addition, each wallet has a unique address which serves as an identifier on the blockchain network. This address can be shared with others to receive funds, but the private key should never be shared. The private key is crucial because it grants access to an individual's cryptocurrencies, regardless of which wallet they use. This means that if one device is compromised, the user can still access their funds on another device as long as they have the corresponding private key or seed phrase. It is important to note that the coins themselves never leave the blockchain; they are simply transferred from one address to another. There are three types of cryptocurrency wallets: software, hardware, and paper wallets. Software wallets are the most common and convenient, while hardware wallets offer the highest level of security. Paper wallets, which consist of a printed copy of the public and private keys, are now considered outdated and unreliable.

Is it obligatory to have a Crypto Wallet to Trade Crypto?

The simple answer is yes. Having a crypto wallet is a requirement for trading cryptocurrencies. Regardless of whether one is a frequent trader or a long-term holder, a wallet address is necessary to store and exchange cryptocurrencies. There are various types of wallets available, including a hot wallet provided by a crypto exchange, a mobile wallet on an Android or IOS smartphone, a browser extension, a computer wallet, or a hardware wallet. The market offers numerous alternatives to choose from. Below are some examples of the different wallet types:

1. **Hot wallet:** Binance exchange.
2. **Mobile crypto wallets:** Trust Wallet, MetaMask.
3. **Browser extension crypto wallets:** MetaMask, MathWallet, Binance Chain Wallet.
4. **Desktop crypto wallets:** Electrum, Exodus.

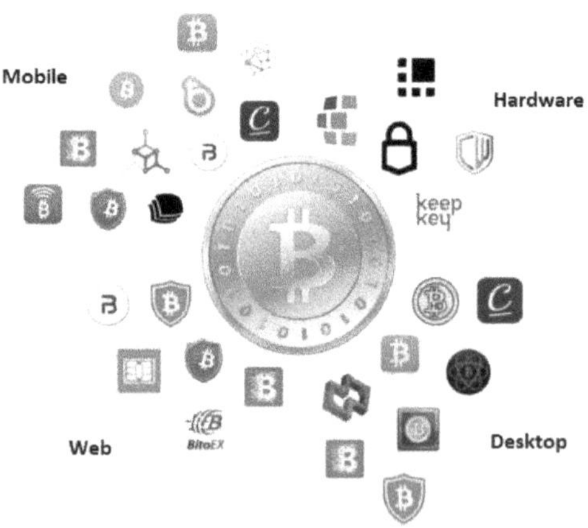

Figure 15.1: Crypto Wallets

Hot vs. Cold Wallets

In the world of cryptocurrency, wallets are essential tools for storing and trading digital assets. However, not all wallets are created equal. Crypto wallets can be classified as either "Hot" or "Cold" based on their mode of operation.

Hot wallets, as the name suggests, are those that are connected to the internet. They are typically provided by crypto exchanges or can be downloaded as software applications for mobile phones or computers. The advantage of hot wallets is their accessibility, which makes them ideal for traders and frequent users who need quick and easy access to their funds. However, the downside of hot wallets is that they are vulnerable to hacking attempts, as they are always connected to the internet.

On the other hand, cold wallets are offline wallets that are not connected to the internet. They use a physical medium to store the private keys, which makes them resistant to online attacks. Cold wallets can be hardware devices, paper wallets, or even offline computers. As a result, they are considered to be a much safer alternative to hot wallets, making them ideal for long-term investors or HODLer's.

It is worth noting that most cryptocurrency exchanges only hold a small percentage of coins in their hot wallets, with the majority kept in cold storage. Cold storage is a technique used by exchanges to store funds offline, away from potential hackers. Decentralized exchanges (DEX) are an alternative for customers who prefer not to keep their funds in a centralized exchange. These exchanges offer a decentralized trading platform that allows users to have complete control over their private keys, while also enabling them to trade directly from their cold storage devices, such as hardware wallets.

Pros & Cons of Crypto Wallets

Pros	Cons
Self-ownership of money	User responsibility
Censorship-resistant transactions	Learning curve
Quick and easy access	Chance of making mistakes

Table 15.1: Pros & Cons of Crypto Wallets

Non-custodial crypto wallets are gaining popularity among crypto users due to their unique benefits. These wallets enable users to hold their private keys, ensuring complete ownership and control over their digital assets. Here are some advantages of using non-custodial crypto wallets:

1. **Self-Ownership of Money:** Non-custodial crypto wallets allow users to have complete control over their funds. Unlike traditional banks, where the money is technically owned by the bank, the digital assets held in non-custodial wallets are entirely owned by the user. This means that users have complete autonomy over how they manage their funds.

2. **Censorship-Resistant Transactions:** Non-custodial crypto wallets provide users with the freedom to send and receive transactions to whomever they like, whenever they like. The decentralized nature of cryptocurrencies ensures that no one controls the network, making it almost impossible for anyone to censor or stop transactions. This feature is particularly valuable for those who live in countries with strict capital controls or for anyone who values their financial privacy.

Criteria	Hot Wallet	Cold Wallet
Definition	Crypto wallets are basically applications connected to the internet that stores private keys is an online environment	A crypto wallet that stores private keys in an offline environment
Working	Hot wallets are connected to the internet, and crypto coins are delivered directly to the wallets through a fast online transaction process	Transactions begin online and then shift to the offline environment for private key signing by the cold wallet. The completed information is then sent back to the online network.

Suitable Users	Cryptocurrency traders or individuals who want to make quick online payments with cryptocurrencies.	People who want to store their cryptocurrencies with the highest levels of security.
Types	Online hot wallets are available in exchanges such as Binance & Coinbase. Two most popular hot wallets are Exodus, Mycelium	Cold wallets are two types, such as paper wallets and hardware wallets. Popular hardware wallets include Trezor, Ledger
Security	Vulnerable to hackers due to connectivity with the internet	Private keys are never exposed online, thereby improving security.

Table 15.2: Hot wallet vs Cold wallet Comparison

Software Wallets

There is a wide range of software wallets available, each with their own distinct features. However, most of them are internet-connected, or "Hot Wallets." Here are some brief descriptions of the most common and necessary types of software wallets:

1.   Web Wallets
2.   PC/Computer Wallets
3.   Mobile Wallets.
4.   Hardware Wallets
5.   Paper Wallets

Web Wallets

Web wallets are a type of software wallet that can be accessed through a web browser without the need to download or install any software. This type of wallet is commonly used for accessing blockchain networks and can be used for exchanging or trading cryptocurrencies. Web wallets can either be exchange wallets or other browser-based wallet providers. With web wallets, users can create a new wallet and set a personal password to access it. However, some providers manage and hold the private keys of their customers, which is a risky practice since it means trusting one's cash to someone else.

To address this issue, many web wallets now offer customers the ability to manage their own private keys, either entirely or through shared control (via multi-signatures). It is important to check the technical approach of each wallet before deciding on the most suitable one for your needs. It is also recommended to make use of the available security tools when using cryptocurrency exchanges. Overall, web wallets provide a convenient way to access blockchain networks but it is important to exercise caution when using them to ensure the safety of one's funds.

PC/Computer Wallets

A desktop wallet, as the name suggests, is a type of software application that a user downloads and installs on their local computer. This software provides complete control over the user's funds and private keys. When a new desktop wallet is created, a file called "wallet.dat" is stored locally on the computer. This file contains the private key information that allows access to the cryptocurrency addresses, so it must be encrypted with a strong password to ensure the security of the wallet.

In order to access a desktop wallet, a user must provide their password each time they open the application. If the wallet is encrypted and the password is lost or forgotten, the user may lose access to their funds. Therefore, it is crucial to back up the wallet.dat file or export the corresponding private key or seed phrase to ensure that the funds can be accessed from another device if the computer stops working or becomes inaccessible.

Desktop wallets are generally considered to be safer than web-based wallets since the private keys are stored locally, but it is essential to ensure that the computer/device running the application is free of viruses and malware before setting up and using a cryptocurrency wallet. It is also advisable to use the latest version of the wallet software and regularly update it to ensure the latest security features and patches are in place.

Mobile Wallets

Mobile wallets are a type of software application designed specifically for smartphones, either Android or IOS. They function similarly to desktop wallets but with a more convenient and user-friendly interface. With mobile wallets, users can send and receive cryptocurrencies easily using QR codes, making them a suitable choice for daily transactions and payments. One of the most prominent examples of a mobile crypto wallet is Trust Wallet. Mobile wallets are generally convenient for users who are always on-the-go, as they can easily access their cryptocurrency funds using their mobile devices. However, just like desktop wallets, users must ensure that their mobile devices are free of viruses and malware before installing and using a mobile crypto wallet.

It is also important to encrypt mobile wallets with a strong password and back up the private keys or seed phrases in case the phone gets lost or broken. Without a backup, users may lose access to their funds forever. Therefore, it is essential to take the necessary precautions and secure mobile wallets properly.

Hardware Wallets

Hardware wallets are specialized physical devices designed to store and secure private keys for cryptocurrency transactions. These wallets use a random number generator (RNG) to generate both public and private keys, which are then saved within the device itself. The keys are never exposed to the internet, meaning that hardware wallets are considered a type of cold wallet and are considered the most secure alternative for storing cryptocurrency. While hardware wallets offer superior levels of security compared to hot wallets, which are connected to the internet, there are still risks to consider. The firmware implementation of the device must be done properly to ensure the device's security. Additionally, hardware wallets can be less user-friendly than other types of wallets, and accessing funds can be more difficult. To overcome this lack of accessibility, hardware wallets can be connected to decentralized exchanges (DEXs) or web wallet service providers to access trading platforms. This method is considered to be a more secure way of accessing funds, as private keys are never exposed to the internet. It is recommended to use a hardware wallet if you plan to hold cryptocurrency for the long term or if you are holding large amounts of cryptocurrency. Most hardware wallets come with security features, such as a unique PIN code and recovery phrase, to protect against loss or theft of the device. *(Ref. No.: 223- 226)*

Paper Wallets

A paper wallet is a physical piece of paper on which a crypto address and its private key are printed in the form of QR codes. This allows users to execute cryptocurrency transactions by scanning the codes with a compatible device. Some paper wallet websites allow users to download code to generate new addresses and keys offline, making these wallets resistant to online hacking attacks and a potential alternative to cold storage. However, the use of paper wallets is currently considered insecure and should be discouraged. It is crucial to understand the risks before using them. The main flaw of paper wallets is that they are not suitable for sending partial funds, only the entire balance at once. For example, if a user generates a paper wallet and funds it with multiple transactions totalling 10 BTC, they cannot spend just 2 BTC. Instead, they must send the entire balance to another wallet type, such as a software or hardware wallet, and then spend the desired amount from there.

If a user imports the private key of their paper wallet into a desktop wallet and spends part of the funds, the remaining coins will be sent to a "change address" generated through the Bitcoin protocol. If the user does not manually set the change address to one, they control, they risk losing their funds. Most software wallets handle the change for their users, sending the remaining coins to an address that is part of the wallet. It is important to remember that after the first transaction, a paper wallet will be empty regardless of the amount sent. It cannot be reused later and must be replaced with a new paper wallet. Therefore, due to these flaws, paper wallets are currently considered an insecure option for storing cryptocurrency and should be avoided. Instead, users should consider alternative wallet types, such as hardware or software wallets, that offer greater security and usability.

The Significance of Backups

It can be very expensive to lose access to a cryptocurrency wallet, which is why it is crucial to regularly back up your wallet. In most cases, this involves backing up wallet.dat files or seed phrases. A seed phrase acts as a root key that generates and provides access to all keys and addresses in a crypto wallet. If a user chooses to encrypt their wallet with a password, it is recommended to back up the password as well. Backing up your wallet is an important step in ensuring that you do not lose access to your cryptocurrency, which could result in significant financial loss. By regularly backing up wallet.dat files or seed phrases, users can ensure that they can recover their wallet in the event of a hardware failure or other unexpected event.

Seed phrases are particularly useful since they allow users to recover their wallet on a different device, and they can be stored securely offline. It is important to remember that backing up your wallet should be done securely. This means using encrypted backups, storing backups in multiple secure locations, and ensuring that backups are regularly updated to reflect any changes to the wallet.

By taking these steps, users can help ensure that they don't lose access to their cryptocurrency and can continue to manage their funds with peace of mind.

Choosing the Right Cryptocurrency Wallet: Factors to Consider

Choosing the right cryptocurrency wallet largely depends on the individual's specific needs and preferences. If someone frequently trades cryptocurrencies using the web, then a software or web wallet can provide quick and easy access to funds for convenient trading. By implementing additional security measures such as two-factor authentication (2FA), users can enhance the security of their accounts and ensure their cryptocurrencies are protected against unauthorized access. However, if someone holds a significant amount of cryptocurrency and does not plan to sell it in the short term, then cold wallets are a more secure alternative. Cold wallets are physical devices that are not connected to the internet, making them highly resistant to online phishing attacks or scams.

By using a cold wallet, users can store their cryptocurrencies offline and significantly reduce the risk of hacking or theft. It is essential to note that cold wallets can be less user-friendly and less convenient to access compared to web wallets. However, for those who prioritize security over convenience, a cold wallet is a better option. Ultimately, the choice of a cryptocurrency wallet should be based on the user's specific needs, priorities, and security preferences. Regardless of the type of wallet one chooses, it is important to implement basic security measures such as strong passwords, regular backups, and the use of 2FA to ensure the safety of their cryptocurrencies. *(Ref. No.: 227- 230)*

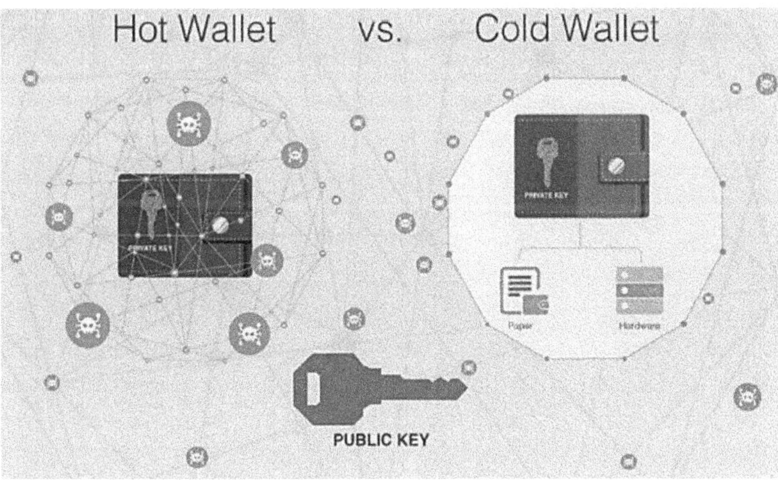

Figure 15.2: Protecting Crypto Assets

Closing Thoughts

Cryptocurrency wallets occupy a pivotal role within the digital currency ecosystem, serving as the primary interface between individuals and blockchain networks. These wallets are the key to secure transactions, enabling users to send and receive funds while maintaining control over their private keys. As our discussion has elucidated, the world of cryptocurrency wallets is diverse, with various options offering their own distinct advantages and limitations. Hence, it becomes paramount for individuals venturing into digital currencies to dedicate time to comprehensive research. Understanding the specific characteristics of different wallet types empowers users to make well-informed choices that align with their unique needs and preferences. Security emerges as the foremost concern when selecting a cryptocurrency wallet. Given the irreversible nature of blockchain transactions and the inherent risks of the digital landscape, safeguarding one's digital assets is of utmost importance. Therefore, users should prioritize wallets that implement robust security features. These features may include strong encryption, two-factor authentication (2FA), and the option to store private keys offline in cold storage solutions like hardware wallets or paper wallets. This precautionary approach helps protect against hacking attempts and unauthorized access.

While security is non-negotiable, user-friendliness plays a vital role in promoting the broader adoption of cryptocurrencies. A user-friendly wallet interface simplifies tasks such as sending and receiving funds, checking balances, and managing transactions, making the experience accessible even to those with limited technical knowledge. Additionally, wallet providers should invest in user education, offering tutorials, guides on identifying phishing attempts, and best practices for securing digital assets. Such resources empower users to navigate the cryptocurrency landscape confidently and securely.

It is essential to recognize that the cryptocurrency wallet landscape is continually evolving. New solutions and technologies emerge, while older ones may become obsolete or susceptible to vulnerabilities over time. Staying informed about the latest developments and updates in the world of cryptocurrency wallets is crucial to maintaining the security and functionality of one's digital assets. In conclusion, selecting the right cryptocurrency wallet is not merely a matter of storage; it is an investment in the security and usability of one's digital financial future. Users should exercise diligence, prudence, and an understanding of their specific needs when making this critical decision.

Chapter 16: Understanding Stablecoins & Coin Burning in Cryptocurrency

"Pretty uniformly, people want the benefits of bitcoin and the blockchain – near-instant transfers, globally available on any Internet-connected device, highly secure, and nearly-free value transfers."

— Jeremy Allaire

Understanding Stable Coins: An Overview of Digital Assets Designed for Price Stability

Stable coins are a type of digital asset that is designed to mimic the value of fiat currencies, such as the US dollar or the euro. These coins enable users to transfer funds across borders quickly and affordably while maintaining price stability. While cryptocurrencies like Bitcoin and Ethereum offer exciting technological advancements in terms of mediums of exchange, they are often highly volatile when priced against fiat currencies. The lack of tethering to any asset means that their value can fluctuate dramatically and unpredictably, rendering them unsuitable for making payments or acting as a reliable store of value. Stable coins, on the other hand, offer a solution to the volatility problem. These digital assets are pegged to an underlying asset or fiat currency, enabling them to maintain a stable price. As such, they function as dependable safe-haven assets in volatile markets. There are several mechanisms used to maintain the stability of stable coins, and this chapter will discuss some of them in detail. By understanding the advantages and limitations of stable coins, users can determine if they are suitable for their needs.

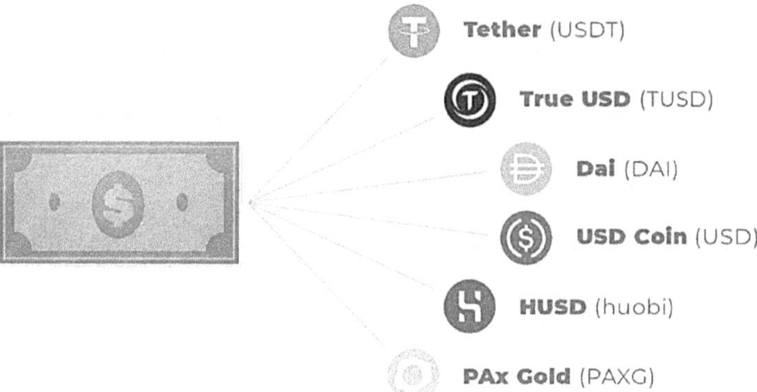

Figure 16.1: Few Popular Stable Coins

Understanding the Mechanisms of Stablecoins

Stablecoins are digital assets that are designed to maintain a stable value, often pegged to a fiat currency such as the US dollar or euro. They provide a reliable way to transfer funds across borders while minimizing the risk of price fluctuations. In contrast to cryptocurrencies like Bitcoin and Ethereum, which are known for their volatility, stablecoins can be an attractive alternative for those seeking a stable store of value.

However, not all stablecoins are created equal. There are several different types of stablecoins, each with their own unique mechanisms for achieving price stability. In this chapter, we will explore some of the most common categories of stablecoins and how they work. By understanding the different approaches that stablecoins take, you can make more informed decisions when it comes to choosing which stablecoin to use.

Fiat-Backed Stable Coins

One of the most popular types of stablecoins is the fiat-collateralized stablecoin. These stablecoins are backed by fiat currency with a 1:1 ratio. A central issuer or bank holds an amount of fiat currency in reserve and issues tokens in a proportionate quantity. For instance, if a company holds a million dollars, they distribute a million tokens worth a dollar each. Users can freely exchange these tokens like cryptocurrencies and redeem them for their equivalent in USD anytime they want. However, there is a high degree of counterparty risk associated with this type of stablecoin because the issuer or company must be trusted. It is difficult for users to determine with confidence whether the company holds funds in reserve. Although the issuing organization can publish audits to increase transparency, the system is far from trustless. For example, Binance, a popular cryptocurrency exchange, offers BUSD, which maintains a peg with the US dollar, and BGBP, which tracks the British Pound, as fiat-backed stablecoins.

Crypto-Backed Stablecoins

Crypto-backed stablecoins are another type of stablecoin that replicates the value of fiat currencies but uses cryptocurrency as collateral instead of fiat currency. The issuance of units is managed by smart contracts, which makes them trustless, but the governance of the financial policy is still determined by the network participants. To acquire a crypto-backed stablecoin, users need to lock their cryptocurrency into a contract, which then issues the equivalent stablecoins. When they want to retrieve their collateral, they need to pay back the stablecoins with some interest into the same contract.

The mechanisms that enforce the peg of crypto-backed stablecoins vary among different systems. In general, game theory and on-chain algorithms are used to incentivize contributors to maintain price stability. These mechanisms can involve the creation of an over-collateralization pool, where users have to put in more cryptocurrency than the stablecoin's value to ensure stability. Alternatively, the system may automatically liquidate the collateral if the value of the cryptocurrency used as collateral falls below a certain threshold. Popular examples of crypto-backed stablecoins include Dai, which is backed by Ether, and USDX, which is backed by Bitcoin. Crypto-backed stablecoins provide a trustless alternative to fiat-backed stablecoins, but they still come with their own set of limitations and risks. It's important to understand how each stablecoin mechanism works and its specific risks before investing or using them for transactions.

Algorithmic Stablecoins & Black Swan Events

Algorithmic stablecoins are a type of stablecoin that differs from fiat-collateralized and crypto-collateralized stablecoins. Rather than being backed by a physical asset like fiat currency or cryptocurrency, algorithmic stablecoins use algorithms and smart contracts to control the supply of tokens and maintain a peg to a specific value. This is similar to the way central banks manage national currencies. In an algorithmic stablecoin system, the token supply is adjusted based on the price of the fiat currency it is tracking. If the price falls below the target value, the system will reduce the token supply, while if the price exceeds the target value, new tokens will be introduced to the market to reduce the value of the stablecoin.

While algorithmic stablecoins are not technically collateralized, they may have a pool of collateral to address unforeseen and highly volatile market movements, which are often referred to as Black Swan events. These events are unexpected and have a significant impact on the market. The term "Black Swan" was originally coined by the Roman poet Juvenal, but it was further developed by statistician and trader Nassim Nicholas Taleb in his book "The Black Swan: The Impact of Highly Improbable." According to Taleb, Black Swan events are outliers beyond regular expectations, carry a significant impact, and are often rationalized after their occurrence. Examples of previous Black Swan events include the rise of the internet, the personal computer, the dissolution of the Soviet Union, and the September 11, 2001 attacks.

Stablecoins: Types & Examples

In practice, collateralized stablecoins are the most widely used. Some well-known examples include USD Tether (USDT), True USD (TUSD), Paxos Standard (PAX), USD Coin (USDC), and Binance USD (BUSD). However, there are also instances of the other categories of stablecoins available on the market. Bitshares USD and DAI are examples of crypto-collateralized coins, while Carbon and the now-defunct Basis are algorithmic versions.

It is worth noting that this list of stablecoins is not exhaustive, as there are hundreds of stablecoin projects currently available in the market. The high number of stablecoin projects is an indication of the growing demand for stable digital currencies, particularly in the context of cryptocurrency trading. These stablecoins provide a means to escape the extreme volatility associated with cryptocurrency trading while still being able to transact using digital currencies. Additionally, stablecoins have other applications, including enabling cross-border payments and remittances and facilitating peer-to-peer lending.

Advantages & Disadvantages of Stablecoins

Stablecoins have become increasingly popular in the cryptocurrency world due to their potential to provide stability and predictability to digital currencies. They offer several advantages, the most significant of which is their ability to serve as a medium of exchange that enhances the usability of cryptocurrencies. This is because stablecoins provide a level of stability that cryptocurrencies currently lack, making them more suitable for everyday use in payment processing. Moreover, by acting as a buffer against volatility, stablecoins can help integrate cryptocurrencies with conventional financial markets. These markets are currently separate ecosystems with few connections. However, with stablecoins available, cryptocurrencies could see increased usage in mortgage and credit markets that have traditionally been dominated by government-issued fiat currencies. *(Ref. No.: 231- 238)*

In addition to their usefulness in financial transactions, stablecoins can be used by traders and investors to hedge their portfolios. By allocating a portion of their portfolios to stablecoins, investors can effectively reduce overall risk. Moreover, stablecoins can be used as a store of value that can be used to buy different cryptocurrencies when prices drop, or as a means to "lock-in" profits made while prices rise, without the need to cash out. However, stablecoins also have their limitations. Fiat-collateralized versions are much less decentralized than typical cryptocurrencies because a central entity is required to maintain the supporting assets. Crypto-collateralized and uncollateralized coins, on the other hand, require users to trust the wider community and the source code to ensure the stability and durability of the systems. Since these are relatively new technologies, they may need some time to mature fully.

Understanding Coin Burning & Binance's BNB Coin Burn Events

Coin burning is a strategy to remove cryptocurrencies permanently from circulation, resulting in a decrease in the overall supply. Binance Coin (BNB

ERC-20) can be used as an example to explain how this technique works. The previous contract for Binance Coin (BNB), while it was on the Ethereum network, is available here. Before Binance Coin migrated to its own blockchain, Binance carried out regular Coin Burn events using a smart contract function called the burn function. These events occur every quarter until 100,000,000 BNB are destroyed, which accounts for 50% of the total BNB ever issued (200,000,000 BNB). The number of BNB coins to be burned is calculated based on the volume of trades executed on the exchange during a three-month period. Thus, after each quarter, Binance burns BNB based on the overall trading volume. Despite the regularity of BNB Coin Burn events, many people still struggle to understand how they are executed. Therefore, this chapter aims to provide important details about the burn function and the quarterly BNB Coin Burn events.

The Process of Token Burning & its Implications on Cryptocurrencies

A token burn event follows a certain sequence of actions:

1. A holder of a cryptocurrency invokes the burn function to destroy a specified number of coins.
2. The smart contract then verifies that the individual indeed possesses the coins in their wallet and that the specified number is valid, as the burning mechanism only accepts positive numbers.
3. If the participant lacks sufficient coins or enters an invalid number (e.g., 0 or -5), the burn function will not be executed.
4. If the holder has enough coins in their wallet, they will be subtracted from the wallet, and the total supply of the coin will be updated, signifying the permanent destruction of the burned coins.
5. Once the burn function is performed, the coins are irreversibly destroyed and can never be recovered. The evidence of coin burning can be easily verified on a blockchain explorer, thanks to blockchain technology.

```
109
110 -   function burn(uint256 _value) returns (bool success) {
111         if (balanceOf[msg.sender] < _value) throw;           // Check if the sender has enough
112         if (_value <= 0) throw;
113         balanceOf[msg.sender] = SafeMath.safeSub(balanceOf[msg.sender], _value);      // Subtract from the sender
114         totalSupply = SafeMath.safeSub(totalSupply,_value);               // Updates totalSupply
115         Burn(msg.sender, _value);
116         return true;
117     }
118
```

The BNB contract offers a burn function that enables participants to remove a specific number of coins from circulation permanently. The blockchain network records all burn transactions, ensuring transparency and allowing any participant to verify that the coins are destroyed. Following each quarterly Coin Burn, Binance makes an official announcement regarding the quantity of BNB coins that were burned based on trading volume. You can review all BNB ERC-20 Coin Burn transactions on an Ethereum blockchain explorer such as Etherscan. The transactions are public, irreversible, and permanently recorded on the blockchain. On Etherscan, details regarding the burning transaction can be found on the Input Data box.

```
Function: burn(uint256 _value)


MethodID: 0x42966c68
[0]:    000000000000000000000000000000000000000000000000000015c20a8ddf4de8407ffff
```

When a token burning event takes place, it is recorded as a transaction on the blockchain network. These transactions are transparent, irreversible, and permanently recorded on the blockchain. To verify the details of a token burning transaction, you can use an Ethereum blockchain explorer, such as Etherscan.

On Etherscan, the information about a burning transaction can be found in the Input Data box. By clicking the "Decode Input Data" button, you can see the amount of the token that was burned. It is important to note that this number includes the token's decimal places, which for most tokens is 18.

For example, if a burning transaction for Binance Coin (BNB) was recorded on the blockchain, the Input Data box on Etherscan would display the details of the transaction. By decoding the input data, you can verify the amount of BNB that was burned. If the decoded input data displays 1,623,818, it means that 1,623,818 BNB were burned in that transaction.

#	Name	Type	Data
0	_value	uint256	1623818000000000000000000

The Contemporary Binance Burn Function

Following the launch of the Binance Chain, BNB ERC-20 tokens were replaced with native Binance Coins (BNB BEP-2), which shifted the Coin Burn events to the Binance Chain from the Ethereum network. All previous BNB ERC-20-coin burns were replicated on the Binance Chain to ensure supply consistency, including the 11,654,397 BNB ERC-20 tokens burned on Ethereum. Users can verify this transaction on the Binance Chain Explorer and check the total supply of BNB.

The new BNB coin burn mechanism relies on a specific command executed on the Binance Chain instead of a smart contract. Further information can be found on the Binance Chain Docs page. As of October 2021, 17 BNB Coin Burn events have been completed, with 33,199,679 BNB coins burned, reducing the Total Supply by 16.59% to 166,801,147 BNB. *(Ref. No.: 239- 246)*

Closing Thoughts

The cryptocurrency market is constantly evolving, and two mechanisms that have been playing a crucial role in its evolution are stablecoins and coin burning. Stablecoins are digital currencies designed to maintain a stable value against a particular asset or basket of assets. They are gaining popularity for various use cases, such as facilitating cross-border payments, providing a stable store of value, and enabling the execution of smart contracts. Stablecoins offer several advantages to users, including stability and predictability that are often lacking in other cryptocurrencies. This makes them attractive to merchants and businesses who can conduct transactions with confidence, knowing that the value of the stablecoin is unlikely to fluctuate widely. Moreover, stablecoins are faster and cheaper to transact with than traditional payment systems, making them a more efficient option. Additionally, stablecoins can be used to provide banking services to unbanked and underbanked populations, promoting greater financial inclusion. Coin burning, on the other hand, is a technique used to regulate the supply and demand dynamics of cryptocurrencies. By reducing the supply of a particular cryptocurrency, coin burning can increase the value of the remaining coins, stimulating demand and driving up prices, which can be more attractive to investors. However, coin burning can also have unintended consequences. For example, if too many coins are burned, it can lead to liquidity problems and reduced trading volumes. Moreover, the process of coin burning is often controlled by centralized entities, which can be seen as arbitrary and opaque. Despite these limitations, both stablecoins and coin burning have important roles to play in the development and growth of the cryptocurrency market. Stablecoins offer a reliable and stable alternative to other cryptocurrencies, while coin burning can help regulate supply and demand dynamics, increase the value of a particular cryptocurrency, and incentivize investors. As the cryptocurrency market continues to mature, it is likely that stablecoins and coin burning will remain essential tools for investors and traders.

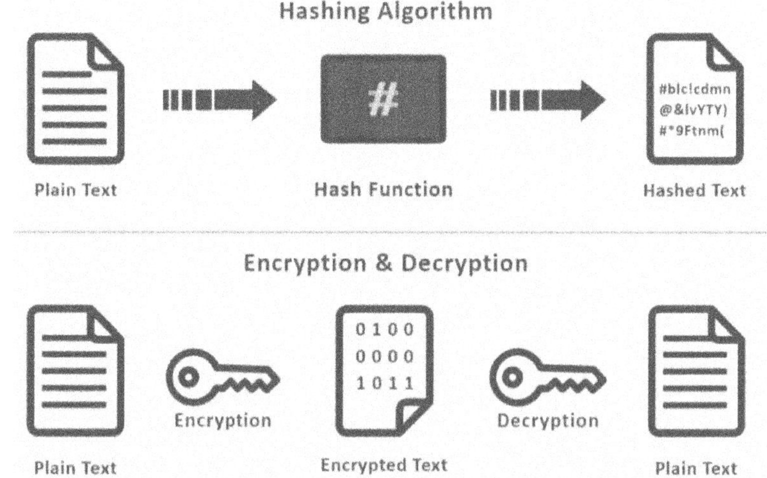

Chapter 17: Hash Functions, Mining & Forks in Blockchain Technology

"The blockchain concept was pioneered within the context of crypto-currency Bitcoin, but engineers have imagined many other ways for distributed ledger technology to streamline the world. Stock exchanges and big banks, for example, are looking at blockchain-type systems as trading settlement platforms."

— Anthony Scaramucci

Importance of Hashing in Cryptocurrencies & Distributed Systems

Hashing is a process that involves generating a fixed-length output, known as a hash, from an input of variable length. This is achieved using mathematical formulas known as hashing functions or algorithms. While some hashing functions do not involve cryptography, cryptographic hashing functions are critical to the functioning of cryptocurrencies. They ensure the integrity and security of information in blockchains and other distributed systems. One of the key features of hashing algorithms, both conventional and cryptographic, is that they are deterministic. This means that given the same input, the algorithm will always produce the same output. The resulting hash is a unique representation of the input and is used to verify the integrity of data.

Cryptographic hashing algorithms are typically designed as one-way functions, which means that it is nearly impossible to recreate the input from the output alone. This property makes them extremely useful for secure data storage and transmission. The harder it is to locate the input given the output, the more secure the hashing algorithm is considered to be. In summary, hashing is a crucial tool in the world of cryptocurrencies and distributed systems. It ensures the integrity and security of data by creating a unique and nearly impossible to reverse hash from an input of variable length.

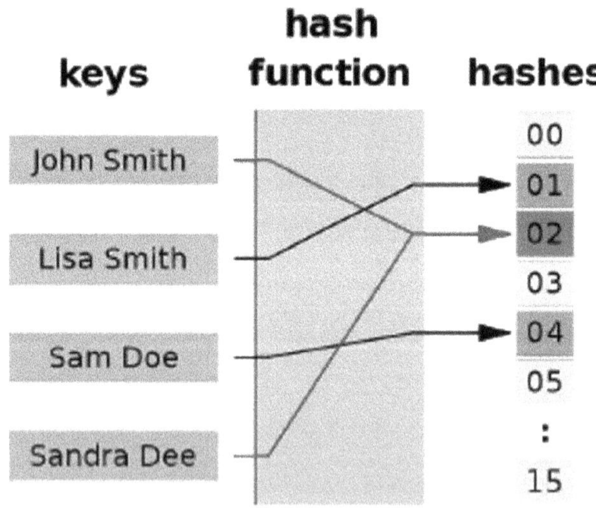

Figure 17.1: Hashing Algorithm

(A hash function that maps names to integers from 0 to 15. There is a collision between keys "Smith" & "Sandra Dee")

Understanding Cryptographic Hashing & its Role in Information Security

Hashing is a process of generating a fixed-length output, called a digest or hash, from a variable-length input. This is done using mathematical formulas known as hashing functions or algorithms. Cryptographic hashing functions are widely used in cryptocurrencies, enabling blockchains and other distributed systems to ensure information security and integrity. It is important to note that different hashing algorithms will produce outputs of varying sizes, but the possible output sizes for each algorithm are always constant. For instance, the SHA-256 algorithm can only produce outputs of 256 bits, while the SHA-1 algorithm will always generate a 160-bit digest. For example, if we run the phrases "Exploring Blockchain" and "exploring blockchain" through the SHA-256 hashing algorithm used in Bitcoin, we will get two completely different hash values. However, since we are using SHA-256, the output will always have a fixed size of 256 bits (or 64 characters), regardless of the input size. Additionally, no matter how many times we run the two phrases through the algorithm, the outputs will remain consistent. On the other hand, if we run the same phrases through the SHA-1 hashing algorithm, we will get different results. The acronym SHA stands for Secure Hash Algorithms, which include a set of cryptographic hash functions such as SHA-0 and SHA-1 algorithms, as well as the SHA-2 and SHA-3 groups. The SHA-256 algorithm belongs to the SHA-2 group, along with SHA-512 and other variants, which are considered secure. It is important to note that most hashing algorithms used in cryptocurrencies are designed as one-way functions, meaning that it is easy to generate the output from the input but very difficult to do the reverse (i.e., generate the input from the output alone). The difficulty in finding the input, also known as the pre-image, is a measure of the security of the hashing algorithm. The harder it is to locate the input, the more secure the hashing algorithm is considered to be.

SHA-256	
Input	**Output (256 bits)**
Exploring Blockchain	0de4a2f06ef4be73667703ccc2aaee2a167122f14a 563d84b98c167bb68f793c

exploring blockchain	d2fb3afbd3630f974e070a9953b02101592fb0f805 64c0f3b6be8f6f658b849d

Table 17.1: Working of SHA256 Hashing Algorithm

SHA-1	
Input	**Output (160 bits)**
Exploring Blockchain	320a227b333ee0939c4d04b6c24b7d1b58ae00d9
exploring blockchain	2c0c8bf82616f2fbe4211b209d8f3e09afb750d0

Table 17.2: Working of SHA-1 Hashing Algorithm

Importance of Hashing in Cryptocurrencies & Blockchain Technology

Hash functions are a crucial tool in both conventional and cryptographic applications. Conventional hash functions find use in database searches, data management, and large-scale data analysis. Cryptographic hash functions are widely employed in information security packages, such as digital fingerprinting and message authentication. In the world of Bitcoin, cryptographic hash functions play a vital role in the mining process, as well as in the generation of new addresses and keys. The real value of hashing lies in its ability to quickly verify the accuracy and integrity of significant amounts of data. For example, an individual can run a large dataset through a hash function and then use the resulting hash output to verify the integrity of the original data. This is possible because of the deterministic nature of hash functions: the input will always produce the same output (hash) if the input does not change. This simplifies the process of data verification and eliminates the need to store and remember large amounts of data.

Hashing is particularly useful in the context of blockchain technology. The Bitcoin blockchain uses hashing extensively in its mining process and to link and solidify groups of transactions into blocks. Almost all cryptocurrency protocols depend on hashing to create blocks that can be linked together effectively, forming a secure and decentralized blockchain. In summary, hashing is a powerful technique that simplifies the process of data verification and integrity checking. Both conventional and cryptographic hash functions are important tools in various applications, including database management, data analysis, and information security. In the context of blockchain technology, hashing is a crucial component in ensuring the security and integrity of the blockchain.

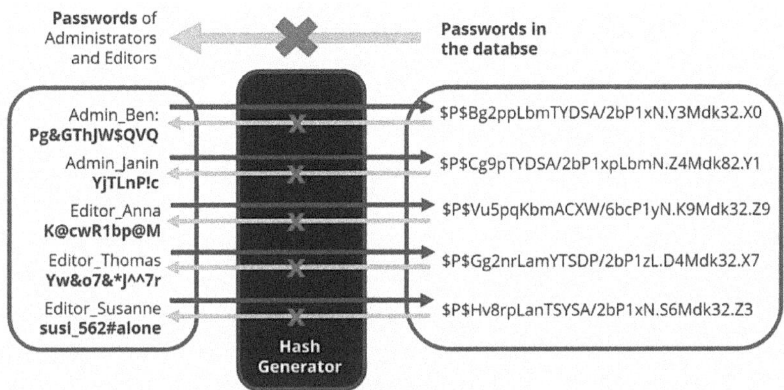

Figure 17.2: Hash Function: Password Encryption

Importance of Cryptographic Hash Functions

A cryptographic hash function is a type of hash function that utilizes cryptographic techniques to secure the output. In contrast to conventional hash functions, cryptographic hash functions are difficult to reverse, and breaking them requires numerous brute-force attempts. A brute-force attack is a trial-and-error method of guessing what the input was until the corresponding output is produced. However, there is also the possibility of different inputs producing the exact identical output, known as a "collision."

For a cryptographic hash function to be considered effectively secure, it must comply with three properties: collision resistance, preimage resistance, and second preimage resistance. Collision resistance means that it is infeasible to locate any distinct inputs that produce the identical hash as output. Preimage resistance means that it is infeasible to "revert" the hash function and locate the input from a given output. Second-preimage resistance means that it is infeasible to discover any second input that collides with a specified input.

These three properties of cryptographic hash functions make them an essential tool in information security and cryptography. They are used to verify the integrity of data, authenticate messages, and provide digital fingerprints. In the context of blockchain technology, cryptographic hash functions play a vital role in securing the Bitcoin blockchain. They are used in the mining process, the generation of new addresses and keys, and linking and solidifying groups of transactions into blocks, creating a blockchain.

Understanding Collision-Resistance in Hash Functions

A collision happens when different inputs result in the same hash output. Therefore, a hash function is deemed collision-resistant until a collision is discovered. It is important to note that collisions are inevitable for any hash function due to the infinite number of possible inputs and the finite number of outputs. In other words, a hash function is considered collision-resistant when the probability of finding a collision is so low that it would take an impractical amount of time to compute it (e.g., hundreds of thousands of years). While there are no hash functions that are entirely collision-free, some are strong enough to resist brute-force attacks, such as SHA-256. Out of the various SHA algorithms, the SHA-0 and SHA-1 groups are no longer secure since collisions have been found. At present, the SHA-2 and SHA-3 groups are deemed to be collision-resistant.

Preimage Resistance in Hash Functions

Preimage resistance is a property that is closely related to the concept of one-way functions. A hash function is considered to have preimage resistance if it is extremely difficult for an attacker to find the input that produced a given output. It is important to note that this property is different from collision resistance, as an attacker attempting to break preimage resistance would be trying to guess the input based on a known output, while a collision occurs when different inputs produce the same output. Preimage resistance is a valuable property for protecting data, as the hash of a message can prove its authenticity without revealing the message itself. In practice, many service providers and web applications store and use hashes generated from passwords instead of the passwords in plaintext. This allows for authentication without the risk of passwords being exposed in the event of a security breach.

Significance of Second-Preimage Resistance in Hash Functions for Information Security.

Second-preimage resistance can be thought of as a middle ground between collision resistance and preimage resistance. When a hash function is second-preimage resistant, it means that it is difficult to find an input that generates the same output as another input that is already known. In contrast to a collision attack, which involves finding any two distinct inputs that produce the same hash, a second-preimage attack involves finding an input that generates the same hash as an already known input. This makes it a more specific and targeted attack. It is important to note that a hash function that is resistant to collisions will also be immune to second-preimage attacks since the latter always implies a collision. However, a collision-resistant function may still be vulnerable to preimage attacks since it only requires finding a single input from a single output. The property of second-preimage resistance is crucial in information security, particularly in the context of digital signatures and message authentication. A hash function that is resistant to second-preimage attacks ensures that a digital signature is securely bound to a message and cannot be used to create a fraudulent signature for another message.

Bitcoin Mining & the Role of Hash Functions in Blockchain Security

Bitcoin mining involves various steps that require the use of hash functions, including checking balances, linking transactions inputs & outputs, and hashing transactions within a block to form a Merkle Tree. However, one of the critical factors contributing to the security of the Bitcoin blockchain is the extensive hashing operations carried out by miners to discover a legitimate key for the next block. In order to validate their block, miners need to generate a hash value for their candidate block by attempting numerous distinctive inputs. The mining difficulty determines the number of zeros required at the start of the output hash, and it changes according to the hash rate committed to the network. The hash rate represents the amount of computer power invested in Bitcoin mining, and if it increases, the mining difficulty automatically adjusts so that the average time needed to mine a block stays close to 10 minutes.

Miners do not have to find collisions as there are multiple hashes that can serve as a valid output starting with a certain number of zeros. Therefore, several possible answers for a certain block exist, and miners only need to find one of them, according to the threshold determined by the mining difficulty. Since Bitcoin mining is a cost-intensive task, miners have no incentive to cheat the system as it could lead to significant financial losses. Moreover, the more miners join a blockchain, the bigger and stronger it becomes, making it even more secure.

Upgrading Cryptocurrency Networks: Hard Forks & Soft Forks

When it comes to updating digital banking apps, people often update them without much thought. It is a necessary process to ensure access to services. However, in open-source cryptocurrencies, the process is very different. Although it is not required to examine every line of code in Bitcoin to use it, having the option to do so is crucial. In a decentralized system, there is no hierarchy or central authority that can push updates and make changes as they please. Thus, implementing new features in blockchain networks can be a challenge. Upgrading cryptocurrency networks is crucial to ensure that the network remains secure and functional. However, doing so in a decentralized system presents a unique set of challenges. To achieve this, blockchain networks use distinct mechanisms known as hard forks and soft forks.

A hard fork occurs when the existing blockchain network splits into two separate chains, with each chain following different rules. This happens when there is a fundamental change to the network's protocol, making older versions of the software incompatible with the new one. Hard forks are often contentious, as they can lead to a split in the community and the creation of a new cryptocurrency.

A soft fork, on the other hand, is a non-contentious upgrade that is backward-compatible with older versions of the software. In a soft fork, the new protocol is a subset of the old protocol, meaning that nodes running older versions can still validate transactions on the network. Soft forks are typically used for minor upgrades and bug fixes. Upgrading a cryptocurrency network is a complex process that requires the consensus of the network's participants. For a hard fork, the majority of the network's nodes and miners must agree to adopt the new protocol, while a soft fork requires only a majority of the network's hash rate. The upgrade process can take several months, and it is essential to ensure that the new software is thoroughly tested before it is implemented to prevent any security vulnerabilities or bugs. *(Ref. No.: 247- 254)*

Participants in Blockchain Governance & Decision-Making

Governance in a blockchain network refers to the process by which participants reach agreement on decisions related to the network, such as updating the protocol or implementing new features. In the case of Bitcoin, the decision-making process is a bit more decentralized than in traditional systems. Developers are responsible for proposing changes to the Bitcoin protocol and creating software to implement those changes. They may introduce updates or new features that improve the network's performance, security, or usability. Miners, on the other hand, are responsible for validating transactions on the network. They compete to solve complex mathematical problems in order to add new blocks to the blockchain and earn rewards. Full node users are individuals who run a copy of the Bitcoin software on their device, allowing them to participate in the network as a whole. They receive and validate transactions, and can even create their own if they wish. While light nodes (such as wallets) can be used to send and receive transactions, they are not considered participants in the decision-making process to the same extent as developers, miners, and full node users. This is because light nodes rely on full nodes to validate transactions and enforce the rules of the network.

In summary, governance in a blockchain network involves a diverse set of participants, each with their own role to play. Developers propose changes, miners validate transactions, full node users run the software and help enforce the rules, and light nodes interact with the network through their full node counterparts.

Role of Developers in a Cryptocurrency Network

Developers play a pivotal role in the cryptocurrency ecosystem as they are the individuals or groups responsible for conceiving, developing, and maintaining the code that forms the foundation of a cryptocurrency network. This code is essentially the software that enables the functioning of the blockchain, including transaction processing, consensus mechanisms, and security protocols.

Role of Miners in a Cryptocurrency Network

Within a blockchain network, miners assume a pivotal role as the guardians of security and the keepers of the ledger. Their responsibilities extend far beyond mere transaction processing, encompassing various facets of network integrity. At the heart of their tasks lies transaction validation—scrutinizing each transaction to ensure it complies with the network's rules, thus mitigating fraudulent or erroneous entries. Yet, the primary function of miners is the creation of new blocks on the blockchain. This involves a competition in cryptographic puzzle-solving, a distinctive feature of the Proof of Work (PoW) consensus algorithm used in networks like Bitcoin. Miners commit substantial computational power to solve complex mathematical puzzles, with the first to succeed granted the privilege of adding a new block to the chain. Moreover, miners contribute to network decentralization. Unlike centralized systems, blockchain networks are open to participation by anyone, and miners exemplify this democratic ethos. Individuals from around the world can participate in mining, contributing their computing resources and strengthening the network's resilience. Miners play a crucial role in maintaining network security, as they invest computational resources to validate and record transactions, fortify the blockchain's integrity, and champion decentralization—a core tenet of blockchain technology.

Role & Decision-Making of Full Node Users in a Cryptocurrency Network

In a cryptocurrency network, several contributors participate in the decision-making process or governance of the network. These contributors include developers, miners, and full node users, each with unique roles and responsibilities. Developers are responsible for creating and updating the code that runs the cryptocurrency network. They allow everyone to contribute to this process through open-source code, which helps developers submit modifications and reviews. Without developers, a software application cannot be successful, and they are also responsible for fixing bugs and adding new features to the network.

Miners are the nodes that secure the blockchain network by running the cryptocurrency code and devoting resources to adding new blocks to the blockchain. In the Bitcoin network, miners accomplish this through Proof of Work and are rewarded for their efforts in the form of a block reward. Full nodes, on the other hand, are the backbone of the cryptocurrency network. They validate, send, and receive blocks and transactions and maintain a copy of the blockchain. Full node users can overlap with developers and miners, or they can be all three or none. However, many cryptocurrency users do not take on any of these roles and instead choose to use light nodes or centralized services. While developers and miners play essential roles in the network, full nodes hold considerable power. This is because the network is opt-in, meaning users can select what software program they are running. Developers cannot coerce anyone to download Bitcoin Core binaries, and if miners force an unwanted change on any user of the network, the users can simply deny the request. It is crucial to note that these parties are not powerful overlords but are simply service providers. The network's value is determined by its users, and if people decide not to use the network, the coin will lose value. This directly impacts miners, whose rewards are worth much less when denominated in dollars. As for developers, they remain unnoticed by users.

Additionally, the software is not proprietary, and anyone can make any updates they want to make. If others execute the modified software application, they can communicate, and in that case, the software will be forked, and a brand-new network will be created.

Decoding the Concept of Forks

A blockchain fork is when the blockchain network splits into two separate chains due to a difference in software code. This is similar to a fork in the road, where the two paths diverge from a common point. Forking is a common occurrence in open-source projects, where developers can copy and modify code to create new projects while building off a common foundation. However, the blockchain space has two distinct types of forks: hard forks and soft forks, which are not found in other software development projects. In a hard fork, the new chain is completely separate from the original one, and the software changes are not compatible with the previous version. This means that a hard fork requires all nodes to update to the new software for the chain to continue to function. Hard forks result in a new cryptocurrency being created, with its own set of rules and features. In contrast, a soft fork is a more backward-compatible version of a hard fork. The new software changes made in a soft fork can still work with the previous version of the software, and nodes that have not updated can still participate in the network. However, the updated nodes will have an advantage in terms of features and capabilities.

Overall, forking in blockchain networks is a way for developers to make changes and add new features to the network, while still allowing the community to choose which direction they want to take. It also serves as a way to resolve conflicts and disagreements within the community.

Hard Forks Vs. Soft Forks

Although Hard Forks & Soft Forks share similar names and objectives, they are fundamentally different. Here, we will examine the characteristics of each type.

Hard Forks in Blockchain Technology

A hard fork in blockchain occurs when there is an update in the software application that is not backward compatible. This means that new rules conflict with old ones, resulting in a split in the blockchain network. The nodes that upload new rules can only communicate with other nodes that operate the new version. Consequently, the blockchain splits into two separate networks: one that operates on the old rules and another that operates on the new rules. After the hard fork, nodes running on the old version turn yellow and reject the blue nodes running the new version. In contrast, blue nodes connect to each other, forming a separate network. Although the two networks continue to propagate blocks and transactions, they operate independently, and the nodes have unique blocks and transactions. Before the hard fork, all nodes had an identical blockchain. However, after the fork, each network has a unique blockchain history. Because of the shared history, if someone held coins before the fork, they would have coins on both networks after the fork. For example, if a person had 5 BTC when a fork occurred at Block 600,000, they could spend those 5 BTC on the old chain in Block 600,001, but not on the new blockchain's Block 600,001. Nonetheless, assuming the cryptography remained unchanged, their private keys would still hold 5 coins on the forked network.

An example of a hard fork was the 2017 fork that split Bitcoin into two separate chains: the original Bitcoin (BTC) and a new one called Bitcoin Cash (BCH). The split occurred after a long-standing dispute over the best approach to scaling. Bitcoin Cash proponents wanted to increase the block size, while Bitcoin proponents opposed the change. Since the nodes could only receive blocks smaller than 1 MB, any 2MB block created, even if valid, would be rejected by other nodes. Therefore, only nodes that had modified their software to accept blocks exceeding 1MB could communicate. This made them incompatible with the previous version of the protocol.

Soft Fork in Blockchain Technology

Soft forks, unlike hard forks, are backward-compatible upgrades that enable upgraded nodes to communicate with non-upgraded nodes. Soft forks introduce a new rule that is not in conflict with the old rules, and therefore, nodes running the old software can still operate on the network. An example of a soft fork could be a decrease in block length. For instance, in Bitcoin, there is no limit to how small a block can be. Therefore, if a user wants to accept only smaller blocks, they can easily reject larger ones. Soft forks do not automatically disconnect users from the network. Although filtered information from nodes not enforcing the rules would have to be done manually. An excellent example of a soft fork is Segregated Witness (SegWit), which was deployed after the Bitcoin/Bitcoin Cash split. SegWit updated the block and transaction layout, but it was designed cleverly. Old nodes could still validate blocks and transactions, but they would not understand them. Only nodes running the new software could parse other data. Although there are benefits to upgrading to the new software, not all nodes have upgraded even two years after SegWit activation. There is no urgency to upgrade since there is no network-breaking change.

Exploring the Differences between Hard Forks & Soft Forks

In the realm of blockchain technology, forks play a pivotal role in implementing updates and alterations to the underlying protocol. There are two primary categories of forks: hard forks and soft forks, each serving unique purposes and exhibiting distinct characteristics. Hard forks represent significant and, at times, transformative software updates that introduce substantial modifications to the blockchain's protocol. These updates are "backward-incompatible," meaning they result in a permanent divergence or split in the blockchain network. Nodes that have not upgraded their software become unable to communicate with those running the upgraded version.

Hard forks can manifest in two forms:

contentious or planned. In contentious hard forks, the blockchain community experiences a lack of consensus, leading to divisions within the network. Such divisions can give rise to entirely new blockchain networks with differing rules and features. Conversely, planned hard forks occur when the community reaches a consensus to modify the software program, with widespread agreement among participants.

Soft Forks: A Gentle Evolution

On the opposite end of the spectrum are soft forks, representing "backward-compatible" upgrades. These upgrades introduce new rules that do not conflict with the existing ones, ensuring that upgraded nodes can continue to communicate with non-upgraded nodes. Soft forks are less disruptive and offer a more conservative approach to network modifications. The key characteristic of soft forks is their compatibility with the existing rules, preventing the fragmentation of the network. Users can implement updates without concern about creating permanent splits, as long as the new changes do not conflict with the old rules.

Choosing Between Hard & Soft Forks: Context Matters

The decision to implement a hard or soft fork hinge on the specific circumstances and objectives of the blockchain community. Hard forks, with their transformative capabilities, offer greater flexibility in modifying the network's rules and features. However, their potential for disruption and the creation of permanent network divisions must be carefully considered. In contrast, soft forks provide a gentler means of evolving the network, preserving compatibility with the existing rules. They are less disruptive and do not lead to permanent splits in the network. The choice between hard and soft forks ultimately depends on the goals, priorities, and consensus within the blockchain community. *(Ref. No.: 255- 262)*

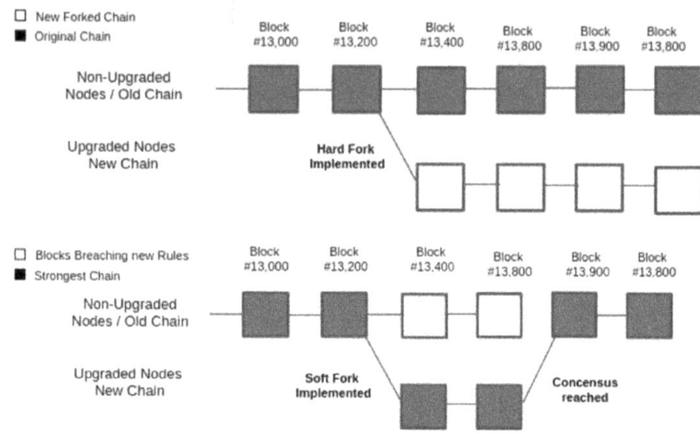

Figure 17.3: Hard forks vs. soft forks

Closing Thoughts

Hash functions are cryptographic tools that transform data into a fixed-size output, making them useful for verifying the integrity and authenticity of data. In the context of cryptocurrencies, hash functions are used to secure and authenticate transactions on the blockchain. When a new block is added to the blockchain, it is first hashed using a cryptographic hash function, and this creates a tamper-evident record of the transaction. Hard forks and soft forks are mechanisms that allow blockchain networks to upgrade and evolve over time. Hard forks create a new blockchain and cryptocurrency, while soft forks are backward-compatible upgrades that do not create a new chain. Forks are crucial to the long-term success of blockchain networks because they allow for updates and enhancements to be made without requiring a centralized authority. This enables the integration of new features and improvements, such as increased transaction speed, reduced fees, and improved privacy. Without forks, blockchain networks would be unable to adapt to changing market conditions or user needs, limiting their potential for growth and innovation.

Chapter 18: Mitigating Blockchain Security Threats

"Blockchain technology has such a wide range of transformational use cases, from recreating the plumbing of Wall Street to creating financial sovereignty in the farthest regions of the world."

— Perianne Boring

Sybil Attacks: Threats to Online Networks & Blockchain Security

A Sybil attack is a type of security threat that targets online networks, where an individual tries to manipulate the network by creating multiple accounts, nodes, or computers. This can be done by a single person who tries to gain control over the network by impersonating multiple users or devices. The attacker can then use these fake accounts or nodes to carry out malicious activities, such as spamming, phishing, or spreading malware.

One common example of a Sybil attack is creating multiple social media accounts to manipulate public opinion or boost the popularity of a particular account or post. In the world of cryptocurrencies, Sybil attacks are more relevant because they can affect the integrity of the blockchain network. For instance, if an attacker creates multiple nodes on a blockchain network, they can gain control over the network and manipulate transactions, double-spend coins, or launch 51% attacks.

The term "Sybil" originates from a case study about a woman named Sybil Dorsett, who had Dissociative Identity Disorder (DID), also known as Multiple Personality Disorder. The term was coined by John R. Douceur in 2002 to describe the vulnerability of peer-to-peer networks to attacks where a single user could create multiple identities. The term has since been used to describe attacks on any type of online network where an attacker can create multiple identities to deceive or manipulate the network.

Sybil Attacks: Impact on Network Integrity, Blockchain Security & Prevention Measures

Sybil attacks can have severe consequences on the integrity of online networks and the security of blockchain systems. Attackers can create multiple fake identities, or Sybil identities, to gain control over the network. This can allow them to out-vote honest nodes, block other users from the network, and refuse to receive or transmit blocks.

In the case of a large-scale Sybil attack, where attackers control a significant portion of the network's computing power or hash rate, they can carry out a 51% attack. This enables them to change the order of transactions and prevent transactions from being confirmed. They may also reverse transactions made while they were in control, resulting in double-spending.

Computer scientists have invested considerable time and effort into developing methods to detect and prevent Sybil attacks. However, there is no fool proof defence against this type of attack. Various prevention measures include using proof-of-work or proof-of-stake algorithms, reputation systems, and identity verification. Nevertheless, attackers continuously develop new techniques to bypass these defences, and it is an ongoing challenge to keep online networks and blockchain systems secure.

Sybil Attack Mitigation in Blockchains: Consensus Algorithms & their Role

Blockchains deploy various consensus algorithms to mitigate the risk of Sybil attacks. These algorithms include Proof of Work (PoW), Proof of Stake (PoS), and Delegated Proof of Stake (DPoS). While these algorithms do not provide fool proof protection against Sybil attacks, they make it impractical for attackers to carry out such attacks effectively. For instance, Bitcoin's blockchain employs a set of rules that dictate the creation of new blocks. One of these rules requires that the ability to create a block should be proportional to the processing power of the PoW mechanism. This means that a participant needs to have the computing power required to create a new block, which can be very difficult and expensive for an attacker to achieve. Since mining Bitcoin requires intensive computing power, miners have a strong incentive to maintain the integrity of the network and avoid Sybil attacks. In PoS and DPoS algorithms, validators are chosen based on the amount of cryptocurrency they hold and stake as collateral. This makes it difficult for attackers to accumulate a significant amount of stake without spending a considerable amount of money. Additionally, PoS and DPoS algorithms enable the network to reach consensus faster than PoW algorithms, making it more efficient and less susceptible to attacks.

While consensus algorithms provide some level of protection against Sybil attacks, attackers continue to develop new techniques to bypass these defences. Therefore, it is crucial to continuously improve and update these algorithms to stay ahead of potential threats and maintain the security of blockchains.

The Evolution & Examples of Sybil Attacks

Sybil Attack, named after Sybil Dorsett, the subject of the book "Sybil," was first introduced in a paper titled "The Sybil Attack" by John R. Douceur at Microsoft Research. Since then, Sybil attacks have evolved and become prevalent in various online platforms and networks. One example of a Sybil attack is the alleged Russian interference in the 2016 United States election, where faux accounts were created on Facebook to sway public opinion. Although the platform itself was not compromised, this falls under the category of a pseudo-Sybil attack. The Tor network, designed for anonymous communication, is also susceptible to Sybil attacks, where attackers can create multiple nodes to control the network and perform various malicious activities. In the context of blockchain, Sybil attacks can lead to the infamous 51% attack, where attackers gain control of the majority of computing power or hash rate, allowing them to manipulate transaction ordering and even reverse confirmed transactions. This can lead to the loss of funds and erode trust in the network. Furthermore, Sybil attacks can also be used for unethical purposes, such as creating multiple fake reviews on e-commerce platforms like Amazon. These attacks can be carried out by leasing mass computing power from locations such as Bangladesh. As Sybil attacks continue to evolve and become more sophisticated, it is crucial for online platforms and networks to stay vigilant and implement robust security measures to protect against such attacks.

Formal Model of Sybil Attack

The Sybil Attack paper presents a simple formal model to describe the components involved in a Sybil attack. The model consists of the following:

1. **E entities:** The entities in the network are divided into two categories, C (correct) entities and F (faulty) entities. The C entities are honest and comply with the protocols and policies set up in the network. Their honesty can be verified. The F entities are dishonest and cannot be predicted. They do not follow the protocols and policies in the network in an honest manner.
2. **Communication Cloud:** The communication cloud is a generalized cloud that serves as a medium through which messages are transmitted between different entities in the network.
3. **Pipe:** A pipe is used to connect an entity with the communication cloud. This facilitates communication between entities.

The model provides a simple framework to understand the Sybil attack. By introducing faulty entities into the network, an attacker can create multiple false identities and gain control over a significant portion of the network. This allows the attacker to manipulate the network and carry out various malicious activities.

While the model used in the Sybil Attack paper is simple, it provides a foundation for more sophisticated models that can be used to study and prevent Sybil attacks in various contexts. As Sybil attacks continue to pose a threat to online networks, it is crucial to develop more robust and effective models to detect and prevent such attacks.

Types of Sybil Attacks & Their Impact

There are two types of Sybil attacks that can be used to compromise honest nodes in a network. The first is a direct attack, where the Sybil node(s) directly influence the honest nodes. This can occur when the Sybil node(s) create false identities to join the network and then use these identities to influence the honest nodes. By controlling multiple identities, the Sybil node(s) can vote or make decisions that favor their interests and disadvantage honest nodes.

The second type of attack is an indirect attack. In this scenario, the Sybil node(s) manipulate a middle node that is communicating with honest nodes. The middle node is compromised because it is under the influence of the Sybil node(s) and is used to spread their malicious influence to the honest nodes. This is a more complex form of attack because it involves manipulating a trusted node to deceive other nodes in the network. However, it can be just as damaging as a direct attack because it can be used to control the network and make decisions that are not in the best interests of the network as a whole.

Using Proof of Work to Prevent Sybil Attacks in the Bitcoin Network

To ensure the authenticity of each block added to the blockchain, the Bitcoin network implements the Proof of Work (PoW) consensus algorithm. PoW requires a significant amount of computing power to perform the work, providing an incentive for miners to perform honest work and earn bitcoin rewards (currently 6.25 bitcoins for every block mined). All nodes verify transactions and reject any faulty transactions included in a block.

Moreover, the Bitcoin network is highly decentralized, with numerous miners around the world. This decentralization makes it very difficult for a single organization or attacker to control 51% of the mining power, which would be required to carry out a 51% attack, a type of Sybil attack. Therefore, it is practically impossible to carry out a 51% attack in the Bitcoin network.

Ways to Prevent Sybil Attack:

1. **Giving exclusive power to different members:** This is based on reputation systems. Members with different power levels are given different reputation levels.
2. **Cost to create an identification:** To prevent multiple faux identities in a network, every identity that aims to join the network should be made chargeable. However, it is important to make it infeasible to operate

multiple fake identities at the same time instead of just creating new identities. Multiple identities can enforce security, anonymity, and censorship prevention.

3. **Validation of identities before joining the network:**
4. **Direct validation:** An already established member verifies the new joiner of the network.
5. **Indirect validation:** An established member verifies some other members who can, in turn, verify other new network joiners. As the members verifying the new joiners are verified and validated by an established entity, the new joiners are trusted to be honest.

Note: While the aforementioned techniques can make it challenging to carry out a Sybil attack on the network, it is important to note that such attacks are not entirely impossible. Attackers can still find ways to create multiple fake identities or manipulate the reputation system to gain control over the network. Therefore, it is essential to continually monitor the network for any suspicious activity and take appropriate measures to prevent and mitigate any potential attacks. Additionally, implementing multiple prevention techniques simultaneously can help strengthen the network's security and decrease the likelihood of a successful Sybil attack.

Understanding the 51% Attack in Blockchain Networks

Blockchain-based systems, such as Bitcoin, operate on the basis of a distributed ledger that is maintained and verified by a network of nodes. One of the strengths of such a decentralized system is that the majority of nodes must regularly reach a consensus on various aspects of the system, including mining, software version, and the validity of transactions. To ensure the validity of a new block of transactions, the network nodes must agree that the block hash provided by the miner is accurate, and that the miner has performed enough computational work to generate a legitimate solution for that block's problem.

Since mining requires a large number of computational resources and electricity, the performance of a miner is based on their hash power or hash rate. This hash power is distributed across many nodes, making it difficult for any single entity to control the network. However, if a single entity is able to control more than 50% of the hash rate, they may be able to launch a 51% attack.

A 51% attack is when an entity or organization gains control of the majority of the hash rate, allowing them to deliberately exclude or modify transactions, reverse transactions they made while in control, and potentially cause a double-spending problem. The attacker may also prevent other miners from mining, leading to a mining monopoly. However, a successful majority attack would not allow the attacker to reverse transactions from other users or prevent transactions from being created and broadcasted to the network.

To prevent a 51% attack, the decentralized nature of the blockchain must be maintained and the hash power must be distributed across many nodes. While it is not impossible for a 51% attack to occur, it is currently unlikely due to the large number of mining nodes in the network.

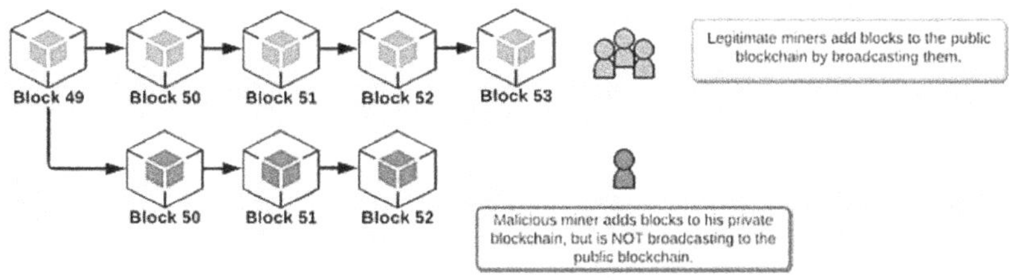

Figure 18.1: 51% Attack in a Blockchain Network

The Unlikelihood of a 51% Attack & Its Impact on Cryptocurrencies

Blockchain networks are highly secure due to their distributed nature. A consensus is reached by a network of nodes working together to maintain the blockchain, making it difficult for any single entity to corrupt or manipulate the data. The larger the network, the more secure it is against attacks and data corruption.

In a Proof of Work-based blockchain, such as Bitcoin, the possibility of a 51% attack is raised with an increased hash rate. Mining involves many hashing attempts, and more computational energy means more trials per second. The competitive state of affairs between miners is one of the reasons why Bitcoin is secure, as miners have no incentive to invest resources unless it is for honest contribution and the chance to receive block rewards. The magnitude of the Bitcoin network makes a 51% attack on the network unlikely. As the blockchain grows, the probability of a single individual or organization acquiring sufficient computing power to outweigh all other contributors rapidly drops to very low levels. *(Ref. No.: 263- 275)*

Additionally, the cryptographic proofs linking the previously confirmed blocks make changing them more difficult as the chain grows. The more confirmations a block has, the higher the cost for altering or reverting transactions. A successful attack would likely only be able to modify a few recent transactions for a short period of time. However, let us imagine a scenario in which a malicious entity is not motivated by profit and decides to attack the Bitcoin network only to break it, irrespective of the costs. Even if the attacker manages to disrupt the network, the Bitcoin software application and protocol would quickly modify and adapt a response to that attack. The other network nodes would need to reach a consensus and agree on these changes, but this would likely happen quickly during an emergency situation. Bitcoin is very resilient to attacks and is considered the most secure and dependable cryptocurrency in existence.

It is more feasible for smaller cryptocurrencies to be vulnerable to a 51% attack as they have a lower amount of hashing power securing their blockchain. Examples of cryptocurrencies that were victims of majority attacks include Monacoin, Bitcoin Gold, and ZenCash. It is essential for smaller cryptocurrencies to take measures to increase their network's security to protect against such attacks.

Dusting Attacks: A New Threat to Cryptocurrency Privacy

A dusting attack is a relatively recent form of cyber-attack aimed at breaching the privacy of Bitcoin or other cryptocurrency users. In this type of attack, scammers or hackers send small amounts of coins to the target's wallet to gain access to their transactional activity. By tracking these transactions, attackers perform a combined analysis of various addresses associated with the wallet to uncover the identity of the individual or company behind it.

The term "dusting" refers to the sending of a small amount of cryptocurrency to the target's wallet. The amount is usually insignificant and often not noticeable to the user. The purpose of sending such small amounts is to establish a link between the wallet and other addresses used in the blockchain network. Once the link is established, the attackers can track the movements of the funds and potentially identify the owner of the wallet. This type of attack is particularly concerning because it can compromise the anonymity and privacy of cryptocurrency users. Blockchain networks are designed to be pseudonymous, meaning that users can transact anonymously without revealing their true identity. However, dusting attacks can reveal the identity of users, making them vulnerable to targeted attacks such as phishing scams, extortion, and other fraudulent activities.

To prevent dusting attacks, cryptocurrency users should be vigilant about any small transactions received in their wallets, especially if they do not recognize the sender. They should also avoid sharing their wallet addresses with anyone they do not trust. Furthermore, users can use tools such as CoinJoin or mixers to obscure their transactional activity, making it more difficult for attackers to track their movements.

The Concept of Dusting Attacks in Cryptocurrency

Cryptocurrency enthusiasts may already be familiar with the term "dust" in the context of digital currencies. But for those who are new to the world of cryptocurrencies, dust refers to the tiny amounts of coins or tokens that are so small that most users do not even notice them. For example, in the case of Bitcoin, the smallest unit of BTC is 1 satoshi (0.00000001 BTC), so we might also use the term dust to refer to multiple hundreds of Satoshis.

Within cryptocurrency exchanges, dust is also the name given to tiny amounts of coins that "get stuck" on users' accounts after trading orders are executed. These balances are not tradeable, but some exchanges like Binance allow users to convert them to their platform token (e.g., BNB).

It is important to note that there is no universal definition for dust in the cryptocurrency world, and each software implementation or client might have a different threshold for what is considered dust. For instance, the Bitcoin Core software defines dust as any transaction output that is lower than the transaction fees, which results in the concept of a dust limit.

Technically speaking, the dust limit is calculated based on the scale of inputs and outputs, and it typically computes to 546 Satoshi's for regular Bitcoin transactions (non-SegWit) and 294 Satoshi's for native SegWit transactions. This means that any regular transaction equal to or smaller than 546 Satoshi's will be considered spam and likely be rejected by the verifying nodes.

Now, dusting attacks refer to a new type of malicious activity in which hackers and scammers try to break the privacy of Bitcoin or other cryptocurrency users by sending tiny amounts of coins to their wallets. After dusting different addresses, the attackers conduct a combined analysis of those addresses in an attempt to identify which ones belong to the same crypto wallet. The goal is to finally link the dusted addresses and wallets to their respective companies or individuals. If successful, the attackers may use this information against their targets either via elaborated phishing attacks or cyber-extortion threats.

Dusting attacks were initially performed on the Bitcoin network, but they are now also arising on other blockchains like Litecoin, BNB, etc. This became possible because most cryptocurrencies are running on top of a traceable and public blockchain. In late October 2018, the Samourai Wallet developers announced that a number of their users were under dusting attacks. The organization sent out a tweet warning user about the attacks and explaining how they may shield themselves. The Samourai Wallet team implemented a real-time alert for dust tracking as well as a "Do Not Spend" feature that lets users mark suspicious funds so that these are not included in future transactions.

Since dusting attacks rely upon a combined analysis of multiple addresses, if a dust fund is not moved, attackers are not able to make the connections they need to "deanonymize" the wallets. Nevertheless, attackers have found ways to bypass the dust limit of 546 Satoshis, and many dusting attacks these days are well above it, typically starting from 1000 to 5000 Satoshis.

Therefore, it is important for cryptocurrency users to be aware of dusting attacks and take appropriate measures to protect themselves. They can use wallets that have implemented measures to counter dusting attacks, such as Samourai Wallet, and avoid sharing their public addresses or using the same address for multiple transactions. Additionally, users can check their wallet for any small incoming transactions and not interact with them, especially if the transaction amount is well above the dust limit.

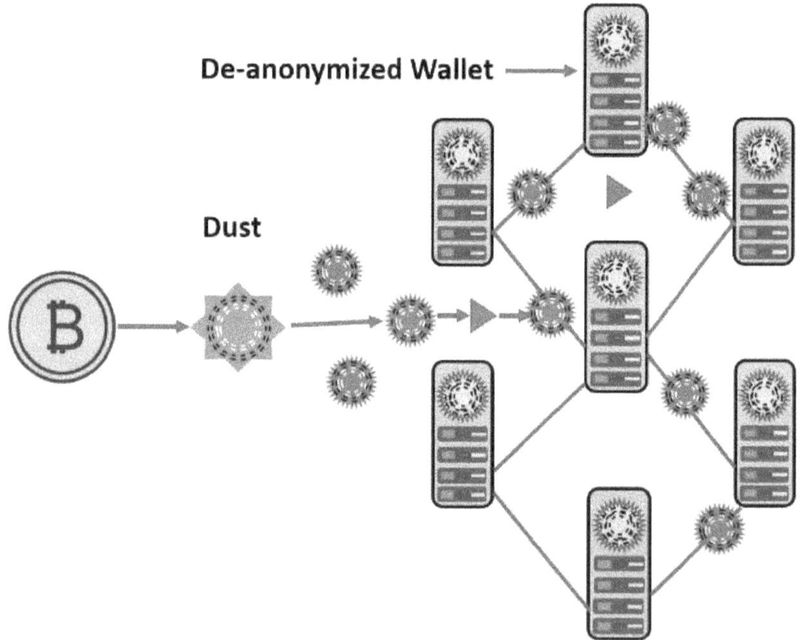

Figure 18.2: How Cryptocurrency Dusting Attack Works

Dusting Attacks on the Binance Chain (BC)

In October 2020, a new form of cyber-attack emerged on the Binance Chain (BC), known as "dusting attack." In this type of attack, scammers send minuscule amounts of BNB to a large number of addresses on the Binance Chain, with the intention of exposing users to their malicious activities. The attackers included a link to a deceitful website in the transaction Memo, enticing users to click on it. The primary goal of dusting attacks is to track user activities, learn their wallet addresses, and link them to their personal identities. Attackers hope to gain access to personal information by tricking users into visiting malicious websites or downloading harmful software.

Dusting attacks on the Binance Chain involve sending tiny amounts of BNB to a large number of addresses with a fraudulent link in the transaction Memo. This form of attack aims to uncover user data and compromise their security by exposing them to malware or phishing scams.

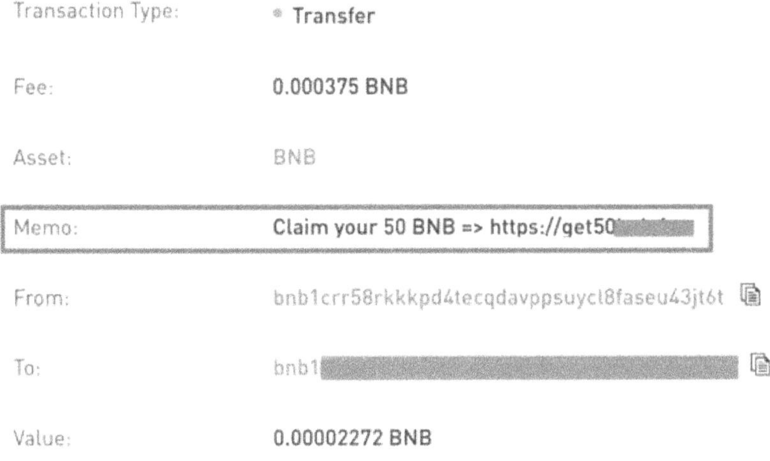

Figure 18.3: An Example of a Binance Chain Dusting Attack.

Bitcoin Pseudonymity & the Importance of Privacy Precautions

Bitcoin provides a certain degree of pseudonymity, which means that while all transactions on the blockchain are public, it can be challenging to determine the real identity behind each address or transaction. This anonymity is because Bitcoin is open and decentralized, allowing anyone to join the network without disclosing personal information.

However, the level of anonymity varies depending on the type of transaction. Peer-to-peer (P2P) transactions are more likely to stay anonymous because they do not involve intermediaries, but when using cryptocurrency exchanges, users are often required to complete a KYC (Know Your Customer) verification process, which can compromise their privacy. When users transfer funds between their private wallets and exchange accounts, they run the risk of being deanonymized.

To maintain anonymity, it is recommended to use a new Bitcoin address for each new receiving transaction or payment request. This practice helps to prevent blockchain analyses from linking transactions to a single user, thus protecting their privacy.

Despite the pseudonymity provided by Bitcoin, it is not entirely anonymous. Some entities such as companies, research labs, and governmental organizations use blockchain analysis to deanonymize the blockchain network. Recent attacks such as dusting attacks further undermine Bitcoin's anonymity. As a result, it is crucial to understand that while Bitcoin provides some level of privacy, it is not entirely anonymous, and users should take necessary precautions to protect their personal information.

Understanding Double Spending in Digital Cash Systems

Double-spending is a potential issue that arises in digital cash systems, where the same funds are spent on two or more recipients at the same time. This problem is particularly dangerous when there are no adequate countermeasures in place, as it undermines the integrity of the entire protocol. Without the ability to verify that the funds they have received have not already been spent elsewhere, users lose confidence in the system, leading to widespread distrust and adoption issues.

In digital cash systems, it is of utmost importance to ensure that specific units of currency cannot be duplicated. The entire system would collapse if a user could receive a certain amount of funds, copy, and paste them multiple times, and end up with an artificially inflated balance. Similarly, the system cannot function if the same funds can be sent to multiple recipients simultaneously, leading to double-spending issues.

To prevent this type of behaviour, digital cash systems must have a mechanism in place to detect and prevent double-spending. The most commonly used method is through a decentralized ledger, such as a blockchain, where every transaction is verified by a network of nodes to ensure that the funds are not already in use.

In summary, double-spending is a serious problem that can undermine the trust and integrity of digital cash systems. Preventing double-spending is essential to ensuring the viability and long-term success of these systems, which require a mechanism in place to detect and prevent such behaviour.

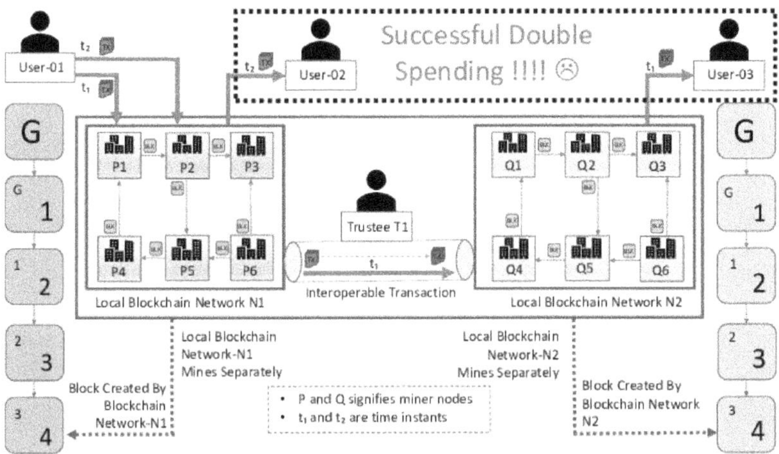

Figure 18.4: Double Spending in a Blockchain Network

Exploring the Centralized Approach to Preventing Double-Spending

Double-spending can be a major problem in digital cash systems. However, there are various approaches to prevent it. One such approach is the centralized method, which involves a single overseer controlling the issuance and distribution of digital cash units. David Chaum's eCash system is a good example of this approach.

To use the eCash system, users must first inform the bank about the digital asset they desire. Assuming that the user, Dan, has the necessary balance in his account, he generates random numbers that are each assigned a cost of $20. To prevent the bank from tracking specific units, Dan obfuscates the random numbers by adding a blinding factor to each one of them.

Dan then submits this information to the bank, which debits his account for $100 and signs messages certifying that each of the five pieces of data is redeemable for $20. Dan can now spend the funds issued by the bank, like when he purchases a meal for $40 at Erin's restaurant.

To prevent double-spending, Dan must reveal the random number associated with each digital cash bill, which serves as a unique identifier for each unit. He reveals this information to Erin, who must redeem them immediately with the bank to prevent Dan from spending them with another merchant. The bank will check that the signatures are legitimate, and if everything appears to be correct, it will credit Erin's account with $40. The bills used are essentially burned, and more must be issued if Erin wishes to spend her new balance in the same way.

While the Chaumian eCash setup may be useful for private transfers, it is not a resilient system due to its dependence on the financial institution. The value of an invoice issued is derived entirely from the financial institution's willingness to exchange it for dollars. Customers are at the mercy of the financial institution and must rely on its goodwill for money to function, which is the exact problem that cryptocurrencies aim to solve.

Double-Spending Prevention in Centralized & Decentralized Approaches

Preventing double-spending in a decentralized ecosystem is a complex task that requires coordination among all participants to uphold a set of regulations and incentives for honest behaviour. In the absence of a central authority, the most significant innovation in the Bitcoin white paper was the introduction of the blockchain as a solution to the double-spending problem. A blockchain is essentially a database that enables all participants in the network (nodes) to synchronize their copy of the database with their peers, creating a public record of all transactions that can be audited by the entire network.

When a user initiates a transaction in a blockchain network, it is not immediately added to the blockchain but is instead included in a block through mining. The recipient must wait until the transaction is confirmed by being included in a block before considering it valid. Otherwise, the sender could spend the same coins elsewhere, and the recipient would risk losing their funds. Once the transaction is confirmed, the coins cannot be double-spent, as the ownership is assigned to a new user, and the whole network can verify this.

Returning to the restaurant scenario, Dan notices a Bitcoin Accepted Here sticker on the restaurant window and decides to pay for his meal using Bitcoin. Erin provides him with a public address to which he must send the funds. Dan broadcasts the transaction, which is essentially a signed message that states that 0.05 BTC that were in Dan's ownership are now transferred to Erin's ownership. Any party presented with Dan's signed transaction can verify that he was indeed in possession of the coins and authorized to send them. However, the transaction is only legitimate if it is confirmed by being added to the blockchain. Therefore, it is recommended that Erin waits for at least six block confirmations (roughly one hour) before accepting Dan's payment. This waiting period is similar to the process of cashing in eCash from the previous scenario with a financial institution. By waiting for multiple confirmations, Erin can ensure that the transaction is genuine and cannot be reversed or double-spent.

Preventing Double-Spending Attacks in Bitcoin: The Role of Confirmations & Unconfirmed Transactions

Bitcoin is a cryptocurrency designed to prevent double-spending attacks, which is one of the most significant challenges in a decentralized financial system. The core idea behind Bitcoin is to ensure that a participant cannot spend the same coins twice. This is achieved through a distributed ledger known as the blockchain, which is a database that keeps a record of all transactions in the network.

When a user broadcasts a transaction, it is not immediately added to the blockchain. Instead, it must be included in a block through a process called mining. Once confirmed, the coins cannot be double-spent, as the ownership is transferred to a new user, and the whole network can verify this. This is why experts recommend waiting for multiple confirmations before accepting payment as valid. Each subsequent block significantly increases the number of attempts required to update or rewrite the chain, making it difficult for attackers to reverse a confirmed transaction.

However, accepting unconfirmed transactions can leave parties vulnerable to double-spending attacks, especially for low-value purchases. For instance, a busy fast-food restaurant might not want to wait for transactions to be included in a block, as it can slow down the payment process and frustrate customers. If a business enables "instant" payments, they open themselves up to double-spends. An attacker could order a burger, pay for it, and immediately send an equal number of coins to their personal wallet. With an increased fee, this new transaction is more likely to be confirmed first, and could consequently invalidate the preceding one. *(Ref. No.: 276- 286)*

Therefore, participants in the Bitcoin network must be careful when accepting unconfirmed transactions, especially for high-value purchases. The more confirmations a transaction has, the more secure it is. It is recommended to wait for at least six confirmations (roughly one hour) before accepting a payment.

This way, parties can be confident that the transaction is legitimate and cannot be reversed without an unrealistic amount of hashing power.

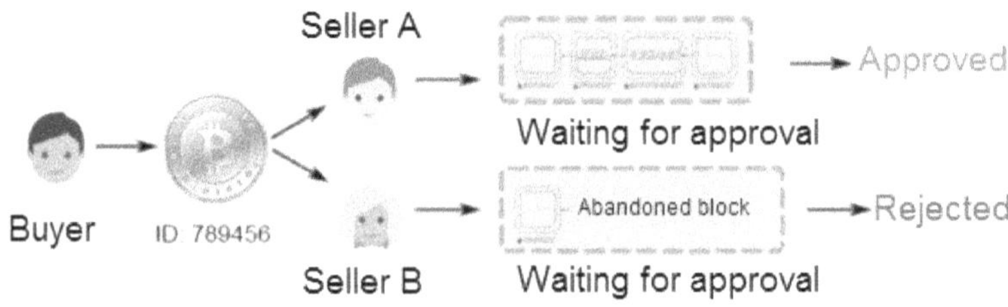

Figure 18.5: Double-Spending in Bitcoin Network

Strategies for Double-Spending in Bitcoin: Understanding the Risks

Double-spending attacks can be a significant concern for participants in blockchain networks. While Bitcoin's design is precisely developed to prevent double-spending attacks, there are still potential risks that need to be addressed. There are three popular strategies for performing a double-spend attack in Bitcoin, each with its own unique characteristics and risks. The first strategy is a 51% attack. This occurs when a single entity or organization controls more than 50% of the hash rate, giving them the ability to exclude or modify the ordering of transactions. While this type of attack is highly unlikely on Bitcoin, it has happened in other blockchain networks. A 51% attack can lead to the creation of a new, separate blockchain, which can cause significant damage to the network. The second strategy is a race attack. In this scenario, two conflicting transactions are broadcast in succession using the same funds, but only one transaction gets confirmed. The attacker intends to invalidate the original transaction by only validating the transaction that benefits them. This is done by sending the same funds to an address that they control. Race attacks require the recipient to accept an unconfirmed transaction as payment, which is why service providers that wait for block confirmations are less likely to fall victim to this type of attack.

The third strategy is a Finney attack. In this type of attack, the attacker pre-mines one transaction into a block immediately, without broadcasting it to the network. Instead, they spend the same coins in another transaction and only then broadcast their previously mined block, which may invalidate the payment. Finney attacks require a specific sequence of events to occur and are also contingent on the recipient's acceptance of unconfirmed transactions. It is crucial for participants to understand these strategies to effectively protect themselves from double-spending attacks. Service providers that wait for block confirmations before accepting payment can significantly reduce the risks of becoming a victim of double-spends.

Closing Thoughts

Blockchain technology has transformed the world of finance by offering novel ways to store and transfer value. However, with the rise of the cryptocurrency industry, it has also brought about new security threats that need to be addressed. One such threat is the Sybil attack, which poses a danger to the blockchain network's security and user privacy. In addition, a 51% attack can provide malicious actors with the power to manipulate transactions and steal funds by controlling a significant portion of the network's computing power. Dusting attacks and double-spending are other security risks that can be exploited by hackers to unmask users and manipulate the transaction history.

Race attacks and Finney attacks are other potential attacks that can compromise the integrity of the blockchain. To address these security threats, it is essential to develop blockchain networks with robust security measures and user verification mechanisms. This includes implementing additional layers of security such as multi-factor authentication and encryption to safeguard user data and funds. Furthermore, educating users about these security threats and promoting best practices for secure cryptocurrency transactions can help mitigate the risks of these attacks. It is crucial to be aware of the various security threats that exist in the cryptocurrency industry and take steps to mitigate them. By implementing careful design and user education, it is possible to make cryptocurrency transactions more secure and reliable for everyone. Users should be vigilant and take proactive measures to safeguard their funds and privacy. In conclusion, while the cryptocurrency industry offers many exciting opportunities, it is essential to prioritize security to ensure its long-term success.

Chapter 19: Decentralized Finance (DeFi): A Comprehensive Guide

"The blockchain cannot be described just as a revolution. It is a tsunami-like phenomenon, slowly advancing and gradually enveloping everything along its way by the force of its progression."

— William Mougayar

Decentralized Finance (DeFi): An Introduction to a New Financial Ecosystem

Decentralized Finance (DeFi) is an emerging financial ecosystem that leverages the power of blockchain technology to create a new, open, and transparent financial service ecosystem. The DeFi system is designed to operate without any central authority, making it accessible to everyone, regardless of their location or financial status. In DeFi, participants have full control over their assets and can interact with the ecosystem through peer-to-peer (P2P) and decentralized applications (dApps). This means that anyone can access financial services without the need for intermediaries or traditional financial institutions. One of the main advantages of DeFi is its ability to provide straightforward access to financial services, particularly for those who are underserved or excluded from the traditional financial system. Another significant advantage of DeFi is its modular framework, which allows interoperable DeFi applications on public blockchains to create entirely new financial markets, products, and services. This chapter introduces DeFi, including its potential applications, benefits, limitations, and much more. By exploring DeFi, we can better understand its potential impact on the financial industry and its role in creating a more accessible, inclusive, and decentralized financial system.

DeFi: Revolutionizing the Traditional Financial System

The traditional financial system is based on intermediaries such as banks and courts to act as facilitators and arbitrators. However, with the emergence of DeFi, financial services can be provided without intermediaries or arbitrators. DeFi applications utilize smart contracts to specify the resolution of any potential conflicts, while users maintain full control over their funds at all times. This eliminates the need for intermediaries and reduces fees associated with traditional financial products, resulting in a more frictionless financial system.

DeFi applications are built on top of blockchain networks, which eliminate single points of failure. Information is recorded on the blockchain and distributed among numerous nodes, making censorship or shutdown of service a challenging task. Furthermore, the frameworks for DeFi programs can be built in advance, making deployment more secure and straightforward.

Another significant advantage of DeFi is its accessibility. The traditional financial system is typically absent in low-income communities due to the intermediaries' profit-driven nature. However, with DeFi, costs are considerably reduced, allowing low-income individuals to benefit from a broader range of financial services. This open ecosystem provides convenience and accessibility to individuals who would otherwise not have access to any financial services.

Potential Use Cases for DeFi: Exploring Decentralized Borrowing & Lending

Decentralized Finance (DeFi) is a transformative force within the financial sector, offering a plethora of innovative applications that extend well beyond its initial use cases. Here, we delve into the diverse applications of DeFi and their potential to revolutionize traditional finance:

1. **Decentralized Borrowing & Lending:** DeFi's cornerstone application enables borrowers and lenders to interact directly on blockchain-based platforms. This eliminates intermediaries, resulting in faster transaction settlement and reduced costs. Importantly, DeFi lending protocols accept digital assets as collateral, eliminating the need for credit score checks and expanding access to financial services. This democratization of lending and borrowing makes financial markets more inclusive and accessible.

2. **Decentralized Exchanges (DEXs):** DeFi empowers users with decentralized exchanges, where peer-to-peer trading occurs directly on blockchain networks. DEXs offer greater privacy, lower fees, and rapid settlement compared to centralized exchanges. They negate the need for third-party custodians, placing users in full control of their assets and trading activities.

3. **Prediction Markets:** DeFi's innovative potential extends to prediction markets, allowing users to speculate on the outcomes of various events, including elections, sports events, and market trends. These markets leverage blockchain's transparency and decentralization to mitigate manipulation and ensure fair outcomes. Participants can make predictions and invest in specific outcomes, providing a novel way to harness collective wisdom.

4. **Yield Farming and Liquidity Provision:** DeFi introduces novel ways to generate passive income through yield farming and liquidity provision. Users can lock up their digital assets in DeFi protocols, earning rewards in the form of interest or tokens. This practice incentivizes liquidity within DeFi platforms, improving overall market efficiency.

5. **Tokenization of Real Assets:** DeFi is increasingly used to tokenize real-world assets, such as real estate, art, and commodities. This process represents physical assets as digital tokens on blockchain networks, making them easier to trade and divide among investors. It enhances accessibility to traditionally illiquid assets, fostering a more liquid market.

DeFi's applications are diverse and continue to evolve. Its decentralized and open-source nature has the potential to reshape traditional finance, making it more inclusive, efficient, and accessible to individuals worldwide. However, as with any emerging technology, DeFi also presents challenges, such as regulatory considerations and security concerns, that must be addressed as the ecosystem matures.

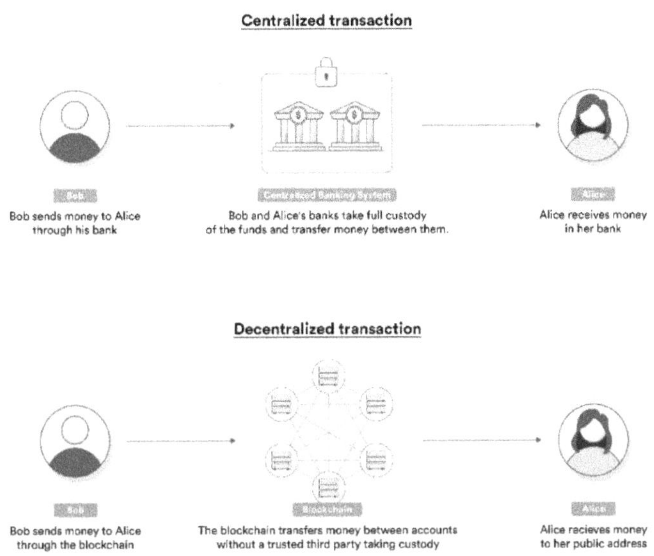

Figure 19.1: Centralized vs Decentralized Transactions

Decentralized Monetary Banking Services: Exploring DeFi's Potential

DeFi applications are inherently financial applications, which makes monetary banking services an obvious use case for them. These services can include the issuance of stablecoins, mortgages, and insurance. One of the most significant advantages of these services in a DeFi ecosystem is their decentralization, which allows for increased transparency and accessibility. Stablecoins are a type of cryptocurrency that is pegged to a real-world asset but can be transferred digitally with relative ease. Decentralized stablecoins could become a digital cash alternative to traditional fiat currencies that are not issued or monitored by a central authority. As cryptocurrency prices can fluctuate rapidly, the use of stablecoins can provide a more stable and reliable option for day-to-day transactions. The process of obtaining a mortgage is typically costly and time-consuming due to the number of intermediaries involved. However, with the use of smart contracts, underwriting and legal charges can be reduced substantially, providing a more streamlined and efficient process. DeFi can potentially democratize access to mortgages and make them available to a wider range of individuals. Another potential use case for DeFi is in the area of insurance. By eliminating intermediaries, insurance on the blockchain could allow for the distribution of risk between many participants, resulting in lower premiums without sacrificing quality of service. This could make insurance more affordable and accessible to people who are currently underserved by the traditional insurance industry.

Decentralized Marketplaces & Their Applications in DeFi

Decentralized marketplaces are a crucial aspect of the DeFi ecosystem, offering a vast array of financial innovations. Decentralized exchanges (DEXes) are a significant part of this category, allowing users to trade digital assets without a central trusted intermediary. Instead, trades are conducted directly between user wallets, facilitated by smart contracts. This eliminates the need for custodial services and reduces the chances of hacking, which is a significant risk in centralized exchanges.

One of the most significant advantages of decentralized exchanges is their low maintenance cost, resulting in reduced transaction fees compared to centralized exchanges. Apart from decentralized exchanges, blockchain technology can also be used to create and manage various conventional financial instruments in a decentralized manner. Security token issuance platforms, for instance, allow the release of tokenized securities on the blockchain with customizable parameters, eliminating the need for intermediaries and single points of failure. Decentralized marketplaces also offer the potential to create and trade derivatives, synthetic assets, decentralized prediction markets, and other innovative financial instruments. As a result, DeFi provides users with a broader range of financial services and the opportunity to participate in a more inclusive and transparent financial system.

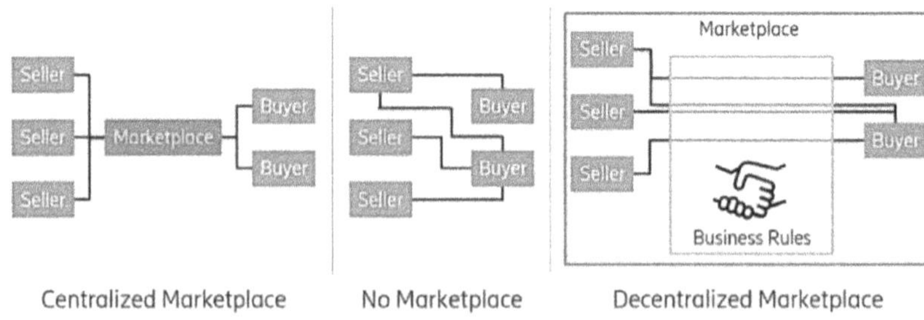

Figure 19.2: Centralized vs Decentralized Marketplace

Significance of Smart Contracts in Decentralized Finance (DeFi)

In the world of DeFi, smart contracts play a critical role in the design and execution of various financial applications. Unlike traditional contracts that rely on legal terminology to specify the terms of the relationship between parties, smart contracts use computerized code to define and enforce the terms of an agreement. With smart contracts, there is no need for intermediaries to oversee the execution of the contract.

One of the most significant benefits of smart contracts is their ability to execute automatically, without the need for manual supervision. This allows for the automation of complex business processes that would otherwise require significant human oversight. Moreover, smart contracts are faster and more efficient than traditional contracts, which can take days or even weeks to finalize.

However, it is important to note that smart contracts also introduce new forms of risk. Since computerized code is vulnerable to bugs and vulnerabilities, any value or private data locked in a smart contract can be at risk if there is a security breach. Additionally, smart contracts are only as good as the code they are written in. If there are errors or vulnerabilities in the code, the contract can fail, resulting in significant financial losses for the parties involved.

Challenges Faced by DeFi

- **Sluggish Performance:** Blockchains are inherently slower than their centralized counterparts, which poses a challenge for DeFi applications. To overcome this, developers need to optimize their products to provide seamless user experience.
- **Increased Risk of User Error:** DeFi applications transfer the responsibility from intermediaries to the user, making it a daunting task for many. To minimize the risk of user error, products should be designed to be user-friendly and deployed on immutable blockchains.
- **Poor User Experience:** Currently, DeFi applications require extra effort on the user's part. To become a vital part of the global financial system, they must provide a clear advantage that motivates users to switch from the conventional system.
- **Complex Ecosystem:** The DeFi ecosystem can be overwhelming, making it difficult for users to find the most suitable application for their specific needs. Building user-friendly applications and considering how they fit into the broader DeFi ecosystem is a significant challenge.

Traditional Finance Vs DeFi

Traditional Finance	Decentralized Finance DeFi
Users must share personal data, which is vulnerable to security breaches.	No need to disclose personal details, just connect a digital wallet.
The unbanked or underbanked cannot access traditional financial services.	Users' custody their own funds, easier to avoid loss of funds.
Payment can be intercepted and market can be shut down.	Trustless – no need to worry if a protocol will do what it says it will do.
The clearing and settling of transactions can take days	Speedy – transactions are completed in near real-time.
The hours of operation are limited	The DeFi markets operate around the clock, 24/7/365.

Table 19.1: Tradition Finance vs DeFi Comparison

Difference between Open Banking & DeFi

Open banking and DeFi are two distinct financial concepts that differ in their approaches to innovation and transformation of the traditional financial system. Open banking refers to a banking system in which third-party financial service providers are granted secure access to financial data through APIs, enabling the networking of accounts and information between banks and non-financial institutions. This allows for the creation of new types of financial services and products within the existing financial system. For instance, open banking can facilitate the management of all traditional financial instruments in one application by securely pulling information from multiple banks and financial institutions. On the other hand, DeFi aims to create a completely new financial system that operates independently of the existing infrastructure. Also known as open finance, DeFi leverages blockchain technology to build a trustless, transparent, and decentralized financial system that is accessible to anyone with an internet connection. Unlike open banking, DeFi enables the creation and management of entirely new financial instruments and provides innovative ways of interacting with them.

For example, DeFi applications allow users to lend and borrow money without intermediaries, invest in decentralized autonomous organizations (DAOs), trade synthetic assets, and more. In contrast, open banking mainly focuses on optimizing and streamlining the existing financial system. While both concepts share some similarities in terms of increasing financial inclusion and access to financial services, DeFi offers a more radical transformation of the traditional financial system. *(Ref. No.: 287- 295)*

Closing Thoughts

The concept of DeFi has been gaining traction and attention in recent years as a potential solution for creating a more open and accessible financial system. DeFi is designed to offer economic services that are independent of the traditional financial system, allowing for a more decentralized and transparent financial landscape. One of the main advantages of DeFi is that it could potentially prevent censorship and discrimination in financial services, making financial products and services available to a broader range of individuals and communities. However, it is important to recognize that not every financial service benefit from decentralization. As DeFi continues to develop, finding the use cases that are most suitable for blockchain technology is essential for building a useful stack of open financial products. DeFi also needs to address challenges such as poor performance, high risk of user error, bad user experience, and a cluttered ecosystem to make it more accessible to mainstream users.

If DeFi is successful, it could potentially shift the power from large, centralized agencies to the open-source community and other interested individuals. However, whether or not DeFi can create a more efficient monetary system will be decided once it is ready for mainstream adoption and has been thoroughly tested in the real world. Overall, DeFi has the potential to bring about significant changes in the financial industry, and it will be interesting to see how it continues to evolve in the coming years.

Chapter 20: Tokenized Networks & Smart Contracts

"On the blockchain, you have a public ledger, which is the form of ownership. That means you are not doing net settlement – just netting everything down, which means that everything has been turned into fungible numbers."

— Patrick M. Byrne

Understanding DAOs: The Innovative Decentralized Autonomous Organizations

A DAO, or Decentralized Autonomous Organization, is a type of organization that operates on a system of hard-coded rules that dictate its actions. Unlike traditional companies, DAOs are based on open-source code and are operated entirely by their community, with no centralization of power or hierarchical management. The DAO model is especially exciting because it utilizes smart contracts to automate many of its functions and work mechanisms, enabling a completely decentralized and automatic business model.

The DAO concept is not new, but its implementation through smart contracts is what makes it unique and innovative. It can be used for various activities, such as automatic fundraising campaigns like Initial Coin Offerings (ICOs), the issuance of digital tokens, tokenization of assets, and decision-making and voting systems. The DAO model also offers the possibility of introducing more efficient systems by reducing the need for human inputs and associated risks, thereby lowering overall operational costs. Just as Bitcoin created a peer-to-peer digital financial system, eliminating the need for banks and other trusted third parties, DAOs have the potential to revolutionize a wide range of industries through the use of decentralized governance models powered by smart contracts. Therefore, the DAO model represents a paradigm shift in the way businesses and organizations operate, and it has the potential to create a more open, transparent, and decentralized economy.

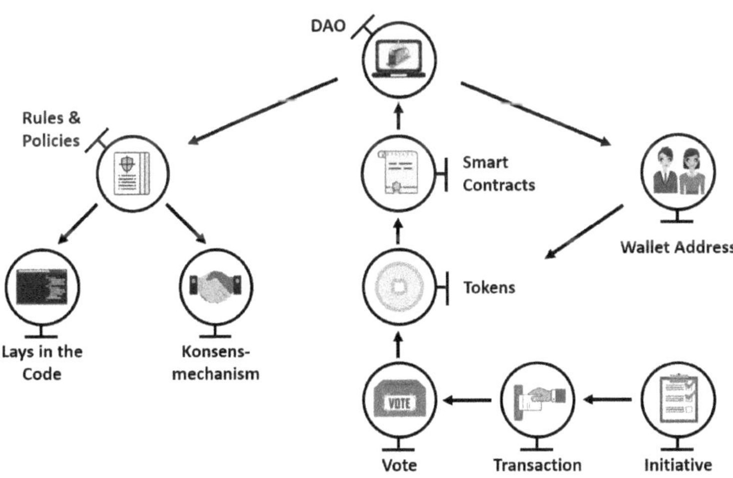

Figure 20.1: Decentralized Autonomous Organization (DAO)

Tokenized Networks & DAOs: Overcoming Principal-Agent Dilemma with Blockchain & Smart Contracts

Blockchain technology and smart contracts have introduced innovative governance systems that offer higher transparency levels and reduce bureaucracy. They also have the potential to minimize the principal-agent dilemma prevalent in traditional institutions, with the help of self-enforcing code. In decentralized networks, tokens play a vital role in providing incentives and aligning interests automatically, without relying on intermediaries.

DAOs, or Decentralized Autonomous Organizations, are designed to address the age-old problem of the principal-agent dilemma. This occurs when an agent has the power to make decisions on behalf of another entity or person, such as a boss acting on behalf of shareholders or a politician representing citizens. In such cases, moral hazard arises when an agent takes more risks because others bear the cost of those risks. Moreover, the boss cannot fully control the agent's activities when the latter prioritizes self-interest over the principal's interests. The principal-agent dilemma is further complicated by information asymmetry.

However, blockchain-based governance models, such as DAOs, mitigate these issues by leveraging smart contracts and token economics. The self-enforcing code of smart contracts ensures transparency and accountability in decision-making, while tokens align interests and incentivize participants to act in the best interest of the network. With DAOs, the principal-agent dilemma can be resolved, leading to a more efficient and transparent governance structure in decentralized networks.

Decentralized Autonomous Organizations (DAOs): Operating Systems for Distributed Internet Tribes

Traditional organizations typically involve representatives of a company who have legal employment contracts to regulate their association with the organization and each other. These contracts contain rights and obligations, and if someone breaches the contract, legal action can be taken against them. However, decentralized autonomous organizations (DAOs) involve a group of people who interact with each other according to an open-source protocol that enforces self-enforcement. Participants in a DAO are incentivized with native network tokens for keeping the network safe and performing tasks. Transactions on DAOs are governed by transparent regulations written in software enforced by machine consensus, and individual behaviour is incentivized with tokens to contribute to a common goal. Members of a DAO are not bound together by a legal entity nor have they entered into any formal legal contracts. Instead, they are steered by incentives linked to the network tokens and fully transparent regulations that are enforced by machine consensus. There are no bilateral agreements, only one governing law - the protocol or smart contract - that regulates the behaviour of all network participants. Unlike traditional organizations that are structured in a top-down manner, DAOs provide an operating system for people and institutions that do not know or trust each other, and who might live in different geographical areas, speak different languages, and therefore be subject to different jurisdictions. In such organizations, all agreements are in the form of open-source code that is self-enforced by the majority consensus of all network actors.

Smart contracts can formalize a DAO, and their use cases range from simple to complex, depending on the number of stakeholders and the number and complexity of functions within the organization. The governance rules of DAOs can have a similarity to companies or nation-states, depending on the purpose of the organization. Proposals take the primary way of making decisions within a DAO, which are voted for by a majority consensus of involved network actors. DAOs are transparent, and in theory, incorruptible since all transactions of the organization are recorded and maintained on a blockchain. The interests of the members of the organization are aligned by the incentive rules affixed to the native token. DAOs can be seen as distributed organisms or distributed Internet tribes that live on the Internet and exist autonomously, but also heavily rely on specialist individuals or smaller organizations to perform certain tasks that cannot be replaced with automation.

Bitcoin Network, the first true decentralized and autonomous organization, is coordinated by a consensus protocol that anyone can adopt. It provides an operating system for money without banks and bank managers. The underlying blockchain protocol enables an incentive network, powered by the governance rules tied to its cryptographic token. These token governance rulesets of the consensus layer allow for automated and transparent coordination of a disparate group of people who do not know or trust each other. The Ethereum Network made the creation of DAOs easily programmable with smart contracts, and "The DAO" in 2016 was an early example of such a complex smart contract on the Ethereum blockchain.

In-Depth Explanation of Smart Contracts & Their Functionality

Smart contracts are a type of digital agreement that is enforced by a specific set of regulations encoded in computer code, which runs on a blockchain network.

First proposed by Nick Szabo in the 1990s, these contracts have gained popularity in the world of cryptocurrencies due to their ability to create trustless protocols. They eliminate the need for intermediaries and reduce operational fees, making them an efficient and secure alternative to traditional contracts.

In this context, a smart contract is an application or program that runs on a blockchain network. The contract's terms and conditions are pre-determined by computer code and are replicated and executed by all network nodes. This ensures that the contract's execution is transparent and cannot be altered or manipulated by any party involved.

Smart contracts have the potential to revolutionize various fields involving contractual agreements, including credit score systems, payment processing, and content rights administration. They allow parties to make commitments via blockchain, without the need to know or trust each other. This eliminates the risk of fraud or manipulation, as the contract will only be executed if all conditions are met.

The Bitcoin protocol has been supporting smart contracts for years, but it was the co-founder of Ethereum, Vitalik Buterin, who popularized them. The Ethereum Virtual Machine (EVM) is an integral part of the Ethereum blockchain and is the primary platform for running smart contracts. It is essential to note that each blockchain may have a different approach to implementing smart contracts. However, this chapter focuses on smart contracts running on the EVM. *(Ref. No.: 303- 314)*

How do Smart Contracts work?

To understand how smart contracts work, it's essential to first recognize that they are not legal contracts, but instead, they are computer programs that execute when certain predefined conditions are met. Typically, these conditions are structured in an "if...then" format.

Smart contracts operate on blockchain networks, such as Ethereum, and are responsible for managing the network's operations. When a user interacts with another user on the blockchain, such as transferring cryptocurrency, the smart contract ensures the transaction's validity and execution.

On the Ethereum network, smart contracts are created using a combination of contract code and public keys. The creator of the contract supplies the first public key, while the second key represents the contract itself, acting as a unique digital identifier.

Once the contract is created, it can only be activated through a blockchain transaction, which must be initiated by an Externally Owned Account (EOA), or another smart contract. An EOA is an account controlled by a user, while smart contracts are entirely managed by computer code.

In summary, smart contracts operate on a "if...then" basis and are created using contract code and public keys. They are activated through blockchain transactions initiated by EOAs and are responsible for managing the network's operations.

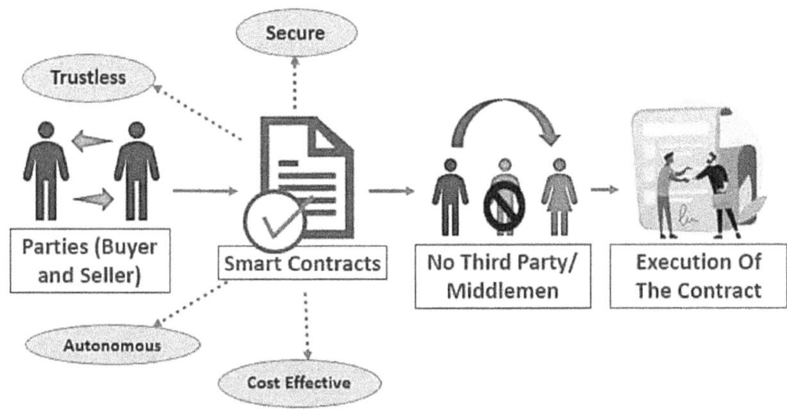

Figure 20.2: Smart Contracts in Blockchain Network

Smart contracts are self-executing computer programs that are designed to automate the execution of an agreement or a task. They are one of the key features of Ethereum, a decentralized blockchain platform. Some of the prominent characteristics of Ethereum smart contracts are:

Distributed: Ethereum smart contracts are replicated and distributed across all nodes of the Ethereum network. This means that there is no central authority controlling the execution of the contract. The decentralized nature of Ethereum ensures that the smart contract can be executed without any interruption, censorship, or downtime.

Deterministic: Smart contracts are programmed to perform specific actions based on pre-defined conditions. They are deterministic, which means that their outcome will be the same every time they are executed, regardless of who initiates the execution.

Autonomous: Smart contracts are self-executing programs that operate without any human intervention once they are deployed. They can automate various tasks and eliminate the need for intermediaries, thereby reducing transaction costs and increasing efficiency.

Immutable: Smart contracts are tamper-proof and cannot be modified once they are deployed on the blockchain. This ensures that the contract code remains transparent and trustworthy.

Customizable: Ethereum smart contracts are highly customizable, and developers can code them in various ways to suit different use cases. The Ethereum blockchain is Turing complete, which means that it can support a wide range of decentralized applications.

Trustless: Ethereum smart contracts enable two or more parties to interact and transact without the need for intermediaries or trusted third parties. The blockchain technology ensures that the data is accurate and transparent.

Transparent: Smart contracts on the Ethereum blockchain are visible to everyone, and their source code is publicly available. This transparency ensures that the contract's behaviour is predictable, and the outcomes are fair and trustworthy.

Modifying & Upgrading Ethereum Smart Contracts

Once an Ethereum smart contract is deployed on the blockchain, its code becomes immutable and cannot be modified. It is designed to execute the instructions programmed into it without any alteration. However, there is a way to replace an existing smart contract with a new one by including a specific function in the code called SELFDESTRUCT. This function can be used to delete the smart contract from the blockchain, freeing up space for a new contract to be deployed.

It is important to note that the SELFDESTRUCT function should be included in the code before the smart contract is deployed. If it is not included, the contract will remain on the blockchain forever. Therefore, smart contract developers need to plan ahead and consider the possibility of needing to update or replace the contract in the future.

To address this challenge, developers have come up with a solution called upgradeable smart contracts. These contracts are designed to be more flexible and allow for changes and updates to be made without compromising the immutability of the blockchain. There are several approaches to creating upgradeable smart contracts, and they can vary in complexity.

One simple example of an upgradeable smart contract is to divide the contract into smaller parts, some of which are designed to be immutable, while others have the SELFDESTRUCT function enabled. This allows the developer to replace certain parts of the code while preserving other functionalities that are still intact.

Overall, while Ethereum smart contracts are designed to be immutable, there are ways to update and replace them when necessary, providing greater flexibility to developers and allowing for the evolution of decentralized applications over time.

Exploring the Versatility & Benefits of Ethereum Smart Contracts for Decentralized Solutions

Smart contracts are programmable code that offer a high degree of customization, allowing developers to create a variety of services and solutions for decentralized applications. As self-executing programs, they can provide increased transparency and reduced operational expenses, leading to increased efficiency, and reduced bureaucratic expenses depending on their implementation.

Smart contracts are especially advantageous in situations that involve the transfer or exchange of funds between two or more parties. They can be designed for a wide range of use cases, including the introduction of tokenized assets, balloting systems, crypto wallets, decentralized exchanges, games, and mobile applications. They can also be deployed in conjunction with other blockchain solutions aimed at tackling fields such as healthcare, charity, supply chain, governance, and decentralized finance (DeFi).

The versatility and benefits of Ethereum smart contracts make them a powerful tool for developers seeking to create decentralized solutions that offer transparency, security, and automation. With their ability to handle complex transactions and enforce contract terms, smart contracts have the potential to revolutionize the way people transact and interact with one another in a trustless, decentralized ecosystem.

Understanding ERC-20 Tokens & their Role in the Ethereum Blockchain

Tokens issued on the Ethereum blockchain are governed by a standard known as ERC-20, which sets out the core capabilities that all Ethereum-based tokens must possess. These digital assets, commonly referred to as ERC-20 tokens, represent a significant portion of the cryptocurrency market.

Many blockchain organizations and startups have leveraged the power of smart contracts to issue their digital tokens on the Ethereum network. In most cases, these companies have distributed their ERC-20 tokens through Initial Coin Offering (ICO) events, taking advantage of the trustless and efficient exchange of funds and distribution of tokens enabled by smart contracts.

ERC-20 tokens offer a wide range of capabilities, including the ability to be traded on cryptocurrency exchanges, used to pay for goods and services, and used as a means of investment. The standard also includes provisions for managing token ownership, enabling developers to design sophisticated decentralized applications that incorporate token ownership and transactional functionality.

Overall, ERC-20 tokens have played a significant role in the growth of the Ethereum ecosystem, enabling developers to create innovative solutions for a wide range of industries and use cases. As the adoption of blockchain technology continues to expand, the role of ERC-20 tokens is expected to become even more important in facilitating secure and efficient transactions in a decentralized environment.

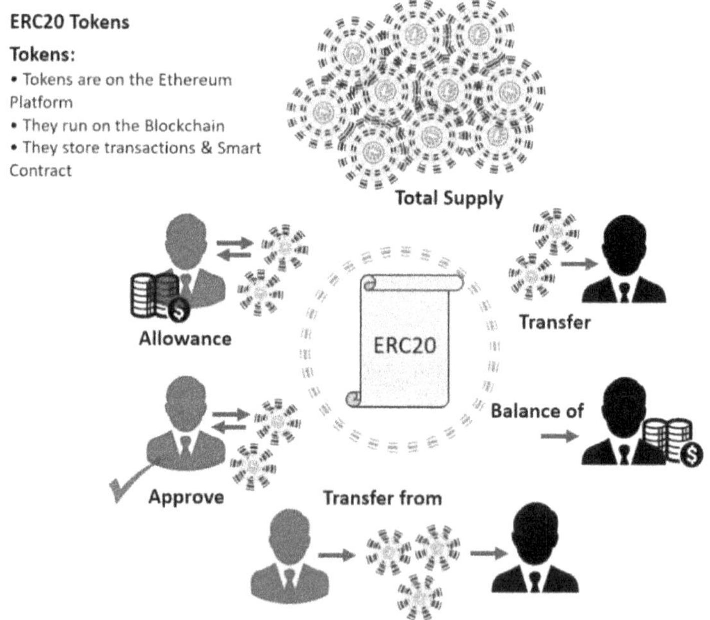

Figure 20.3: ERC20 Tokens

Smart Contracts Limitations & Risks

Smart contracts offer a lot of potential in terms of reducing transaction costs, increasing efficiency, and improving transparency in a trustless manner. However, they also have some limitations and risks. One of the main concerns with smart contracts is that they are written by humans, and therefore, they are subject to vulnerabilities and bugs. This is a significant risk as smart contracts often involve sensitive data or large amounts of money. Ideally, they should be written and deployed by skilled programmers who can ensure their security.

Additionally, some people argue that centralized systems can provide most of the same solutions and functionalities as smart contracts. The primary difference is that smart contracts run on a distributed peer-to-peer (P2P) network instead of a centralized server. However, the immutability of smart contracts can also be a double-edged sword. While it can be beneficial in some cases, it can be catastrophic in others. For instance, in 2016, a DAO on the Ethereum network was hacked, and millions of Ethers were stolen due to flaws in their smart contract code. Since smart contracts are immutable, developers were unable to fix the code, leading to a hard fork of the Ethereum network and the creation of a second Ethereum chain (Ethereum Classic).

Another limitation of smart contracts is their uncertain legal status. Smart contracts are in a Gray area in most countries, and they do not always fit within the current legal framework. For example, many contracts require both parties to be properly identified and over 18 years old. The pseudonymity provided by blockchain technology, combined with the lack of intermediaries, may threaten those requirements. While potential solutions exist, the legal enforceability of smart contracts is a real challenge, especially in borderless, distributed networks.

In conclusion, while smart contracts offer many benefits, they also have limitations and risks that must be carefully considered. To mitigate these risks, skilled programmers should develop and deploy smart contracts, and the legal status of smart contracts must be clarified to ensure their enforceability.

Applications & Limitations of Smart Contracts on the Ethereum Blockchain

There are some who view smart contracts as a revolutionary technology that has the potential to disrupt many industries. However, others criticize their usefulness in certain scenarios. Despite their potential benefits, smart contracts can be less attractive in some situations due to their distributed, deterministic, transparent, and immutable nature.

One of the primary criticisms of smart contracts is that they may not be the most appropriate solution for all real-world problems. For instance, some organizations may find that traditional server-based alternatives offer better efficiency in terms of speed and cross-network communication. Additionally, centralized servers are often less complex and less expensive to maintain. In some cases, it may be more practical and cost-effective to use centralized systems over smart contracts. Another criticism of smart contracts is related to their complexity and the potential for bugs or vulnerabilities in the code. Given that smart contracts are written by humans, the code is subject to the same risks as any other software application. Moreover, since smart contracts are immutable, fixing any issues that arise can be challenging, if not impossible. In situations where a bug or vulnerability is discovered, developers must often create a new smart contract to address the issue, which can be time-consuming and costly.

Furthermore, smart contracts' legal status remains uncertain in many jurisdictions. While smart contracts may offer many benefits, such as pseudonymity and lack of intermediaries, these same features may pose a challenge to the enforceability of traditional contract law. Smart contracts may not be suitable for all types of contracts, and their legal status needs to be addressed to ensure their widespread adoption.

Blockchain Scalability: Sidechains & Payment Channels

Scalability is an important factor in any system's ability to handle growth and increasing demands. In the realm of computing, upgrading hardware can often improve performance and efficiency for handling various tasks. In the blockchain world, scalability refers to the ability to handle more transactions and data. However, not all blockchain protocols are designed to be scalable. Bitcoin, for example, has its own strengths but scalability is not one of them. Unlike a centralized database where an administrator can easily increase speed and throughput, Bitcoin's decentralized nature and value propositions, such as censorship-resistance, require many participants to sync a duplicate of the blockchain.

Exploring the Issue of Blockchain Scalability: Sidechains & Payment Channels

In the world of computing, scalability refers to a system's ability to expand and adapt to meet increasing demands. When it comes to blockchain technology, scalability pertains to the capacity to handle a growing number of transactions. While blockchain protocols like Bitcoin have many strengths, they are often criticized for their lack of scalability. If Bitcoin operated on a centralized database, it would be relatively simple for an administrator to increase its speed and processing capabilities. However, the decentralized nature of the blockchain, along with its value proposition of censorship-resistance, requires thousands of participants to sync a copy of the blockchain, making scalability a challenging issue.

Running a Bitcoin node is relatively inexpensive, and even basic computing devices can do it. However, because many nodes need to stay in sync, there are certain limitations to their capacity. Caps are placed on the number of transactions that can be processed on-chain to prevent the database from becoming too unwieldy. If blocks are too big, they cannot be rapidly relayed around the network, causing nodes to fall behind. This bottleneck creates a situation where the blockchain operates like a train service that departs at regular intervals, with limited seats in each carriage. If everyone tries to board the train at the same time, prices will increase. Similarly, a network clogged with pending transactions will require users to pay higher fees to ensure that their transactions are included in a timely manner. One possible solution to this problem is to increase the size of the carriages, which would mean more seats, higher throughput, and cheaper ticket prices. However, there is no guarantee that the additional seats will not quickly fill up. Blocks or block gas limits cannot scale indefinitely, making it more expensive for nodes to remain on the network, as they will need more expensive hardware to remain in sync.

Ethereum creator Vitalik Buterin developed the Scalability Trilemma to explain the challenges faced by blockchains. He theorizes that protocols must make trade-offs between scalability, security, and decentralization, which are often at odds with each other. Focusing too much on scalability could compromise decentralization, which is a core value of blockchain technology.

To address this issue, many in the blockchain community are exploring off-chain solutions, such as sidechains and payment channels, to improve scalability. These solutions allow for a higher volume of transactions to be processed without clogging the main blockchain network. By maximizing security and decentralization on the blockchain itself, while achieving scalability off-chain, the blockchain community hopes to find a way forward to meet the growing demands of its users.

Off-Chain Scaling Solutions: Sidechains & Payment Channels

Off-chain scaling solutions are methods that enable transactions to be conducted without overburdening the blockchain network. Instead of executing all transactions on the primary chain, protocols can be integrated into the system, allowing users to send and receive funds without causing congestion on the main chain. This technique has become increasingly popular, as it provides a way to address the scalability issue of blockchain networks.

Two notable off-chain scaling solutions are sidechains and payment channels. Sidechains are essentially separate blockchains that are attached to the primary chain and operate independently. They allow transactions to be conducted without adding to the primary chain's load, and the resulting data can be returned to the main chain as proof-of-work. Payment channels, on the other hand, are a type of sidechain that facilitates the execution of numerous transactions between two parties off-chain. These channels are established using smart contracts, which enable the parties involved to send transactions to one another without recording them on the primary chain. Instead, the transactions are aggregated and only the final state of the channel is recorded on the main chain. Off-chain scaling solutions offer several benefits, including faster transaction processing times, lower transaction fees, and reduced network congestion. However, they also introduce new challenges, such as increased complexity and potential security risks. Nonetheless, these solutions represent a promising way forward for blockchain networks looking to overcome scalability limitations while maintaining the security and decentralization benefits of distributed ledger technology.

Sidechains: A Solution to Blockchain Scalability

Sidechains are an off-chain scaling solution that allows for the creation of separate blockchains that are interoperable with the primary chain. In simple terms, a sidechain is a separate blockchain that is pegged to the primary chain, allowing assets to move between the two networks. This approach has gained popularity due to its ability to address the scalability problem faced by primary blockchain networks like Bitcoin and Ethereum.

One of the essential features of sidechains is their interoperability with the primary chain. This means that assets can move freely between the two chains. There are several ways to facilitate this, with the most popular methods being depositing assets into a special address on the primary chain or sending funds to a custodian who exchanges the deposit for funds on the sidechain. The primary chain then locks the assets, and a corresponding amount is issued on the sidechain.

Sidechains offer several advantages over traditional on-chain transactions. They reduce network congestion, improve transaction speed, and lower fees. They also provide greater flexibility, allowing for the development of custom applications and features that may not be feasible on the primary chain.

However, sidechains also have their limitations. One significant challenge is ensuring the security of the funds being transferred between the two chains. There is always a risk of a potential security breach that could result in the loss of funds. Therefore, implementing robust security measures is crucial.

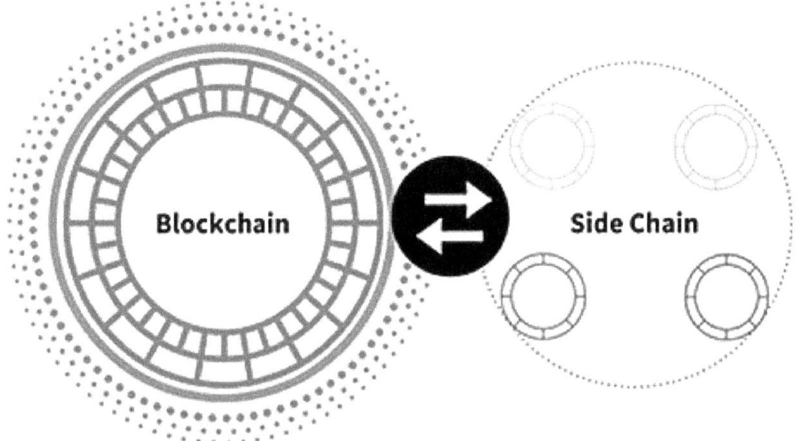

Figure 20.4: Blockchain Sidechains

How Sidechains Work: An Overview of Two-Way Peg System & Transactions

Let us assume that Alice possesses five bitcoins and intends to exchange them for five equivalent units on a Bitcoin sidechain, known as sidecoins. The sidechain employs a Two-Way Peg system, which means that users can move their assets from the primary chain to the sidechain and vice versa.

It is essential to remember that the sidechain operates as a distinct blockchain with its own unique blocks, nodes, and validation mechanisms. To obtain her sidecoins, Alice would need to send her five bitcoins to another address. This address may be held by a person who will then credit Alice's sidechain address with five sidecoins once they receive the bitcoins. Alternatively, the address may have a trust-minimized configuration where the software program automatically credits the sidecoins after detecting a payment.

Now that Alice has exchanged her coins for sidecoins, she can always revert the process to recover her bitcoins. After entering the sidechain, she is free to conduct transactions on this separate blockchain. Alice can transfer sidecoins to other users or receive them, just like she would on the primary chain.

For example, Alice could buy a Binance hoodie from Bob for one sidecoin. When she wants to return to Bitcoin, she can send her remaining four sidecoins to a designated address. Following confirmation of the transaction, four bitcoins will be unlocked and delivered to an address that Alice controls on the primary chain.

Why Use Sidechains: Advantages Over Bitcoin Blockchain & Potential for Experimentation

Some may question the rationale behind utilizing sidechains. Why can't Alice just utilize the Bitcoin blockchain? The reason is that sidechains are capable of achieving things that the Bitcoin network cannot accomplish. Blockchains are meticulously engineered systems with trade-offs. Although Bitcoin is the most secure and decentralized cryptocurrency, it lacks throughput. Bitcoin transactions are faster than conventional methods, but they remain slow in comparison to other blockchain systems. Blocks are mined every ten minutes, and fees can skyrocket when the network is congested. Admittedly, this level of security is probably not required for small daily transactions. Alice will not wait for the transaction to confirm if she is paying for a coffee. She would hold up the queue, and her drink would be cold when she got it. Sidechains are not subject to the same constraints. In fact, they do not even have to employ Proof of Work to function. Any consensus mechanism can be utilized, a single validator can be trusted, and any number of parameters can be tweaked. It is even possible to incorporate upgrades that do not exist on the primary chain, generate larger blocks, and enforce rapid settlements. Sidechains can also have critical bugs without affecting the underlying chain. This enables them to be employed as experimentation platforms and to introduce features that would otherwise necessitate consensus from the majority of the network. If users are content with the trade-offs, sidechains could be a crucial step toward effective scaling. Main chain nodes do not have to store every transaction from the sidechain. Alice could enter the sidechain with a single Bitcoin transaction, execute hundreds of sidecoin transactions, and then leave the sidechain; according to the Bitcoin blockchain, she only performed two transactions, one to enter and one to exit. Ethereum's Plasma is comparable to sidechains but has some significant variations.

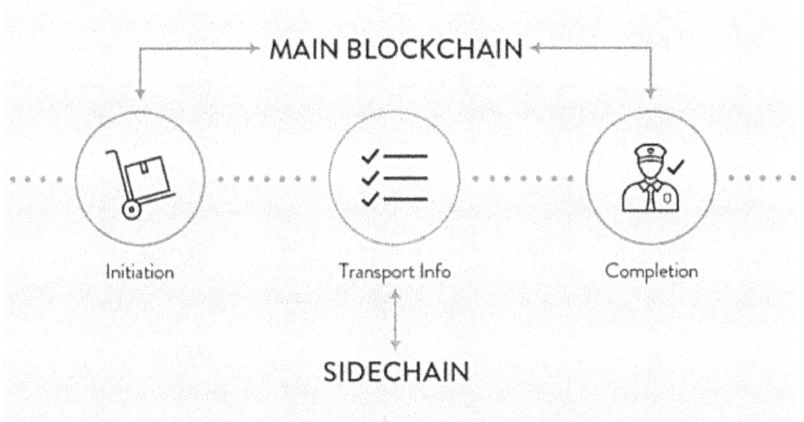

Figure 20.5: Use case of Sidechains

Exploring Payment Channels: A Smart Way to Transact Off-Chain

Payment channels, similar to sidechains, aim to address the issue of scalability on the blockchain. However, they differ in their approach. Payment channels do not require a separate blockchain, unlike sidechains, to function. Rather, they leverage smart contracts to enable users to transact without publishing their transactions to the blockchain. A payment channel is a two-way communication channel between two participants that uses a smart contract to enforce a set of rules for conducting transactions. It allows users to transact back and forth without the need to interact with the main chain every time they initiate a transaction. This off-chain transaction technique can greatly reduce the transaction fees and processing time that users would otherwise have to pay for each on-chain transaction.

To initiate a payment channel, the two participants, let us say Alice and Bob, first create a funding transaction on the blockchain, locking a certain amount of cryptocurrency into a multisig wallet. This wallet requires both Alice and Bob's approval to release funds. Once the funds are locked, the payment channel is established, and the two parties can transact back and forth, updating the balance between themselves. The updates are done using a set of off-chain transactions, which are signed by both parties and exchanged between them. The new balance is updated and recorded in the multisig wallet at the end of each update. The final balance is then broadcasted to the blockchain as a single transaction, settling the balance between the two parties.

Payment channels provide a secure and efficient way for users to transact off-chain without compromising on the security of the underlying blockchain. They are especially useful for small or frequent transactions, as they significantly reduce transaction costs and time while maintaining a high level of security.

Comparing Sidechains & Payment Channels for Blockchain Scalability

Payment channels provide a solution to the scalability problem that blockchain technology faces. The Lightning Network is a popular example of a payment channel that enables users to conduct transactions without the need for each transaction to be recorded on the blockchain. Instead, users deposit funds into a multi-signature address, which requires both parties' signatures for funds to be spent. Once Alice and Bob have deposited 10 BTC each into a jointly owned multi-signature address, they can create a balance sheet to keep track of their funds. If Alice wants to send a coin to Bob, they can update the balance sheet to indicate that Alice has nine BTC, and Bob has eleven BTC. These updates do not need to be published to the blockchain, so there are no miner fees to pay and no block confirmations to wait for.

When the payment channel comes to an end, Alice and Bob can create a transaction that sends their respective balances to their own addresses. This transaction will need to be signed by both parties and broadcasted to the blockchain. Alice and Bob could have recorded multiple transactions on their balance sheet, but as far as the blockchain is concerned, they have only performed two on-chain operations: the initial funding transaction and the final reallocation transaction. However, the above example assumes that both parties will cooperate. In practice, payment channels require mechanisms to prevent cheating. If either party tries to cheat, the other party can use the information stored in the balance sheet to prove their case and get their fair share of the funds. These mechanisms ensure that parties can safely interact with each other without having to trust one another.

Payment Routing & the Advantages of a Network of Payment Channels

Payment routing within a network of payment channels is a dynamic process that facilitates secure and efficient fund transfers between parties, even when a direct payment channel is not established between them. While payment channels are a valuable means for conducting high-volume transactions directly between two parties, the concept of a payment channel network elevates the advantages of this technology. Within a payment channel network, participants connect to multiple peers, forming an intricate and distributed web of connections. This web-like structure creates numerous possible routes for transmitting funds to a desired destination. To illustrate, suppose Alice intends to send funds to Carol, but they lack a direct channel between them. In this case, they can leverage the network to find an indirect route. For instance, Alice can transmit her funds to Bob, a participant with an open channel to Carol. Bob, in turn, forwards these funds to Carol via their channel connection. If Carol has another connection, say with Dan, this payment can be further routed through Dan. *(Ref. No.: 315- 333)*

This network-based approach offers a range of significant advantages:

Speed and Efficiency: Payment routing within a network of channels enables swift and cost-effective transactions. Participants can conduct transfers off-chain, avoiding the need for time-consuming block confirmations typically associated with on-chain transactions. Expanded Connectivity: Participants can transact with a more extensive network of peers, even those with whom they lack a direct channel. This expanded connectivity increases the potential number of payment partners, fostering a more robust and flexible ecosystem. Optimized Routes: The network allows participants to select the most efficient route for their transactions. By identifying the path with minimal fees and costs, users can optimize their payments and minimize transaction expenses.

In essence, payment routing within a network of payment channels enhances the utility and practicality of this technology, making it a compelling solution for fast, low-cost, and widely accessible digital transactions. It represents a crucial advancement in the realm of blockchain-based financial systems, offering both convenience and cost-effectiveness to users.

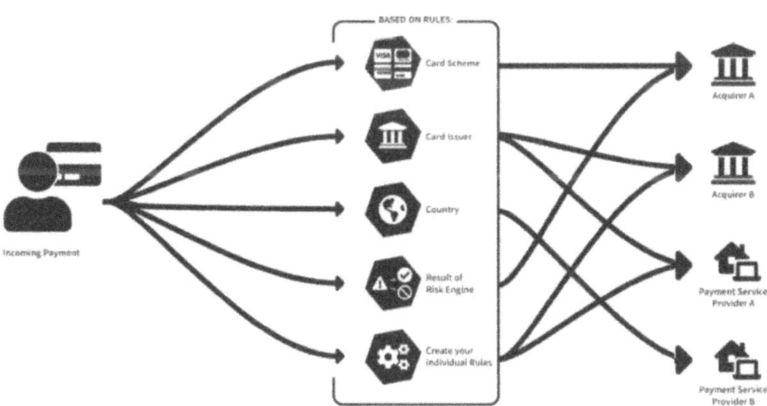

Figure 20.6: Routing in Smart Contract

Closing Thoughts

The advent of blockchain technology and smart contracts has ushered in a wave of innovation with the potential to reshape various facets of society. These technologies introduce secure and decentralized methods for handling transactions and data, offering a promising alternative to traditional governance and financial systems. However, while blockchain and smart contracts hold tremendous potential, their widespread adoption faces several substantial challenges, including scalability, interoperability, and regulatory concerns.

Scalability stands as a foremost challenge in the path of blockchain technology. As these networks expand, the strain on their underlying infrastructure can lead to slower transaction processing times and higher costs. To tackle this issue, researchers and developers are actively exploring solutions like sidechains and payment channels. These innovations aim to offload a significant portion of transaction processing from the main blockchain, thus alleviating congestion and enhancing overall scalability. While these solutions are still in their developmental stages, they hold immense promise for the future of blockchain technology.

Decentralization represents a core tenet of blockchain systems, underpinning trust, and transparency in the absence of central authorities. However, as blockchain networks grow, maintaining this decentralization becomes increasingly challenging. To address this, it becomes essential to implement mechanisms that limit the rapid expansion of the blockchain network, allowing new nodes to join seamlessly. Proponents of off-chain scalability solutions suggest that, in the future, the primary blockchain might predominantly serve high-value transactions or act as the bridge for connecting to sidechains and initiating and concluding payment channels. This strategic approach can help preserve the essential aspect of decentralization in blockchain networks.

In essence, while blockchain and smart contracts offer a revolutionary shift in how transactions and data are managed, overcoming challenges such as scalability and preserving decentralization are pivotal for their long-term success. Researchers, developers, and the blockchain community at large are actively working on these issues, and as solutions emerge, blockchain technology is likely to continue evolving and expanding its influence across a wide array of industries and applications.

Glossary of Terms

Term	Definition
51% Attack	51% attack: When a majority of miners in a blockchain conspire to attack and steal assets or engage in double spending.
Address	Blockchain address: Similar to a URL, it serves as the destination or source for transactions within the blockchain.
Alt-coin	Altcoin: Any cryptocurrency or token other than Bitcoin.
Attestation Ledger	Attestation ledger: Register or account book providing evidence and support for individual transactions, often used for verification or authentication.
Bitcoin	Bitcoin: The first and most popular cryptocurrency based on DLT technology, created by Satoshi Nakamoto in 2008.
Block	Block: A group of transactions in a blockchain, analogous to a page in a ledger.
Blockchain	Blockchain: Immutable, decentralized digital ledger storing digital transactions or data, linked via cryptographic signatures for security and reliability.

Blockchain (Private a.k.a. Permissioned)	Private blockchain: Blockchain residing on a private network accessible only to authorized individuals.
Blockchain (Public i.e., Permissionless)	Public blockchain: Blockchain residing on a global network accessible to everyone.
Byzantine Fault Tolerance (BFT)	Fault tolerance: Property of distributed systems to resist failure even with nodes acting maliciously or failing.
Centralized	Centralized system: System or process with a single source of authority, control, or truth.
Chain of Custody	Supply chain: Entire chain of ownership documentation for a product from raw materials to the end user.
Chain code	Smart contract: Another term for a smart contract, which is a self-executing contract with predefined rules.
Consensus Mechanism - Proof of Authority (PoA)	Proof of Authority (PoA): Validators stake their identity to gain the right to validate blocks, with their actions tied to their identity. Rigorous KYC process is required for participant verification.

Consensus Mechanism - Proof of Burn (PoB)	Proof of Burn (PoB): Miners destroy cryptocurrency to earn the right to add blocks, obtaining virtual mining rigs. Burning currency determines rig size. PoB is considered an energy-efficient version of PoW.
Consensus Mechanism - Proof of Capacity (PoC)	Proof of Capacity (PoC): Miners use available hard drive space to determine mining rights instead of computational power or stake.
Consensus Mechanism - Proof of Stake (PoS)	Proof of Stake (PoS): Miners stake the blockchain's cryptocurrency to improve their chances of validating a block. Staked amount serves as a deposit and can be burned for rule violations.
Consensus Mechanism - Proof of Work (PoW)	Proof of Work (PoW): Miners use computational resources to repeatedly apply transaction data and random strings to a mathematical formula until desired outcome is achieved. Requires significant computational power.
Consensus Mechanism (a.k.a. Consensus Protocol)	Byzantine Fault Tolerance (BFT): Validation process across a distributed blockchain network designed to achieve consensus despite potential malicious actions.

Cryptocurrency	Cryptocurrency: Digital money using encryption and consensus algorithms for generation and transfer of funds, generally decentralized.
Cryptography	Cryptography: Science of securing communication through personalized codes for private message access.
DAO (Decentralized Autonomous Organization)	Decentralized governance: Governance structure rewarding good behaviour and penalizing bad behaviour based on predefined rules, with changes made through voting and stake involvement.
DApp	Decentralized software: Enables information sharing among users without relying on a central system or database, often utilizing a decentralized database like a blockchain.
Decentralization/ Decentralized	Decentralized decision-making: A system where each node independently makes decisions, resulting in collective behaviour.
Digital Identity (a.k.a. Self-Sovereign Identity)	Blockchain-based digital identity: An Internet equivalent of a real identity that empowers individuals by giving them ownership and control over their personal information.

Digital Signature	Digital signatures: Mathematical schemes that verify the authenticity and integrity of digital messages or documents.
Digital Signature - Multi-signature	Multisig addresses: Addresses that require multiple digital signatures to authorize a transaction, enhancing security.
Digital Signature - Ring	Ring signatures: Digital signatures that make it impossible to determine which member of a group signed a transaction.
Distributed	Distributed systems: Systems where processing and/or data are shared across multiple nodes, although decision-making may still be centralized.
Distributed Ledger Technology (DLT)	Distributed ledger technology (DLT): A technology class that includes blockchain, recording asset transactions across multiple copies simultaneously without a central data store.

Double Spending	Double spending: The act of spending or trading the same cryptocurrency coins or tokens twice.
Ethereum	Public smart contract blockchain: A public blockchain that supports smart contracts.
Fiat	Legal tender: Government-backed currency, such as US dollars or Euros.
Fork	Hard fork: An agreed-upon software update by all nodes in a distributed network, sometimes coexisting with the previous version.
Fungible	Fungibility: The property of being exchangeable with other similar items, such as currencies.
Gas	Gas fee: The fee charged for writing a transaction to a public blockchain, rewarding the validating miner.
Genesis Block	Genesis block: The first block or initial blocks in a blockchain.

Governance	Governance: Establishing and monitoring policies for proper implementation in an organization or system.
Hash Function	Hash function: A function that converts input of any size into a unique string of uniform length.
Hyperledger Fabric	IBM Blockchain: IBM's private blockchain toolset for permissioned networks.
Identity	Entity identifier: Information used by computer systems to uniquely represent a person, organization, application, or device.
Immutable/Immutability	Immutability: The property of being unchangeable once a transaction is added to a block and written to a blockchain.
Initial Coin Offering (ICO)	Initial coin offering (ICO): The first sale of a blockchain coin or token.
Interoperability	Blockchain interoperability: The ability of multiple systems to communicate and exchange data, with ongoing efforts to connect different blockchains.

IPFS (Interplanetary File System)	IPFS: A peer-to-peer protocol for storing and sharing data in a distributed file system using content-addressing to uniquely identify files.
Know Your Customer (KYC)	KYC (Know Your Customer): The legal process of a business verifying the identity of its clients, with requirements varying across jurisdictions.
Liquidity	Liquidity: The ease of converting an asset or cryptocurrency into cash (fiat).
Mainnet	Mainnet: The production version of a blockchain.
Merkle Tree/Hash Tree	Merkle tree: A tree structure where leaf nodes contain hash values of data blocks, and non-leaf nodes contain hashes of their child nodes.
Mining	Mining: The process of verifying and adding transactions to the public blockchain, with miners rewarded in the blockchain's cryptocurrency.
Node	Node: A computer that holds a copy of the blockchain ledger.
Non-Fungible	Non-fungible: The property of an item not being exchangeable with similar items, such as unique artworks or collectibles.
Off-chain	Off-chain data: Data stored externally to the blockchain.

On-chain	On-chain data: Data stored within the blockchain.
Open Source	Open source: Software products that allow permission to use, modify, and enhance their source code, design documents, or content.
Oracle	Blockchain integration: An application that connects blockchain applications to legacy systems.
Peer-to-Peer (P2P)	Peer-to-peer: A direct connection between participants in a system, either computer-to-computer or person-to-person.
Provenance	Product provenance: The entire history of a product, including its lifecycle, chain of custody, and value-added services.
Public/Private Key	Public key cryptography: A cryptographic system using a unique public key derived from a private key for encryption and decryption.
Satoshi Nakamoto	Satoshi Nakamoto: The pseudonymous name associated with the development of Bitcoin and the creation of the first blockchain database.
Seed Phrase	Recovery phrase: A random sequence of words used to restore a lost wallet.

Sharding	Sharding: Database partitioning technique that divides large databases into smaller, more manageable parts to potentially improve blockchain performance.
Sidechain	Sidechain: A discrete blockchain linked to a main blockchain via two-way pegs, allowing assets to be interchanged and enabling scaling and faster transactions.
Smart Contract	Smart contract: Self-executing computer code deployed on a blockchain to perform various functions, often involving value exchange.
Solidity	Solidity: JavaScript-like programming language for implementing smart contracts on the Ethereum blockchain.
Stablecoin	Stablecoin: Cryptocurrency backed by assets like fiat currency or commodities, designed to minimize price volatility.
State Channel	Off-chain execution: Process of executing blockchain transactions outside the main chain to enhance performance and reduce costs.
Testnet	Testnet: Staging blockchain environment for testing applications before deployment on the production network.

Token	Cryptographic tokens: Programmable assets or access rights managed by smart contracts and distributed ledgers, accessible only through private keys.
Token Generation Event	Initial coin offering (ICO): Creation and initial sale of a blockchain coin or token.
Token Type ERC-20	ERC-20: Fungible token standard on Ethereum, defining functions for total supply retrieval, wallet-to-wallet transfer, and transaction approval.
Token Type - ERC-721	ERC-721: Non-fungible token standard on Ethereum, defining functions for unique token management and wallet-to-wallet transfer.
Token/Coin Exchange	Cryptocurrency exchange: Application for buying, selling, and trading cryptocurrencies.
Token less Ledger	Native currency-free ledger: Ledger operating without a native currency requirement.
Tokenomics	Monetary governance: Study, design, and implementation of monetary management and distribution using blockchain technology.
Transactions Per Second (TPS)	Transactions per second (TPS): Measurement of blockchain speed, where low TPS can limit business and financial applications.

Transparency	Transparency: Public blockchain property allowing any participant to view transactions.
Trust	Trust: Confidence in the integrity of an entity or system.
Trustless	Trustless: Elimination of trust from transactions in a blockchain system, relying on properties like digital signatures and cryptography.
Vyper	Vyper: Python-like programming language for Ethereum, prioritizing security, simplicity, and auditability.
Wallet	Wallet: Digital file holding coins and tokens, with a blockchain address for receiving transactions.
Wallet (Cold)	Cold wallet: Offline wallet disconnected from the internet.
Wallet (Hot)	Hot wallet: Online wallet connected to the internet.
Wallet (Multisignature)	Multi-signature wallet: Wallet requiring multiple digital signatures for transaction execution.
Zeppelin/Open Zeppelin	Developer community: Community of like-minded smart contract developers.

References

1) Perez, J., 2021. 5 Courses to Beef Up Your Knowledge of Blockchain Technology. IEEE Spectrum. Retrieved from: https://spectrum.ieee.org/5-courses-on-blockchain-technology#toggle-gdpr

2) Zhao, W., 2021. Blockchain Technology: Principles and Applications. IEEE Access. Retrieved from: https://ieeeaccess.ieee.org/closed-special-sections/blockchain-technology-principles-and-applications/

3) Simplilearn., 2022. What is Blockchain Technology and How Does It Work? Retrieved from: https://www.simplilearn.com/tutorials/blockchain-tutorial/blockchain-technology

4) Schmidt, J., 2022. What is Blockchain? Forbes Advisor. Retrieved from: https://www.forbes.com/advisor/in/investing/what-is-blockchain/

5) Little, K., 2022. What Is Blockchain and How Does It Work? Time. Retrieved from: https://time.com/nextadvisor/investing/cryptocurrency/what-is-blockchain/

6) Iansiti, M., Lakhani, K. R., 2017. The Truth About Blockchain. Harvard Business Review. Retrieved from: https://hbr.org/2017/01/the-truth-about-blockchain

7) Voigt, K., Rosen, A., 2022. What is Blockchain? Blockchain Technology, Explained. NerdWallet. Retrieved from: https://www.nerdwallet.com/article/investin/blockchain

8) Darlington, N., 2021. Blockchain For Beginners: What Is Blockchain Technology a Step-by-Step Guide. Block geeks. Retrieved from: https://blockgeeks.com/guides/what-is-blockchain-technology/

9) Hayes, A., 2022. What Is a Blockchain? Investopedia. Retrieved from: https://www.investopedia.com/terms/b/blockchain.asp

10) Pratt, M. K., Gillis, A. S., 2021. Blockchain. TechTarget. Retrieved from: https://www.techtarget.com/searchcio/definition/blockchain

11) NDTV Business Desk., 2022. What Is Blockchain Technology and How It Works. NDTV. Retrieved from: https://www.ndtv.com/business/what-is-blockchain-technology-that-powers-the-cryptocurrencies-2975865

12) Norton, S. 2016. What Is Blockchain? The Wall Street Journal. Retrieved from:
https://www.wsj.com/articles/BL-CIOB-8993

13) Duggan, W. 2022. All you need to know about blockchain technology and how it enables cryptocurrency transactions and other services. U.S. News & World Report. Retrieved from:
https://money.usnews.com/investing/term/blockchain

14) Chaturvedi, A., 2022. What is Blockchain? How is it used in cryptocurrency. Business Today. Retrieved from:
https://www.businesstoday.in/crypto/story/what-is-blockchain-how-is-it-used-in-cryptocurrency-332411-2022-05-05

15) Iansiti, M., Lakhani, K. R., 2015. Learn More About blockchain. Merriam-Webster Dictionary. Retrieved from:
https://www.merriam-webster.com/dictionary/blockchain

16) Mansur, R., 2021. What is blockchain? Blocks, distributed ledgers and nodes explained in simple terms. Your Story. Retrieved from:
https://yourstory.com/the-decrypting-story/what-is-blockchain-technology-blocks-distributed-ledgers-nodes/amp

17) Binance Academy., 2019. What Makes a Blockchain Secure? Binance Academy. Retrieved from:
https://academy.binance.com/en/articles/what-makes-a-blockchain-secure

18) Sahu, M., 2021. What Makes Blockchain Secure: Key Characteristics & Security Architecture. UpGrad Blog. Retrieved from:
https://www.upgrad.com/blog/what-makes-blockchain-secure/

19) Orcutt, M., 2018. How secure is blockchain really? MIT Technology Review. Retrieved from:
https://www.technologyreview.com/2018/04/25/143246/how-secure-is-blockchain-really/

20) Gadgets 360 News desk., 2021. How Secure Is the Bitcoin Blockchain, and Is Your Cryptocurrency Safe? Gadgets 360. Retrieved from:
https://gadgets360.com/cryptocurrency/features/how-secure-is-bitcoin-blockchain-cryptocurrency-technology-2516405

21) Sajan, K. S., 2022. What makes blockchain secure? Tutorials point. Retrieved from:
https://www.tutorialspoint.com/what-makes-blockchain-secure

22) Stouffer, C., 2022. What is blockchain security? An overview. Norton. Retrieved from:
https://us.norton.com/internetsecurity-privacy-how-to-blockchain-security.html

23) Stephen, R., Alex, A., 2018. A Review on Blockchain Security. IOP Conference Series: Materials Science and Engineering, 396 (1), 012030. Retrieved from:
https://iopscience.iop.org/article/10.1088/1757-899X/396/1/012030/pdf

24) Liu, Z., Luong, N. C., & Wang, W., 2019. A Survey on Applications of Game Theory in Blockchain. arXiv preprint arXiv:1902.10865. Retrieved from:
https://arxiv.org/pdf/1902.10865.pdf

25) Zephyr, O., 2022. Game Theories of Cryptocurrency Investing I. Medium. Retrieved from:
https://medium.datadriveninvestor.com/game-theories-of-cryptocurrency-investing-i-1ff0aa15a55d

26) Binance Academy., 2019. Game Theory and Cryptocurrencies. Binance Academy. Retrieved from:
https://academy.binance.com/en/articles/game-theory-and-cryptocurrencies

27) Caginalp, C., Caginalp, G., 2019. Establishing Cryptocurrency Equilibria Through Game Theory. ResearchGate. AIMS Mathematics, 4 (3), 420–436. Retrieved from:
https://www.researchgate.net/publication/332981760_EstablishingCryptocurrency_Equilibria_Through_Game_Theory

28) Garcia, J., 2021. A Look at The Game Theory of Bitcoin. Nasdaq. Retrieved from:
https://www.nasdaq.com/articles/a-look-at-the-game-theory-of-bitcoin-2021-09-13

29) Rosic, A., 2020. What is Cryptocurrency Game Theory: A Basic introduction. Blockgeeks. Retrieved from:
https://blockgeeks.com/guides/cryptocurrency-game-theory/

30) Buntinx, J., 2021. What Is Game Theory and Why Is It Relevant to Cryptocurrency? Vaultoro. Retrieved from:
https://vaultoro.com/what-is-game-theory-and-why-is-it-relevant-to-cryptocurrency/

31) Finestone, M., 2018. Game Theory and Blockchain. Matthew Finestone on Medium. Retrieved from:
https://matthewfinestone.medium.com/game-theory-and-blockchain-db46e67933d7

32) Geeks for Geeks., 2022. Practical Byzantine Fault Tolerance (pBFT). Retrieved from:
https://www.geeksforgeeks.org/practical-byzantine-fault-tolerancepbft/

33) Binance Academy., 2018. Byzantine Fault Tolerance Explained. Binance Academy. Retrieved from: https://academy.binance.com/en/articles/byzantine-fault-tolerance-explained

34) Daly, L., 2022. What Is Byzantine Fault Tolerance? The Motley Fool. Retrieved from: https://www.fool.com/investing/stock-market/market-sectors/financials/cryptocurrency-stocks/byzantine-fault-tolerance/

35) Iredale, G., 2021. Byzantine Fault Tolerance - A Complete Guide. 101 Blockchains. Retrieved from: https://101blockchains.com/byzantine-fault-tolerance/

36) Konstantopoulos, G., 2017. Understanding Blockchain Fundamentals, Part 1: Byzantine Fault Tolerance. Loom Network on Medium. Retrieved from: https://medium.com/loom-network/understanding-blockchain-fundamentals-part-1-byzantine-fault-tolerance-245f46fe8419

37) Patel, R., 2020. Byzantine Fault Tolerance (BFT) and its significance in the Blockchain world. HCL Technologies. Retrieved from: https://www.hcltech.com/blogs/byzantine-fault-tolerance-bft-and-its-significance-blockchain-world

38) Castro, M., Liskov, B., 1999. Practical Byzantine Fault Tolerance. Proceedings of the Third Symposium on Operating Systems Design and Implementation, 173-186. Retrieved from: https://pmg.csail.mit.edu/papers/osdi99.pdf

39) Datta, R., Marchang, N., 2012. Byzantine fault detection. In Computer Science Review (Vol. 6, No. 3, pp. 179-202). Elsevier. Retrieved from: https://www.sciencedirect.com/topics/computer-science/byzantine-fault

40) Sharma, T. K., 2020. Types of Blockchains Explained- Public Vs. Private Vs. Consortium. Blockchain Council. Retrieved from: https://www.blockchain-council.org/blockchain/types-of-blockchains-explained-public-vs-private-vs-consortium/

41) Singh, J., 2021. Private, public and consortium blockchains: The differences explained. Coin telegraph. Retrieved from: https://cointelegraph.com/explained/private-public-and-consortium-blockchains-the-differences-explained

42) Mueller, T., 2018. Public vs Private Vs Consortium Blockchains — what is best for enterprises? Evan Network on Medium. Retrieved from:

https://medium.com/evan-network/public-vs-private-vs-consortium-blockchains-3ad180d1e74

43) CNBCTV18.com., 2022. Explained | What are private, public, and consortium blockchains? CNBCTV18.com. Retrieved from: https://www.cnbctv18.com/cryptocurrency/explained-what-are-private-public-and-consortium-blockchains-12543932.htm

44) Obafemi, F., 2021. Blockchain Applications in Business: Public vs Private. Flexiple. Retrieved from: https://flexiple.com/blockchain/blockchain-applications-in-business-public-vs-private/

45) Buterin, V., 2015. On Public and Private Blockchains. Ethereum Blog. Retrieved from: https://blog.ethereum.org/2015/08/07/on-public-and-private-blockchains/

46) Vatsraj, D., 2021. Difference between private and public blockchain. iPleaders. Retrieved from: https://blog.ipleaders.in/difference-between-private-and-public-blockchain/

47) Joshi, N., 2020. Public vs private blockchain: Who wins? Retrieved from: https://www.allerin.com/blog/public-vs-private-blockchain-who-wins

48) Times of India., 2021. What are Blockchain Nodes? The Times of India. Retrieved from: https://timesofindia.indiatimes.com/business/cryptocurrency/blockchain/what-are-blockchain-nodes/articleshow/88116308.cms

49) S., J., 2018. Blockchain: What are nodes and master nodes? Coin monks on Medium. Retrieved from: https://medium.com/coinmonks/blockchain-what-is-a-node-or-masternode-and-what-does-it-do-4d9a4200938f

50) Abrol, A., 2016. What are Blockchain nodes? Detailed Guide. Blockchain Council. Retrieved from: https://www.blockchain-council.org/blockchain/blockchain-nodes/

51) Gadgets 360 News desk., 2021. What Is a Blockchain Node and How Is It Used in Cryptocurrency? Gadgets 360. Retrieved from: https://gadgets360.com/cryptocurrency/features/what-is-a-blockchain-node-how-does-cryptocurrency-work-2515427

52) Newman, D., 2021. Blockchain Node Providers and How They Work. InfoQ. Retrieved from: https://www.infoq.com/articles/blockchain-as-a-service-get-block/

53) Nibley, B., 2021. What Are Nodes? 7 Types of Blockchain Nodes. SoFi. Retrieved from:

https://www.sofi.com/learn/content/what-are-nodes/

54) Vermaak, W., 2021. What Is a Node? Coin Market Cap. Retrieved from:
https://coinmarketcap.com/alexandria/article/what-is-a-node

55) NDTV Business., 2022. Cryptocurrency Mining: It's More Than Just Creating New Coins. Read About It Here. NDTV. Retrieved from:
https://www.ndtv.com/business/what-is-cryptocurrency-mining-know-all-about-it-here-2956751#:~:text=Mining%20cryptocurrency%20is%20a%20method,well%20as%20validate%20existing%20transactions

56) Hong, E., 2022. How Does Bitcoin Mining Work? Investopedia. Retrieved from:
https://www.investopedia.com/tech/how-does-bitcoin-mining-work/

57) Ingalls, S., 2021. Cryptocurrency Mining. Webopedia. Retrieved from:
https://www.webopedia.com/definitions/cryptocurrency-mining/

58) Infosec Insights, Sectigo Store., 2020. What Is Crypto Mining? How Cryptocurrency Mining Works. Retrieved from:
https://sectigostore.com/blog/what-is-crypto-mining-how-cryptocurrency-mining-works/

59) Baker, B., 2022. What is Bitcoin mining and how does it work? Bankrate. Retrieved from:
https://www.bankrate.com/investing/what-is-bitcoin-mining/

60) N26., 2022. What is crypto mining? Retrieved from:
https://n26.com/en-eu/blog/what-is-crypto-mining

61) Seaton, J., 2021. Cryptomining. Techopedia. Retrieved from:
https://www.techopedia.com/definition/33729/cryptomining

62) Binance Academy., 2018. What Is Cryptocurrency Mining? Binance Academy. Retrieved from:
https://academy.binance.com/en/articles/what-is-cryptocurrency-mining

63) Jain, A., 2022. What is cryptocurrency and how does it work? Forbes. Retrieved from:
https://www.forbes.com/advisor/in/investing/what-is-cryptocurrency-and-how-does-it-work/

64) Voigt, K., Rosen, A., 2022. What Is Cryptocurrency? Here's What Investors Should Know. NerdWallet. Retrieved from:
https://www.nerdwallet.com/article/investing/cryptocurrency

65) Little, K., 2022. What is Cryptocurrency? Time. Retrieved from:
https://time.com/nextadvisor/investing/cryptocurrency/what-is-cryptocurrency/

66) Arora, S., 2022. What Is Cryptocurrency: Types, Benefits, History and More. Simplilearn. Retrieved from: https://www.simplilearn.com/tutorials/blockchain-tutorial/what-is-cryptocurrency

67) Kerner, S. M., 2021. Cryptocurrency. TechTarget. Retrieved from: https://www.techtarget.com/whatis/definition/cryptocurrency

68) CoinDesk., 2021. What Is Cryptocurrency? CoinDesk. Retrieved from: https://www.coindesk.com/learn/what-is-cryptocurrency/

69) Ramsey Solutions., 2022. What Is Cryptocurrency and Should I Invest in It? Ramsey Solutions. Retrieved from: https://www.coindesk.com/learn/what-is-cryptocurrency/

70) Reynolds, M., 2019. What is Cryptocurrency and How Does it Benefit Me? Blackhawk Bank. Retrieved from: https://www.blackhawkbank.com/what-is-cryptocurrency-and-how-does-it-benefit-me

71) Cope, J., 2002. What is a Peer-to-Peer (P2P) Network? Computerworld. Retrieved from: https://www.computerworld.com/article/2588287/networking-peer-to-peernetwork.html#:~:text=In%20its%20simplest%20form%2C%20a,Serial%20Bus%20to%20transfer%20files

72) Techopedia., 2016. What Does Peer-To-Peer Network (P2P Network) Mean? Retrieved from: https://www.techopedia.com/definition/25777/peer-to-peer-network-p2p-network

73) Binance Academy., 2019. Peer-to-Peer Networks Explained. Binance Academy. Retrieved from: https://academy.binance.com/en/articles/peer-to-peer-networks-explained

74) Geeks for Geeks., 2020. What is P2P (Peer-to-peer process)? Retrieved from: https://www.geeksforgeeks.org/what-is-p2ppeer-to-peer-process/

75) XcelToken Plus., 2019. Peer-To-Peer Network Explained. Medium. Retrieved from: https://medium.com/xcellab-magazine/peer-to-peer-network-explained-c5038f6e8366

76) Indeed Editorial Team., 2021. What Is a Peer-to-Peer Network? Indeed. Retrieved from: https://in.indeed.com/career-advice/career-development/peer-to-peer-network

77) Alpern, N. J., Shimonski, R. J., 2010. Peer to Peer Networks. ScienceDirect. Retrieved from: https://www.sciencedirect.com/topics/computer-science/peer-to-peer-networks

78) Iverson, D., 2018. What is P2P? Rampecdn. Retrieved from: https://www.rampecdn.com/enterprise/what-is-p2p/?gclid=EAIaIQobChMIkqbN7rj69wIVEDUrCh3v0QGDEAMYASAAEgIJCvD_BwE

79) Werner Vermaak., 2021. What Is Peer-to-Peer (P2P) Networks? Coin Market Cap. Retrieved from: https://coinmarketcap.com/alexandria/article/what-is-peer-to-peer-p2p

80) 101 Blockchains., 2020. Understanding Peer to Peer Network. 101 Blockchains. Retrieved from: https://101blockchains.com/peer-to-peer-network

81) Bay Bit Learn., 2022. Peer-to-Peer Blockchain Networks: The Rise of P2P Crypto Exchanges. Bay Bit Learn. Retrieved from: https://learn.bybit.com/bybit-p2p-guide/peer-to-peer-blockchain-network

82) Alden Baldwin., 2021. Peer-to-Peer in Blockchain: how it works. Cryptopolitan. Retrieved from: https://www.cryptopolitan.com/peer-to-peer-in-blockchain-how-it-works

83) Shermin Voshmgir., 2020. What is Blockchain? Retrieved from: https://blockchainhub.net/blockchain-intro

84) Laura Gonzalez., 2019. Blockchain's Role in Peer-To-Peer Lending. CSULB Graduate Programs. Retrieved from: https://www.csulb.edu/cob-graduate-programs/faculty-and-research/article/blockchain%E2%80%99s-role-peer-to-peer-lending

85) Suman Chouhan., 2021. A quick rundown of the peer-to-peer (P2P) blockchain. LinkedIn. Retrieved from: https://www.linkedin.com/pulse/quick-rundown-peer-to-peer-p2p-blockchain-suman-chauhan

86) L. J. García Villalba., 2018. Cryptocurrency Networks: A New P2P Paradigm. Hindawi. Retrieved from: https://www.hindawi.com/journals/misy/2018/2159082

87) Chirag., 2021. A Guide to Understand Blockchain Consensus Algorithms? Appinventiv. Retrieved from: https://appinventiv.com/blog/blockchain-consensus-algorithms-guide

88) Vishal Chawla., 2020. What Are the Top Blockchain Consensus Algorithms? Analytics India Magazine. Retrieved from: https://analyticsindiamag.com/blockchain-consensus-algorithms

89) Academy Binance., 2018. What Is a Blockchain Consensus Algorithm? Binance Academy. Retrieved from:
https://academy.binance.com/en/articles/what-is-a-blockchain-consensus-algorithm

90) TechTarget Contributor., 2020. Consensus Algorithm? TechTarget. Retrieved from:
https://www.techtarget.com/whatis/definition/consensus-algorithm

91) Hasib Anwar., 2018. Consensus Algorithms: The Root of Blockchain Technology. 101 Blockchains. Retrieved from:
https://101blockchains.com/consensus-algorithms-blockchain

92) Azeem Irshad., 2021. Implementation of Blockchain Consensus Algorithm on Embedded Architecture. Hindawi. Retrieved from:
https://www.hindawi.com/journals/scn/2021/9918697

93) Lalithnaryan C., 2021. An Overview of Consensus Protocols in Blockchain. Section.io Engineering Education. Retrieved from:
https://www.section.io/engineering-education/blockchain-consensus-protocols

94) Pethuru Raj., 2021. The Blockchain Technology for Secure and Smart Applications across Industry Verticals. ScienceDirect. Retrieved from:
https://www.sciencedirect.com/topics/computer-science/consensus-algorithm

95) Academy Binance., 2019. Hybrid PoW/PoS Consensus Explained. Binance Academy. Retrieved from:
https://academy.binance.com/en/articles/hybrid-pow-pos-consensus-explained

96) Toshendra Kumar Sharma., 2020. A Brief Introduction to Hybrid PoW PoS Consensus Mechanism. Blockchain Council. Retrieved from:
https://www.blockchain-council.org/blockchain/a-brief-introduction-to-hybrid-powpos-consensus-mechanism

97) Kishor Datta Gupta, Abdur Rahman, Mohammad Nurul, Subash Poudyal., 2019. A Hybrid POW-POS Implementation Against 51% Attack in Cryptocurrency System. ResearchGate. Retrieved from:
https://www.researchgate.net/figure/Hybrid-PoW-PoS-Mechanism_fig3_337831342

98) Cryptopedia Staff., 2022. Varieties of Proof of Stake: LPoS, PPoS, HPoS, PoV. Gemini. Retrieved from:
https://www.gemini.com/cryptopedia/proof-of-stake-delegated-proof-of-stake-consensus-mechanism

99) Max Thake., 2018. What is Proof of Stake? (PoS). Medium. Retrieved from:

https://maxthake.medium.com/what-is-proof-of-stake-pos-479a04581f3a

100) Jake Frankenfield., 2021. Proof-of-Stake (PoS). Investopedia. Retrieved from:
https://www.investopedia.com/terms/p/proof-stake-pos.asp

101) BitFury Group., 2015. Proof of Stake versus Proof of Work. BitFury. Retrieved from:
https://bitfury.com/content/downloads/pos-vs-pow-1.0.2.pdf

102) Kishor Datta Gupta., 2019. Hybrid pow-pos-based-system against majority attack-in-cryptocurrency system compact version. SlideShare. Retrieved from:
https://www.slideshare.net/KishorDattaGupta/hybrid-powposbasedsystem-against-majority-attackincryptocurrency-system-compact-version

103) GeeksforGeeks., 2022. Delegated Proof of Stake (DPoS). GeeksforGeeks. Retrieved from:
https://www.geeksforgeeks.org/delegated-proof-of-stake/

104) Academy Binance., 2018. Delegated Proof of Stake Explained. Binance Academy. Retrieved from:
https://academy.binance.com/en/articles/delegated-proof-of-stake-explained

105) BybitLearn., 2021. What is Delegated Proof of Stake DPoS? Bybit Learn. Retrieved from:
https://learn.bybit.com/blockchain/delegated-proof-of-stake-dpos/

106) Himanshi., 2022. Delegated Proof of Stake (DPoS). Naukri Learning. Retrieved from:
https://www.naukri.com/learning/articles/delegated-proof-of-stake-dpos/

107) Delton Rhodes., 2020. What is Delegated Proof of Stake? An Overview of DPoS Blockchains. Komodo Platform. Retrieved from:
https://komodoplatform.com/en/academy/delegated-proof-of-stake/

108) Dan Schmidt., 2020. Delegated Proof of Stake. Benzinga. Retrieved from:
https://www.benzinga.com/money/delegated-proof-of-stake/

109) Shaan Ray., 2018. What is Delegated Proof of Stake? Hackernoon. Retrieved from:
https://hackernoon.com/what-is-delegated-proof-of-stake-e184f0c17c6f

110) Explaining Cardano's Proof-of-Stake (PoS) vs. Delegated Proof-of-Stake (DPoS) Blockchain., English Knowledge., Feb 20, 2020., Retrieved from:

https://emurgo.io/explain-proof-of-stake-pos-dpos/

111) Proof of Burn Explained., academy Binance., 2018. Retrieved from: https://academy.binance.com/en/articles/proof-of-burn-explained

112) Proof of Burn Consensus Algorithm in Blockchain., Geeks for Geeks., 2022. Retrieved from: https://www.geeksforgeeks.org/proof-of-burn-consensus-algorithm-in-blockchain/

113) What is Proof of Burn (PoB)? TheLuWizz., 2021. Retrieved from: https://medium.datadriveninvestor.com/what-is-proof-of-burn-pob-e8f7e7dfbbfa

114) Explained | Coin Burning: How is it done & why? CNBCTV18.com., 2022. Retrieved from: https://www.cnbctv18.com/cryptocurrency/explained--coin-burning-how-is-it-done--why-13245532.htm

115) What is Proof of Burn (PoB)? bit2academy., 2020. Retrieved from: https://academy.bit2me.com/en/que-es-proof-of-burn-pob/

116) Proof of Burn Explained – An Alternative Crypto Consensus Algorithm., Steve Walters., 2018. Retrieved from: https://www.coinbureau.com/education/proof-of-burn-explained/

117) Proof of Burn (PoB)., Jargon Various., 2016. Retrieved from: https://dyor-crypto.fandom.com/wiki/Proof-of-Burn_(PoB)

118) Proof of Burn., Kostis Karantias., 2020. Retrieved from: https://link.springer.com/chapter/10.1007/978-3-030-51280-4_28

119) Proof of Authority Explained., academy Binance., 2020. Retrieved from: https://academy.binance.com/en/articles/proof-of-authority-explained

120) Proof of Authority consensus., GitHub., 2019. Retrieved from: https://apla.readthedocs.io/en/latest/concepts/consensus.html

121) Proof of Authority: consensus model with Identity at Stake., POA Network., 2017. Retrieved from: https://medium.com/poa-network/proof-of-authority-consensus-model-with-identity-at-stake-d5bd15463256

122) Proof of Authority Explained., Dimitar Bogdanov., 2021. Retrieved from: https://limechain.tech/blog/proof-of-authority-explained/

123) Proof of Authority (PoA) in Blockchain., Himanshi., 2022. Retrieved from: https://www.naukri.com/learning/articles/proof-of-authority-in-blockchain/

124) What is Proof-of-Authority consensus? Narayan Prusty., 2020. Retrieved from:

https://www.oreilly.com/library/view/building-blockchain-projects/9781787122147/827f4856-1b32-4a10-a53a-e02050f74a15.xhtml

125) How does the Proof of Authority algorithm work? Christina Comben., 2019. Retrieved from: https://finance.yahoo.com/news/does-proof-authority-algorithm-090006261.html

126) A Complete Guide to the Proof of Authority (PoA) Algorithm., Mariia Rousey., 2016. Retrieved from: https://changelly.com/blog/what-is-proof-of-authority-poa/

127) Leased Proof of Stake Consensus Explained., academy Binance., 2019. Retrieved from: https://academy.binance.com/en/articles/leased-proof-of-stake-consensus-explained

128) A Complete Guide to Leased Proof-of-Stake (LPOS)., Andy Amora., 2020. Retrieved from: https://www.mycointainer.com/insight/a-complete-guide-to-leased-proof-of-stake-lpos/

129) Varieties of Proof of Stake: LPoS, PPoS, HPoS, PoV., Cryptopedia Staff., 2022. Retrieved from: https://www.gemini.com/cryptopedia/proof-of-stake-delegated-proof-of-stake-consensus-mechanism

130) Techskill Brew., 2022. Leased Proof of Stake or LPoS in Blockchain (Part 13- Blockchain Series). Retrieved from: https://medium.com/techskill-brew/leased-proof-of-stake-or-lpos-in-blockchain-part-13-blockchain-series-e1b373580cd1

131) Leased Proof of Stake., Exbase.Io., 2021. Retrieved from: https://exbase.io/en/wiki/leased-proof-of-stake

132) The Proof-of-Stake Guidebook: PoS, DPoS, LPoS, BPoS, Kézako? Finyear., 2019. Retrieved from: https://www.finyear.com/The-Proof-of-Stake-Guidebook-PoS-DPoS-LPoS-BPoS-Kezako_a40716.html

133) Leased Proof of Stake., Gleb Kostarev., 2018. Retrieved from: https://wiki.p2pfoundation.net/Leased_Proof_of_Stake

134) Proof of Stake (PoS) in Blockchain., Geeks for geeks., 2022. Retrieved from: https://www.geeksforgeeks.org/proof-of-stake-pos-in-blockchain/

135) Delayed Proof of Work (dPoW)., Consensus., 2019. Retrieved from: https://tokens-economy.gitbook.io/consensus/chain-based-proof-of-work/delayed-proof-of-work-dPoW#:~:text=consensus-

,Delayed%20Proof%20of%20Work%20(dPoW),power%20of%20a%20
20secondary%20blockchain.

136) Delayed Proof of Work Explained., academy Binance., 2018.
Retrieved from:
https://academy.binance.com/en/articles/delayed-proof-of-work-
explained

137) Delayed Proof of Work., Komodo., 2022. Retrieved from:
https://developers.komodoplatform.com/basic-docs/start-here/core-
technology-discussions/delayed-proof-of-work.html#a-foundational-
discussion-of-blockchain-security

138) Delayed Proof of Work: the multi-blockchain consensus algorithm.,
Emanuele Pagliari., 2019. Retrieved from:
https://en.cryptonomist.ch/2019/06/22/delayed-proof-of-work-
blockchain/

139) Proof of Work (dPoW)., Jake Frankenfield., 2022. Retrieved from:
https://www.investopedia.com/terms/p/proof-work.asp

140) What is Proof-of-Work (PoW)? Max Thake., 2018. Retrieved from:
https://maxthake.medium.com/what-is-proof-of-work-pow-
2574ddebf916

141) Delayed Proof of Work (dPoW) Security., Explained., Satinder
Grewal., 2018., Retrieved from:
https://satindergrewal.medium.com/delayed-proof-of-work-explained-
9a74250dbb86

142) Proof of Work Explained: Pros and Cons of PoW., Explained., Crypto
geek., 2020. Retrieved from:
https://cryptogeek.info/en/blog/proof-of-work-explained

143) Proof-Of-Work (Pow)., @minimalsm., 2022. Retrieved from:
https://ethereum.org/en/developers/docs/consensus-mechanisms/pow/

144) What Is Proof of Work (PoW) in Crypto? 2022. Retrieved from:
https://www.fool.com/investing/stock-market/market-
sectors/financials/cryptocurrency-stocks/proof-of-work/

145) Proof of Work Explained., Andrew Tar., 2018. Retrieved from:
https://cointelegraph.com/explained/proof-of-work-explained

146) Proof of Work (PoW)., technopedia., 2019. Retrieved from:
https://www.techopedia.com/definition/33594/proof-of-work-pow

147) Proof of Work Explained., Andrew Tar., 2018. Retrieved from:
https://cointelegraph.com/explained/proof-of-work-explained

148) Blockchain Proof of work., Java point., 2021. Retrieved from:
https://www.javatpoint.com/blockchain-proof-of-work

149) What Is Proof of Work (PoW)? Alex Lielacher., 2022. Retrieved
from:

https://www.axi.com/int/blog/education/blockchain/proof-of-work

150) What Is Proof of Work (PoW)? Osman., 2021. Retrieved from: https://learn.eqonex.com/news/what-proof-work-pow

151) Proof of stake Explained., E. Napoletano, Benjamin Curry., 2022. Retrieved from: https://www.forbes.com/advisor/investing/cryptocurrency/proof-of-stake/

152) What Is Proof of Stake (PoS) in Crypto? Lyle Daly., 2022. Retrieved from: https://www.fool.com/investing/stock-market/market-sectors/financials/cryptocurrency-stocks/proof-of-stake/

153) What Is Proof-of-Stake? Alyssa Hertig., 2022. Retrieved from: https://www.coindesk.com/learn/2020/12/30/what-is-proof-of-stake/

154) What Is Proof-of-Stake, and Why Is Ethereum Adopting It? Ian Bezek., 2021. Retrieved from: https://money.usnews.com/investing/cryptocurrency/articles/what-is-proof-of-stake-and-why-is-ethereum-adopting-it

155) What Is Proof-of-Stake? Everett Muzzy., 2020. Retrieved from: https://consensys.net/blog/blockchain-explained/what-is-proof-of-stake/

156) Proof of Stake: Verifying the Blocks in Blockchain., Zeb pay., 2021. Retrieved from: https://zebpay.com/blog/overview-on-proof-of-stake-pos/

157) Proof of Stake: A process used to validate crypto transactions through staking., Brian Nibley., 2021. Retrieved from: https://www.businessinsider.com/personal-finance/proof-of-stake?IR=T

158) Proof-of-Work vs. Proof-of-Stake: Which Is Better? Luke Conway., 2022. Retrieved from: https://blockworks.co/proof-of-work-vs-proof-of-stake-whats-the-difference/

159) A Brief History of Cryptography., Huzaifa Sidhpurwala., 2013. Retrieved from: https://www.redhat.com/en/blog/brief-history-cryptography

160) History of Cryptography., academy binance., 2021. Retrieved from: https://academy.binance.com/en/articles/history-of-cryptography

161) History of Cryptography., Gustavus J. Simmons., 2021. Retrieved from: https://www.britannica.com/topic/cryptology/The-Data-Encryption-Standard-and-the-Advanced-Encryption-Standard

162) A Brief History of Cryptography., Tony M. Damico., 2009. Retrieved from:
http://www.inquiriesjournal.com/articles/1698/a-brief-history-of-cryptography

163) A Brief History of Encryption (and Cryptography)., Thales Group., 2021., Retrieved from:
https://www.thalesgroup.com/en/markets/digital-identity-and-security/magazine/brief-history-encryption

164) The Story of Cryptography: History., Caesar Box., 2019. Retrieved from:
https://ghostvolt.com/articles/cryptography_history.html

165) The History of Cryptography., History Computer Staff., 2021. Retrieved from:
https://history-computer.com/the-history-of-cryptography/

166) History of Cryptography., Usama Azad., 2021. Retrieved from:
https://linuxhint.com/history-of-cryptography/

167) Public key cryptography., IBM., 2021. Retrieved from:
https://www.ibm.com/docs/en/ztpf/1.1.0.14?topic=concepts-public-key-cryptography

168) What is Public Key Cryptography? Nate Lord., 2018. Retrieved from:
https://digitalguardian.com/blog/what-public-key-cryptography

169) Public Key Cryptography (PKC)., techopedia., 2018. Retrieved from:
https://www.techopedia.com/definition/9021/public-key-cryptography-pkc

170) Public Key Encryption., geeksforgeeks., 2021. Retrieved from:
https://www.geeksforgeeks.org/public-key-encryption/

171) How does public key encryption work? | Public key cryptography and SSL., cloudflare., 2020. Retrieved from:
https://www.cloudflare.com/en-in/learning/ssl/how-does-public-key-encryption-work/

172) Public key., TechTarget., 2021. Retrieved from:
https://www.techtarget.com/searchsecurity/definition/public-key

173) Digital signature algorithm., Shubhani Aggarwal, Neeraj Kumar., 2021. Retrieved from:
https://www.sciencedirect.com/topics/computer-science/public-key-cryptosystems

174) What is Public Key Cryptography? Kelley Robinson., 2018. Retrieved from:
https://www.twilio.com/blog/what-is-public-key-cryptography

175) End-to-End encryption (E2EE)., Ben Lutkevich, Madelyn Bacon., 2021. Retrieved from:

https://www.techtarget.com/searchsecurity/definition/end-to-end-encryption-E2EE

176) What is End-to-End Encryption (E2EE) and How Does it Work? Cisco., 2021. Retrieved from:
https://www.cisco.com/c/en/us/products/security/what-is-end-to-end-encryption.html

177) End-to-End Encryption: An Overview., techtarget., March 2021. Retrieved from:
https://searchsecurity.techtarget.com/definition/end-to-end-encryption-E2EE

178) End-to-End Encryption: The Complete Guide., cybersecurity., 2023. Retrieved from:
https://cybersecurity.att.com/blogs/security-essentials/end-to-end-encryption

179) What is End-to-End Encryption (E2EE)? academy Binance., 2020., Retrieved from:
https://academy.binance.com/en/articles/what-is-end-to-end-encryption-e2ee

180) What is end-to-end encryption (E2EE)? atsign., 2021. Retrieved from:
https://atsign.com/resources/articles/what-is-end-to-end-encryption-e2ee/

181) What is end-to-end encryption (E2EE)? Baris Yeldiren., 2016. Retrieved from:
https://www.gbnews.ch/what-is-end-to-end-encryption-e2ee/

182) End-to-End Encryption (E2EE) Explained., Earl Potters., 2020. Retrieved from:
https://www.section.io/engineering-education/end-to-end-encryption/

183) The growing list of applications and use cases of blockchain technology in business and life., Insider Intelligence., 2022. Retrieved from:
https://www.insiderintelligence.com/insights/blockchain-technology-applications-use-cases/

184) Research leading blockchain use cases., IBM., 2020., Retrieved from:
https://www.ibm.com/blockchain/use-cases/

185) Blockchain Applications and Real-World Use Cases Disrupting the Status Quo., Sam Daley., 2021. Retrieved from:
https://builtin.com/blockchain/blockchain-applications

186) Top 17 Blockchain Applications & Use Cases in 2022., Cem Dilmegani., 2020. Retrieved from:
https://research.aimultiple.com/blockchain-applications/

187) Today's blockchain use cases and industry applications., Pam Baker., 2021. Retrieved from: https://www.techtarget.com/searchcio/feature/Todays-blockchain-use-cases-and-industry-applications

188) Blockchain Technology Use Cases You Should Know in 2022., Mayank Sahu., 2021. Retrieved from: https://www.upgrad.com/blog/blockchain-technology-use-cases/

189) 11 Real Blockchain Use Cases for Business., Aygun Suleymanova., 2021. Retrieved from: https://www.salesforce.com/blog/real-use-cases-enterprise-blockchain/

190) 15+ Practical Blockchain Use Cases in 2022., 2021. Retrieved from: https://101blockchains.com/practical-blockchain-use-cases/

191) 5 blockchain healthcare use cases in digital health., stlpartners., 2022. Retrieved from: https://stlpartners.com/articles/digital-health/5-blockchain-healthcare-use-cases/

192) 14 Major Real Use Cases of Blockchain in Healthcare., Bryan Weinberg., 2019. Retrieved from: https://openledger.info/insights/blockchain-healthcare-use-cases/

193) 3 Promising Use Cases for Blockchain Technology in Healthcare., innovation at work., 2021. Retrieved from: https://innovationatwork.ieee.org/3-promising-use-cases-for-blockchain-technology-in-healthcare/

194) Use Cases of Blockchain in Healthcare., Leeway Hertz., 2021. Retrieved from: https://www.leewayhertz.com/blockchain-in-healthcare/

195) Blockchain Technology Use Cases in Healthcare., Peng Zhang., Douglas C. Schmidt., Jules White., Gunther Len., 2018. Retrieved from: https://www.sciencedirect.com/science/article/abs/pii/S0065245818300196

196) 10 Possible Use Cases of Blockchain in Healthcare., apriorit., 2020. Retrieved from: https://www.apriorit.com/dev-blog/677-blockchain-blockchain-in-healthcare-possible-use-cases

197) Blockchain in healthcare: use cases for your practice., Nadejda Alkhaldi., 2020. Retrieved from: https://www.itransition.com/blog/blockchain-in-healthcare

198) How Using Blockchain in Healthcare Is Reviving the Industry's Capabilities., Sam Daley. 2021. Retrieved from:

https://builtin.com/blockchain/blockchain-healthcare-applications-companies

199) Blockchain for supply chain management: use cases and applications., 2022. Retrieved from:
https://www.r3.com/customers/supply-chain/#:~:text=Blockchain%20use%20cases%20for%20supply,physical%20and%20financial%20supply%20chains.

200) Top 5 use cases of Blockchain in the supply chain in 2021., Mark Morley., 2020. Retrieved from:
https://blogs.opentext.com/blockchain-in-the-supply-chain/

201) Building a Transparent Supply Chain., Vishal Gaur and Abhinav Gaiha., 2020. Retrieved from:
https://hbr.org/2020/05/building-a-transparent-supply-chain

202) 7 real-life blockchain in the supply chain use cases and examples., Linda Rosencrance., 2020. Retrieved from:
https://www.techtarget.com/searcherp/feature/4-key-blockchain-in-supply-chain-use-cases-and-examples

203) Blockchain in the Supply Chain: 10 Real-Life Use Cases and Examples., Darya Yafimava., 2019. Retrieved from:
https://openledger.info/insights/blockchain-in-the-supply-chain-use-cases-examples/

204) Use Cases and Benefits of Blockchain in Supply Chain Management., Diego Geroni., 2021. Retrieved from:
https://101blockchains.com/benefits-of-blockchain-in-supply-chain-management/

205) Use Cases for Blockchain in Supply Chain Management., Stuart Rauch., 2021. Retrieved from:
https://www.simplilearn.com/use-cases-for-blockchain-in-supply-chain-management-article

206) Blockchain in Supply Chain | Use Cases, Advantages, and Features., Mudit Kumar., 2020. Retrieved from:
https://blockchain.oodles.io/blog/blockchain-supply-chain-management-use-cases/

207) Blockchain in Digital Identity., ConsenSys., 2021. Retrieved from:
https://consensys.net/blockchain-use-cases/digital-identity

208) Blockchain for Digital Identity., Nec., 2020. Retrieved from:
https://www.nec.com/en/global/solutions/blockchain/blockchain-for-digital-identity.html

209) Digital Identity and the Blockchain: Universal Identity Management and the Concept of the "Self-Sovereign" Individual., Andrej J. Zwitter1, Oskar J. Gstein2 and Evan Yap3., 2020., Retrieved from:

https://www.frontiersin.org/articles/10.3389/fbloc.2020.00026/full

210) Blockchain Identity Management: The Definitive Guide., 2021. Retrieved from:
https://tykn.tech/identity-management-blockchain/

211) Practical Thoughts on Blockchain and Identity., 2020., Retrieved from:
https://www.okta.com/resources/whitepaper/practical-thoughts-on-blockchain-and-identity/

212) What Are the Benefits of Blockchain Identity Management., Aran Davies., 2021. Retrieved from:
https://www.devteam.space/blog/benefits-of-blockchain-for-identity-management/

213) How Blockchain Is Used in Digital Identity Management., Pixelplex., 2021. Retrieved from:
https://pixelplex.io/blog/blockchain-digital-identity/

214) Blockchains: Use Cases in Cybersecurity., Satish Mohan., 2019. Retrieved from:
https://www.forbes.com/sites/forbestechcouncil/2022/02/02/blockchains-use-cases-in-cybersecurity/?sh=6b1a9fc25989

215) Digital Signature., Ben Lutkevich., 2021. Retrieved from:
https://www.techtarget.com/searchsecurity/definition/digital-signature

216) Understanding digital signatures., DocuSign., 2022. Retrieved from:
https://www.docusign.com/how-it-works/electronic-signature/digital-signature/digital-signature-faq

217) Understanding digital signatures., cisa., 2020. Retrieved from:
https://www.cisa.gov/uscert/ncas/tips/ST04-018

218) Digital signature overview., IBM., 2021-03-05., Retrieved from:
https://www.ibm.com/docs/en/b2badv-communication/1.0.0?topic=overview-digital-signature

219) Digital Signatures: What They Are & How They Work., Sectigo., 2020. Retrieved from:
https://sectigo.com/resource-library/how-digital-signatures-work

220) What is a Digital Signature? John Carl Villanueva., 2022. Retrieved from:
https://www.jscape.com/blog/what-is-a-digital-signature

221) What Is Digital Signature: How It Works, Benefits, Objectives, Concept., Eliza Paul., 2017. Retrieved from:
https://www.emptrust.com/blog/benefits-of-using-digital-signatures/

222) Digital Signature., Jason Andress., 2014. Retrieved from:
https://www.sciencedirect.com/topics/computer-science/digital-signature

223) Blockchain Wallet., Jake Frankenfield., 2022. Retrieved from:
https://www.investopedia.com/terms/b/blockchain-wallet.asp

224) What is a Crypto Wallet? A Beginner's Guide., Crypto., 2022. Retrieved from:
https://crypto.com/university/crypto-wallets

225) What is a crypto wallet? Understanding the software that allows you to store and transfer crypto securely., Brian Nibley., 2021. Retrieved from:
https://www.businessinsider.com/personal-finance/crypto-wallet?IR=T

226) What is Blockchain Wallet and How Does It Work? Simplilearn., 2022. Retrieved from:
https://www.simplilearn.com/tutorials/blockchain-tutorial/blockchain-wallet

227) What is a Crypto Wallet., n26., 2021? Retrieved from:
https://n26.com/en-eu/blog/what-is-a-crypto-wallet

228) Your First Crypto Wallet: What Is a Crypto Wallet and How to Use It., Robert Stevens., 2022. Retrieved from:
https://www.coindesk.com/learn/your-first-crypto-wallet-how-to-use-it-and-why-you-need-one

229) A Crypto Wallet Can Help Keep Your Coins Safe. Here's How to Decide If You Need One., Kendall Little., 2021., Retrieved from:
https://time.com/nextadvisor/investing/cryptocurrency/best-bitcoin-cryptocurrency-wallet/

230) How Crypto Wallets Work: Key Things to Know., NDTV Business., 2021. Retrieved from:
https://www.ndtv.com/business/here-s-what-you-need-to-know-about-how-crypto-wallets-work-2565966

231) Stablecoin., Adam Hayes., 2022. Retrieved from:
https://www.investopedia.com/terms/s/stablecoin.asp

232) What Is a Stablecoin? Alyssa Hertig., 2022. Retrieved from:
https://www.coindesk.com/learn/what-is-a-stablecoin/

233) What Are Stablecoins? cbinsights., 2022. Retrieved from:
https://www.cbinsights.com/research/report/what-are-stablecoins/

234) What are stablecoins and how do they affect the cryptocurrency market? James Royal., 2022. Retrieved from:
https://www.bankrate.com/investing/stablecoin-cryptocurrency/

235) What Are Stablecoins? Cryptopedia Staff., May 10., 2022. Retrieved from:
https://www.gemini.com/cryptopedia/what-are-stablecoins-how-do-they-work

236) What Are Stablecoins and How Do They Work? Crypto., 2022. Retrieved from:
https://crypto.com/university/what-are-stablecoins

237) How Stablecoins Can Help Crypto Investors, and Why the U.S. Government Has Taken Notice., Alex Gailey., 2022. Retrieved from:
https://time.com/nextadvisor/investing/cryptocurrency/what-are-stablecoins/

238) What are Stablecoins, and why are they not falling amid the crypto market crash? business today., 2022. Retrieved from:
https://www.businesstoday.in/crypto/story/what-are-stablecoins-and-why-are-they-not-falling-amid-crypto-market-crash

239) Explained | Coin Burning: How is it done & why? CNBCTV18., 2022. Retrieved from:
https://www.cnbctv18.com/cryptocurrency/explained--coin-burning-how-is-it-done--why-13245532.htm

240) Cryptocurrency Burning., Nathan Reiff., 2022. Retrieved from:
https://www.investopedia.com/tech/cryptocurrency-burning-can-it-manage-inflation/

241) What Does It Mean to Burn Crypto? Benedict George., 2022. Retrieved from:
https://www.coindesk.com/learn/what-does-it-mean-to-burn-crypto/

242) What Is Cryptocurrency Coin Burning? Lyle Daly., Nov 10, 2021. Retrieved from:
https://www.fool.com/investing/stock-market/market-sectors/financials/cryptocurrency-stocks/coin-burn/

243) Cryptocurrency: What is coin burning and why it should be a part of every investor's calculus? Rajeev Kumar., 2021. Retrieved from:
https://www.financialexpress.com/money/cryptocurrency-what-is-coin-burning-and-why-it-should-be-a-part-of-every-investors-calculus/2323339/

244) What Is a Coin Burn? Binance Academy., 2023. Retrieved from:
https://academy.binance.com/en/articles/what-is-a-coin-burn

245) Explained: What is coin burning in the cryptocurrency universe? Pawan Nahar., 2021. Retrieved from:
https://economictimes.indiatimes.com/markets/cryptocurrency/explained-what-is-coin-burning-in-the-cryptocurrency-universe/articleshow/85658968.cms

246) Burn/Burned? Alexandria. 2019. Retrieved from:
https://coinmarketcap.com/alexandria/glossary/burned

247) Explained: What Is Hashing in Blockchain? Bybit Learn., 2020. Retrieved from:

https://learn.bybit.com/blockchain/what-is-hashing-in-blockchain/

248) What Is Hashing? [Step-by-Step Guide-Under Hood of Blockchain] Blockchain., Ameer Rosic., 2020. Retrieved from: https://blockgeeks.com/guides/what-is-hashing/

249) What Is Hashing in Blockchain? dcx Learn., 2021. Retrieved from: https://dcxlearn.com/blockchain/what-is-hashing/

250) Hash functions in blockchain., Howard Poston., September 29, 2020. Retrieved from: https://resources.infosecinstitute.com/topic/hash-functions-in-blockchain/

251) Hashing in Blockchain: Explained., Leslie Chally., 2021. Retrieved from: https://blog.ineuron.ai/Hashing-in-Blockchain-Explained-zEfewYMCtx

252) Hash., Henrique Erhardt., 2020. Retrieved from: https://academy.binance.com/en/glossary/hash

253) Cryptographic Hashing: A Complete Overview., Toshendra Kumar Sharma., 2021. Retrieved from: https://www.blockchain-council.org/blockchain/cryptographic-hashing-a-complete-overview/

254) Hashing In Cryptography., 101 Blockchains., 2021. Retrieved from: https://101blockchains.com/hashing-in-cryptography/

255) Explained | What are hard forks and soft forks in blockchain? CNBCTV18., 2022. Retrieved from: https://www.cnbctv18.com/cryptocurrency/explained-what-are-hard-forks-and-soft-forks-in-blockchain-12406512.htm

256) Hard Fork (Blockchain)., Jake Frankenfield., 2022. https://www.investopedia.com/terms/h/hard-fork.asp

257) Hards Forks and Soft Forks Explained., academy Binance., 2018. Retrieved from: https://academy.binance.com/en/articles/hard-forks-and-soft-forks

258) How do Hard and Soft Forks work? bitpanda., 2020. Retrieved from: https://www.bitpanda.com/academy/en/lessons/how-do-hard-and-soft-forks-work/

259) Soft Fork vs Hard Fork: What Are the Differences? Finance Magnates Staff., 2022., Retrieved from: https://www.financemagnates.com/cryptocurrency/education-centre/soft-fork-vs-hard-fork-what-are-the-differences/

260) Hard and Soft Forks - What is the difference between them? Pragya Soni., 2021., Retrieved from:

https://www.analyticssteps.com/blogs/hard-and-soft-forks-what-difference-between-them

261) Bitcoin Forks Explained: Hard Fork vs. Soft Fork., Nicolas Tang., 2021. Retrieved from:
https://phemex.com/academy/bitcoin-hard-fork-vs-soft-fork

262) Hard Fork Versus Soft Fork: Understanding the Difference., centrum coin., 2019. Retrieved from:
https://medium.com/centrumcoin/hard-fork-versus-soft-fork-understanding-the-difference-81219a8353b2

263) Jake Frankenfield., 2022. 51% Attack: Definition, Who Is at Risk, Example, and Cost…. Retrieved from:
https://www.investopedia.com/terms/1/51-attack.asp

264) Griffin McShane., 2020. What Is a 51% Attack? Retrieved from:
https://www.coindesk.com/learn/what-is-a-51-attack/

265) What are 51% attacks in cryptocurrencies? Pawan Nahar., 2021., Retrieved from:
https://economictimes.indiatimes.com/markets/cryptocurrency/what-are-51-attacks-in-cryptocurrencies/articleshow/85802504.cms?from=mdr

266) What Is a 51% Attack? academy Binance., 2018., Retrieved from:
https://academy.binance.com/en/articles/what-is-a-51-percent-attack

267) Understanding a 51% Attack on the Blockchain., Ephraim Njoroge., 2021., Retrieved from:
https://www.section.io/engineering-education/understanding-the-51-attack-on-blockchain/

268) Jimi S., 2018. Blockchain explained: how a 51% attack works (Double Spend Attack)., Retrieved from:
https://blog.goodaudience.com/what-is-a-51-attack-or-double-spend-attack-aa108db63474?gi=4bd44e89ed4f

269) What Is a 51% Attack? Eric Rosenberg., October 28, 2021., Retrieved from:
https://www.thebalance.com/what-is-a-51-attack-5207550

270) What Is a 51% Attack? Sofi Learn. 2021., Retrieved from:
https://www.sofi.com/learn/content/51-attack/

271) Academy Binance., 2018. What Is a Dusting Attack? Retrieved from:
https://academy.binance.com/en/articles/what-is-a-dusting-attack

272) Cryptopedia Staff., 2022. What Is a Crypto Dusting Attack? Retrieved from:
https://www.gemini.com/cryptopedia/crypto-dusting-attack-bitcoin

273) Nambiampurath, R., 2021. What Is a Dusting Attack? Retrieved from:
https://beincrypto.com/learn/dusting-attack/

274) Mihajlović, M., 2022. What Is a Dusting Attack? Crypto Exploit Explained. Retrieved from: https://academy.shrimpy.io/post/what-is-a-dusting-attack-crypto-exploit-explained

275) Ledger Support., 2022. What Is a Dusting Attack? Retrieved from: https://support.ledger.com/hc/en-us/articles/4410970438929-What-is-a-dusting-attack-?support=true

276) Bit2meacademy., 2022. What Is a Dusting Attack? Retrieved from: https://academy.bit2me.com/en/que-es-dusting-attack/

277) Batt, S., 2021. What Is a Cryptocurrency Dusting Attack? Retrieved from: https://www.makeuseof.com/what-is-a-cryptocurrency-dusting-attack/

278) Team Luno., 2020. What Is a Dusting Attack? Retrieved from: https://discover.luno.com/what-is-a-dusting-attack/

279) Frankenfield, J., 2022. Double-Spending. Retrieved from: https://www.investopedia.com/terms/d/doublespending.asp

280) Academy Binance., 2020. Double Spending Explained. Retrieved from: https://academy.binance.com/en/articles/double-spending-explained

281) Bit panda., 2020. What is double-spending and why is it such a problem? Retrieved from: https://www.bitpanda.com/academy/en/lessons/what-is-double-spending-and-why-is-it-such-a-problem/

282) Warholak, M., 2021. Bitcoin's Double Spending Problem: Definition, Solution, and Future Outlook. Retrieved from: https://www.sofi.com/learn/content/double-spending/

283) Digital Assets., 2022. What Is Double-Spending and How to Prevent It in Crypto Asset Transactions. Retrieved from: https://zipmex.com/learn/double-spending/

284) Cryptopedia Staff., 2021. What Is Double-Spending? Retrieved from: https://www.gemini.com/cryptopedia/double-spending-problem-crypto

285) Mark., 2018. Double-Spending Explained. Retrieved from: https://www.mycryptopedia.com/double-spending-explained/

286) Rhodes, D., 2020. The Double Spending Problem, Explained. Retrieved from: https://komodoplatform.com/en/academy/double-spending-problem/

287) Academy Binance., 2019. The Complete Beginner's Guide to Decentralized Finance (DeFi). Retrieved from https://academy.binance.com/en/articles/the-complete-beginners-guide-to-decentralized-finance-defi

288) Coelho-Prabhu, S., 2020. A Beginner's Guide to Decentralized Finance (DeFi). Retrieved from: https://blog.coinbase.com/a-beginners-guide-to-decentralized-finance-defi-574c68ff43c4

289) Stevens, R., 2021. DeFi: The Ultimate Beginner's Guide to Decentralized Finance. Retrieved from: https://decrypt.co/resources/defi-ultimate-beginners-guide-decentralized-finance

290) Coin Telegraph., 2020. Decentralized finance (DeFi): A beginner's guide. Retrieved from: https://cointelegraph.com/defi-101/defi-a-comprehensive-guide-to-decentralized-finance

291) Chirag., 2021. A Beginner's Guide to What is Decentralized Finance (DeFi). Retrieved from: https://appinventiv.com/blog/decentralized-finance-defi-guide/

292) N26., 2022. What is DeFi? A beginner's guide to decentralized finance. Retrieved from: https://n26.com/en-eu/blog/what-is-defi

293) Forbes., 2022. A beginner's guide to DeFi. Retrieved from: https://www.forbesindia.com/article/brand-connect/a-beginners-guide-to-defi/74925/1

294) Royal, J., 2022. What is DeFi? A beginner's guide to decentralized finance. Retrieved from: https://www.bankrate.com/investing/what-is-decentralized-finance-defi-crypto/

295) Daley, S., 2019. Tastier coffee, hurricane prediction and fighting the opioid crisis: 26 ways blockchain & AI make a powerful pair. Retrieved from: https://builtin.com/artificial-intelligence/blockchain-ai-examples

296) IBM., 2022. Defining blockchain and AI. Retrieved from: https://www.ibm.com/topics/blockchain-ai

297) Geeks for geeks., 2021. Integration of Blockchain and AI. Retrieved from: https://www.geeksforgeeks.org/integration-of-blockchain-and-ai/

298) Choudhury, A., 2020. Top 8 AI-Blockchain Projects One Must Know. Retrieved from: https://analyticsindiamag.com/top-8-ai-blockchain-projects-one-must-know/

299) Salah, K., Rehman, M. H. U., Nizamuddin, N., Al-Fuqaha, A., 2021. Blockchain for AI: Review and Open Research Challenges. Retrieved from:

https://ieeexplore.ieee.org/abstract/document/8598784%2

300) Lavrov, B., 2017. Blockchain Vs. Artificial Intelligence. Retrieved from:
https://towardsdatascience.com/blockchain-vs-artificial-intelligence-3e1c50908ebb

301) Banafa, A., 2019. Blockchain and AI: A Perfect Match? Retrieved from:
https://www.bbvaopenmind.com/en/technology/artificial-intelligence/blockchain-and-ai-a-perfect-match/

302) Oleksiuk, A., 2022. Five Benefits of Combining AI and Blockchain. Retrieved from:
https://intellias.com/five-benefits-of-combining-ai-and-blockchain/

303) academy Binance., 2019. How Blockchain Technology Will Impact the Banking Industry. Retrieved from:
https://academy.binance.com/en/articles/how-blockchain-technology-will-impact-the-banking-industry

304) Hosanagar, K., Werbach, K., 2018. How the Blockchain Will Impact the Financial Sector. Retrieved from:
https://knowledge.wharton.upenn.edu/article/blockchain-will-impact-financial-sector/

305) cm-alliance., 2022. How Blockchain Is Changing the Banking System. Retrieved from:
https://www.cm-alliance.com/cybersecurity-blog/how-blockchain-is-changing-the-banking-system

306) Kulkarni, S., 2021. 7 Ways How Blockchain Could Disrupt Banking and Finance Industry. Retrieved from:
https://www.goodreturns.in/classroom/7-ways-how-blockchain-could-disrupt-banking-and-finance-industry-1214096.html

307) Hexacta., 2021. How Blockchain Is Transforming the Banking Industry. Retrieved from:
https://www.hexacta.com/blockchain-in-banking-transforming-the-industry/

308) Dior., 2021. How Blockchain Can Affect Traditional Banking? Retrieved from:
https://diro.io/how-does-blockchain-affect-banking-industry/

309) Jain, P., 2020. Impact of Block-Chain Technology on Financial Services. Retrieved from:
https://techdayhq.com/community/articles/impact-of-block-chain-technology-on-financial-services

310) Reiff, N., 2021. Decentralized Autonomous Organization (DAO). Retrieved from:

https://www.investopedia.com/tech/what-dao/

311) Blockchain hub., 2015. Tokenized Networks: What is a DAO? Retrieved from:
https://blockchainhub.net/dao-decentralized-autonomous-organization/

312) Coin Telegraph., 2021. What is a decentralized autonomous organization, and how does a DAO work? Retrieved from:
https://cointelegraph.com/decentralized-automated-organizations-daos-guide-for-beginners/what-is-decentralized-autonomous-organization-and-how-does-a-dao-work

313) Shuttleworth, D., 2021. What Is A DAO and How Do They Work? Retrieved from:
https://consensys.net/blog/blockchain-explained/what-is-a-dao-and-how-do-they-work/

314) Hackl, C., 2021. What Are DAOs And Why You Should Pay Attention. Retrieved from:
https://www.forbes.com/sites/cathyhackl/2021/06/01/what-are-daos-and-why-you-should-pay-attention/?sh=12d05de97305

315) Barber, K. 2022. WTF is a decentralized autonomous organization (DAO). Retrieved from:
https://digiday.com/media/wtf-is-a-dao/

316) Bove, T., 2022. What is a DAO, and how it could change the future of business. Retrieved from:
https://fortune.com/2022/02/15/what-is-a-dao-explaining-decentralized-autonomous-organizations/

317) Bogna, J., 2022. What Is a DAO? Decentralized Autonomous Organizations Explained. Retrieved from:
https://www.pcmag.com/how-to/what-is-a-dao-decentralized-autonomous-organizations-explained

318) Frankenfield, J., 2022. Smart Contracts. Retrieved from:
https://www.investopedia.com/terms/s/smart-contracts.asp

319) Arora, S., 2022. What is a Smart Contract in Blockchain and How Does it Work? Retrieved from:
https://www.simplilearn.com/tutorials/blockchain-tutorial/what-is-smart-contract

320) Lawton, G., 2020. smart contract. Retrieved from:
https://www.techtarget.com/searchcio/definition/smart-contract

321) Geeks for Geeks., 2022. Blockchain | Smart Contracts. Retrieved from:
https://www.geeksforgeeks.org/smart-contracts/

322) Chain-link., 2021. What Is a Smart Contract? Retrieved from:

https://chain.link/education/smart-contracts?utm_medium=paid-search&utm_source=google

323) Mearian, L., 2019. What is a smart contract (and how does it work)? Retrieved from: https://www.computerworld.com/article/3412140/whats-a-smart-contract-and-how-does-it-work.html

324) Iberdrola., 2022. Smart contracts: contracts to formalise agreements in the digital age. Retrieved from: https://www.iberdrola.com/innovation/smart-contracts

325) Hussey, M., Phillips, D., 2022. What Are Smart Contracts and How Do They Work? Retrieved from: https://decrypt.co/resources/smart-contracts

326) Academy Binance., 2020. Blockchain Scalability - Sidechains and Payment Channels. Retrieved from: https://academy.binance.com/en/articles/blockchain-scalability-sidechains-and-payment-channels

327) Kathy Nguyenn., 2021. Can Sidechains and Payment Channels Solve Blockchain Scalability Issues? Retrieved from: https://holdingb.com/can-sidechains-and-payment-channels-solve-blockchain-scalability-issues/

328) Dollar cost bitcoin., 2021. Blockchain Scalability - Sidechains and Payment Channels. Retrieved from: https://dollarcostbitcoin.com/blog/blockchain-scalability-sidechains-and-payment-channels/

329) Techskill Brew., 2021. Layer 2 Blockchain scaling solutions: Channels, Sidechains, Rollups and Plasma (Part 16). Retrieved from: https://medium.com/techskill-brew/layer-2-blockchain-scaling-solutions-channels-sidechains-rollups-and-plasma-part-16-79819e058ef6

330) Cryptopedia Staff., 2022. Layer-1 and Layer-2 Blockchain Scaling Solutions. Retrieved from: https://www.gemini.com/cryptopedia/blockchain-layer-2-network-layer-1-network

331) Hive Power., 2020. Blockchain Second Layer Solutions: State Channels vs Side Chains. Retrieved from: https://www.hivepower.tech/blog/blockchain-second-layer-solutions-state-channels-vs-side-chains

332) Blockchain Hub., 2019. Blockchain Scalability: State Channels, Sidechains & more. Retrieved from: https://blockchainhub.net/blog/blog/blockchain-scalability-statchannels-sidechains/

333) Davids, M., 2021. Blockchain Scalability. Retrieved from: https://nominex.io/blog/education/blockchain-scalability/

334) Bandyopadhyay, L., Chaulya, S., Mishra, P., 2010. Wireless communication in underground mines. RFID-Based Sensor Network. 1-477.

335) Mandal, R., Maity, T., Chaulya, S.K., Prasad, G.M., Verma, R.P., 2016. Automation of underground coal mines using PLC. J. Mines Met. Fuels, 174-181.

336) Kumar, S., Kumar, D., Donta, P.K., Amgoth, T., 2022. Land subsidence prediction using recurrent neural networks. Stoch. Environ. Research Risk Assess. 36, 373–388. https://link.springer.com/article/10.1007/s00477-021-02138-2

337) Kumar, A., Kumar, D., Singh, A. K., Ram, S., Kumar, R., 2021. Development of empirical model for strength estimation of irregular-shaped-heightened-rib/snook for mechanised depillaring. Int. J. of Rock Mech. and Min. Sciences. 148, 104969. https://www.sciencedirect.com/science/article/pii/S136516092100351 8?via%3Dihub

338) Pranjal, P., Kadiyan, N., Chatterjee, R.S. et al. 2021. Interpreting land subsidence impacts due to groundwater depletion using remote sensing-based GRACE gravity anomaly and DInSAR technique: a study on north-western parts of India. Environ Earth Sci 80, 596. https://link.springer.com/article/10.1007/s12665-021-09905-y

339) Vikram, S., Kumar, D., Duvvuri, S., 2021. Support System Design for Deep Coal Mining by Numerical Modelling and a Case Study. Theory and Practice of Tunnel Engineering. 1-11. https://www.intechopen.com/chapters/76814

340) T, Sivasankar., Kumar, D., S, Hari., Patel, P., 2018. Advances in Radar Remote Sensing of Agricultural Crops: A Review. International Journal on Advanced Science, Engineering, and Information Technology. Vol. 8 (4), 1126-1137. https://ijaseit.insightsociety.org/index.php?option=com_content&view =article&id=9&Itemid=1&article_id=5797.

341) Singh, A., Kumar, D., Hötzel, Jürgen., 2018. IoT based Information and Communication System for Enhancing Underground Mines Safety and Productivity: Genesis, Taxonomy and Open Issues. Ad Hoc Networks. 78, 115-129. https://www.sciencedirect.com/science/article/pii/S157087051830352 4?via%3Dihub

342) Singh, A., Singh, U., Kumar, D., 2018. IoT in mining for sensing, monitoring and prediction of underground mines roof support.

International Conference on Recent Advances in Information Technology (RAIT).
https://ieeexplore.ieee.org/document/8389041

343) Kumar, A., Kumar, D., Verma, A., Singh, A., Ram, S., Kumar, R., 2018. Influence of overlying roof strata on rib design in mechanised depillaring. J. Geo. Society of Ind. 91, 341-347. https://link.springer.com/article/10.1007/s12594-018-0860-7

344) Kumar, D., 2015. Application of Modern Tools and Techniques for Mine Safety & Disaster Management. J. Inst. Eng. India Ser. D 97(1), 77–85
https://link.springer.com/article/10.1007/s40033-015-0071-y

345) Ram, S., Kumar, D., Singh, R., 2014. Roof bolt-based breaker-line design during underground coal pillar extraction. International Colloquium on Geomechanics and Geophysics, 69-71. https://www.researchgate.net/publication/311776053_Roof_bolt_base d_breakerline_design_during_underground_coal_pillar_extraction.

346) Shankar, V., Kumar, D., Narasimha, A. R., 2013. Application of Ground Based Lidar for Monitoring & Stability Analysis Of Dump Slope In Open pit Coal Mine: A Case Study. MEJ, Vol. 14, No. 9. https://www.researchgate.net/publication/337223456